Exploring and Modeling the Magma-Hydrothermal Regime

Exploring and Modeling the Magma-Hydrothermal Regime

Editors

John Eichelberger
Alexey Kiryukhin
Silvio Mollo
Noriyoshi Tsuchiya
Marlène Villeneuve

MDPI • Basel • Beijing • Wuhan • Barcelona • Belgrade • Manchester • Tokyo • Cluj • Tianjin

Editors

John Eichelberger
University of Alaska Fairbanks
USA

Noriyoshi Tsuchiya
Tohoku University
Japan

Alexey Kiryukhin
Institute of Volcanology and Seismology
Russia

Marlène Villeneuve
Montanu Universität Leoben
Austria

Silvio Mollo
Sapienza University of Rome
Italy

Editorial Office
MDPI
St. Alban-Anlage 66
4052 Basel, Switzerland

This is a reprint of articles from the Special Issue published online in the open access journal *Geosciences* (ISSN 2076-3263) (available at: https://www.mdpi.com/journal/geosciences/special_issues/magma_regime#Research).

For citation purposes, cite each article independently as indicated on the article page online and as indicated below:

LastName, A.A.; LastName, B.B.; LastName, C.C. Article Title. *Journal Name* **Year**, *Article Number*, Page Range.

ISBN 978-3-03936-636-1 (Hbk)
ISBN 978-3-03936-637-8 (PDF)

Cover image courtesy of Yevhenii Chulovskyi on Shutterstock.com License for reproduction purchased from Shutterstock.com by John Eichelberger.

© 2020 by the authors. Articles in this book are Open Access and distributed under the Creative Commons Attribution (CC BY) license, which allows users to download, copy and build upon published articles, as long as the author and publisher are properly credited, which ensures maximum dissemination and a wider impact of our publications.

The book as a whole is distributed by MDPI under the terms and conditions of the Creative Commons license CC BY-NC-ND.

Contents

About the Editors vii

John Eichelberger, Alexey Kiryukhin, Silvio Mollo, Noriyoshi Tsuchiya and Marlène Villeneuve
Exploring and Modeling the Magma–Hydrothermal Regime
Reprinted from: *geosciences* **2020**, *10*, 234, doi:10.3390/geosciences10060234 1

Giancarlo Tamburello, Séverine Moune, Patrick Allard, Swetha Venugopal, Vincent Robert, Marina Rosas-Carbajal, Sébastien Deroussi, Gaëtan-Thierry Kitou, Tristan Didier, Jean-Christophe Komorowski, François Beauducel, Jean-Bernard De Chabalier, Arnaud Le Marchand, Anne Le Friant, Magali Bonifacie, Céline Dessert and Roberto Moretti
Spatio-Temporal Relationships between Fumarolic Activity, Hydrothermal Fluid Circulation and Geophysical Signals at an Arc Volcano in Degassing Unrest: La Soufrière of Guadeloupe (French West Indies)
Reprinted from: *geosciences* **2019**, *9*, 480, doi:10.3390/geosciences9110480 7

Yasuhisa Tajima, Setsuya Nakada, Fukashi Maeno, Toshio Huruzono, Masaaki Takahashi, Akihiko Inamura, Takeshi Matsushima, Masashi Nagai and Jun Funasaki
Shallow Magmatic Hydrothermal Eruption in April 2018 on Ebinokogen Ioyama Volcano in Kirishima Volcano Group, Kyushu, Japan
Reprinted from: *geosciences* **2020**, *10*, 183, doi:10.3390/geosciences10050183 33

Arthur Jolly, Corentin Caudron, Társilo Girona, Bruce Christenson and Roberto Carniel
'Silent' Dome Emplacement into a Wet Volcano: Observations from an Effusive Eruption at White Island (Whakaari), New Zealand in Late 2012
Reprinted from: *geosciences* **2020**, *10*, 142, doi:10.3390/geosciences10040142 59

Ben M. Kennedy, Aaron Farquhar, Robin Hilderman, Marlène C. Villeneuve, Michael J. Heap, Stan Mordensky, Geoffrey Kilgour, Art. Jolly, Bruce Christenson and Thierry Reuschlé
Pressure Controlled Permeability in a Conduit Filled with Fractured Hydrothermal Breccia Reconstructed from Ballistics from Whakaari (White Island), New Zealand
Reprinted from: *geosciences* **2020**, *10*, 138, doi:10.3390/geosciences10040138 73

Alexey V. Kiryukhin and Gennady Karpov
A CO_2-Driven Gas Lift Mechanism in Geyser Cycling (Uzon Caldera, Kamchatka)
Reprinted from: *geosciences* **2020**, *10*, 180, doi:10.3390/geosciences10050180 91

Alexey Kiryukhin, Evgenia Chernykh, Andrey Polyakov and Alexey Solomatin
Magma Fracking Beneath Active Volcanoes Based on Seismic Data and Hydrothermal Activity Observations
Reprinted from: *geosciences* **2020**, *10*, 52, doi:10.3390/geosciences10020052 115

Geri Agroli, Atsushi Okamoto, Masaoki Uno and Noriyoshi Tsuchiya
Transport and Evolution of Supercritical Fluids During the Formation of the Erdenet Cu–Mo Deposit, Mongolia
Reprinted from: *geosciences* **2020**, *10*, 201, doi:10.3390/geosciences9110480 131

Michael A. Stearns, John M. Bartley, John R. Bowman, Clayton W. Forster, Carl J. Beno, Daniel D. Riddle, Samuel J. Callis and Nicholas D. Udy
Simultaneous Magmatic and Hydrothermal Regimes in Alta–Little Cottonwood Stocks, Utah, USA, Recorded Using Multiphase U-Pb Petrochronology
Reprinted from: *geosciences* **2020**, *10*, 129, doi:10.3390/geosciences10040129 . . . . **151**

John M. Bartley, Allen F. Glazner, Michael A. Stearns and Drew S. Coleman
The Granite Aqueduct and Autometamorphism of Plutons
Reprinted from: *geosciences* **2020**, *10*, 136, doi:10.3390/geosciences10040136 . . . . **173**

Allen F. Glazner
Climate and the Development of Magma Chambers
Reprinted from: *geosciences* **2020**, *10*, 93, doi:10.3390/geosciences10030093 . . . . **185**

Knútur Árnason
New Conceptual Model for the Magma-Hydrothermal-Tectonic System of Krafla, NE Iceland
Reprinted from: *geosciences* **2020**, *10*, 34, doi:10.3390/geosciences10010034 . . . . **199**

John Eichelberger
Distribution and Transport of Thermal Energy within Magma–Hydrothermal Systems
Reprinted from: *geosciences* **2020**, *10*, 212, doi:10.3390/geosciences10060212 . . . . **227**

About the Editors

John Eichelberger's career spans volcanology, geothermal energy, natural hazards, and international Arctic education. Educated at MIT and Stanford, he was on the research staff at the Los Alamos National Laboratory 1974–1979 and Sandia National Laboratories 1979–1991, New Mexico. He then moved to University of Alaska Fairbanks (UAF) when he was appointed Professor of Volcanology and led the Alaska Volcano Observatory (AVO). Sixteen years later, he became Program Coordinator for the Volcano Hazards Program of the U.S. Geological Survey, Reston, VA. He returned to UAF in 2012 as Dean of the Graduate School and Vice President Academic of University of the Arctic. John is best known for discoveries concerning mixing and degassing of magmas, leadership in scientific drilling, and advocacy of international collaboration in natural hazards. In 2015, this advocacy was formally recognized when he received the Sergey Soloviev Medal from the European Geosciences Union. He is now Professor Emeritus at UAF, based at the International Arctic Research Center.

Alexey Kiryukhin Education: 1974–1979: St. Petersburg Mining Institute (Candidate of Science, Hydrology, 1984); 1977–1981: St. Petersburg University, Mathematics; 1993: Institute of Earth Crust, SB RAS, Doctor of Science, Hydrogeology and Professor in 2001. Positions: Institute of Volcanology (IV) Far East Branch Russia Academy of Sciences, Engineer, Staff Scientist of the Geothermal and Geochemistry Department (1979–1995), Deputy Director (Science) (1996–2004), Chief Scientist, Head of the Laboratory of the Heat and Mass Transfer (2004–); 1991, 1998, 2001–2002: Geological Scientist, Earth Sciences Division, Lawrence Berkeley National Laboratory (LBNL). Research: Focused on heat and mass transfer, geofluids mechanics studies in volcanic areas. Developed methods of analysis for conditions of hydrothermal systems for exploitation. Proposed geomechanical models of magma injections beneath active volcanoes and methods of identification of their magma feeding systems based on seismic data. Established a relationship between strong earthquakes and thermohydrodynamic perturbations in hydrothermal systems. Demonstrated the CO_2 gas-lifting mechanism of geyser activity. Piip Award 2010 (volcanology and seismology) of the FEB RAS for "Modeling and Experimental Study of Heat and Mass Transfer Processes in Volcanic Areas".

Silvio Mollo has served as Professor of Petrology at the Department of Earth Sciences of Sapienza University of Rome in Italy since 2015. He studied at the Department of Sciences of the University of Roma Tre, where he concluded his Ph.D. in 2008. Afterward, Dr. Silvio Mollo was employed as a researcher at the National Institute of Volcanology and Geophysics in Rome until 2015. His research concerns experimental petrology, magma dynamics related to fractional crystallization, assimilation and mixing, trace element partitioning, isotope geochemistry, equilibrium and disequilibrium cation exchanges between crystals and melts, thermobarometric and hygrometric modeling, and the physicochemical properties of rocks and fluids.

Noriyoshi Tsuchiya has served as Professor at the Graduate School of Environmental Studies, Tohoku University, since 2004. He studied at the Department of Resources Engineering at Tohoku University and became a research assistant after having finished his Ph.D. in 1988. He participated in the 31st Japanese Antarctic Research Expedition (1989–1990) and in the 35th Japanese Antarctic Research Expedition (199–1994). In 2009–2010, Professor N. Tsuchiya was expedition leader of Earth Scientific Research Party of the Sør Rondane Mountains, East Antarctica, as part of the 51st Japanese Antarctic Research Expedition. He studies water–rock interactions involving geothermal and metamorphic fluids under various crustal situations. He also carries out hydrothermal experiments for understanding the interactions between rock and subcritical and supercritical geofluids. He is now a board member of the Japanese Association for Petroleum Technology, Geothermal Research Society of Japan, Society of Resource Geology, and Japanese Association of Mineralogical Sciences. His current fields of study are geothermal and resource geology as well as environmental geology.

Marlène Villeneuve is a rock engineer with experience with tunneling, geomechanics, and drilling in a variety of settings ranging from the European Alps to New Zealand volcanoes as well as to deep geothermal systems in New Zealand and Iceland. She received her education at Queen's University, Canada, where she was awarded her PhD in Geological Engineering in 2008. She was Associate Professor at Canterbury University, New Zealand, becoming Chair of Subsurface Engineering at Montanuniversität Leoben, Austria, in 2020. Her particular focus is on exploring the nexus between geological history, properties, and rock mass behavior. She uses engineering tools to help geologists describe and understand geological processes while also using geological tools to help engineers better predict rock mass behavior for engineering projects. This has allowed Dr. Villeneuve to work from the tops of 2700 m tall volcanoes to 2500 m deep tunnels and everywhere in between.

Editorial

Exploring and Modeling the Magma–Hydrothermal Regime

John Eichelberger [1],*, Alexey Kiryukhin [2], Silvio Mollo [3], Noriyoshi Tsuchiya [4] and Marlène Villeneuve [5]

1. International Arctic Research Center, University of Alaska Fairbanks, Fairbanks, AK 99775, USA
2. Institute of Volcanology and Seismology, FEB RAS, 683006 Petropavlovsk-Kamchatsky, Russia; avkiryukhin2@mail.ru
3. Department of Earth Sciences, Sapienza University of Rome, 00185 Roma, Italy; silvio.mollo@uniroma1.it
4. Graduate School of Environmental Studies, Geomaterial and Energy Lab., Tohoku University, Sendai 980-8579, Japan; noriyoshi.tsuchiya.e6@tohoku.ac.jp
5. Montanu Niversität Leoben, Erzherzog Johann-Straße 3, A-8700 Leoben, Austria; marlene.villeneuve@unileoben.ac.at
* Correspondence: jceichelberger@alaska.edu; Tel.: +1-907-888-0204

Received: 16 June 2020; Accepted: 16 June 2020; Published: 18 June 2020

Abstract: This special issue comprises 12 papers from authors in 10 countries with new insights on the close coupling between magma as an energy and fluid source with hydrothermal systems as a primary control of magmatic behavior. Data and interpretation are provided on the rise of magma through a hydrothermal system, the relative timing of magmatic and hydrothermal events, the temporal evolution of supercritical aqueous fluids associated with ore formation, the magmatic and meteoric contributions of water to the systems, the big picture for the highly active Krafla Caldera, Iceland, as well as the implications of results from drilling at Krafla concerning the magma–hydrothermal boundary. Some of the more provocative concepts are that magma can intrude a hydrothermal system silently, that coplanar and coeval seismic events signal "magma fracking" beneath active volcanoes, that intrusive accumulations may far outlast volcanism, that arid climate favors formation of large magma chambers, and that even relatively dry rhyolite magma can convect rapidly and so lack a crystallizing mush roof. A shared theme is that hydrothermal and magmatic reservoirs need to be treated as a single system.

Keywords: magma–hydrothermal; geothermal energy; volcanology; magma convection; heat transport; gas and fluid geochemistry; phreatic eruption; volcano monitoring; geophysical imaging; drilling

1. Introduction

The purpose of this special issue is to report new results bearing on the close connection between magma and hydrothermal reservoirs. The editors' intent is igniting broader discussion and a more holistic view wherein magma and hydrothermal regimes are viewed as an integrated system of mass and heat transport from Earth's mantle to the upper crust, surface, and atmosphere. A challenge is that people use different kinds of data to understand limited portions and/or processes of these huge systems, in parts of which the fluid phase is silicate melt and in another, cooler part, aqueous liquid. Throughout there can exist a water-dominated vapor phase. Geologists have long studied the dead exposed remnants of magma–hydrothermal systems. Microanalytical techniques, including chemistry of fluid inclusions and radiometric dating of individual minerals and zones within them have produced great advances in understanding their evolution. Volcanologists are largely surface-bound, and rely on remote geophysical and geochemical sensing from surface networks, space, or the atmosphere between, as well as analyses of rock and gas samples from eruptions. As an example of resulting differences, those who study volcanism are more convinced that magmas convect than those who study cold exposed igneous intrusion. This may be just a matter of stage. Exposed plutons are the end product of magma–hydrothermal activity whereas volcanoes erupt only magmas that are in a relatively early stage of crystallization and hence are mobile.

The greater disconnect is between magmatic and geothermal investigations. The former is usually undertaken for basic science or for mitigating the disaster risk from eruptions, the latter for the economic incentive of electric power and/or heat production. The data acquired and the models for interpreting data pertain on the one hand to mass and heat transport in a multi-phase suspension dominated by silicate melt and on the other to porous flow of aqueous fluid through a solid medium of variable physical properties. Given the economic incentive, geothermal data in the private sector are often not as accessible as data from magmatic systems, typically produced by academic or government scientists. However, accidental geothermal drilling encounters with magma are bringing these communities together. Magma is no longer an abstract concept to geothermalists, and those who theorize about magma must deal with the subsolidus complexities above it and accept the realities revealed by samples of magma quenched at depth.

Until recent drilling encounters with magma, a magma chamber was essentially a black box to scientists. Mass and heat output could reasonably be inferred by observations within hydrothermal systems or at the surface, plus volume measurements of volcanic deposits. Input from below to the box is more difficult to estimate, though CO_2 emission rate and surface deformation are helpful.

It is not too extreme to assert that hydrothermal circulation largely controls what happens inside the magma box [1,2]. This is because advective heat transport in the hydrothermal reservoir is orders of magnitude faster than simple conduction without hydrothermal circulation. The rate of release of thermal energy from the magma is controlled by such parameters as porosity, permeability, and water supply in the overlying rock.

2. Phreatomagmatic Activity

We can begin at the surface where the clearest evidence of direct magma–hydrothermal interaction are steam-rich explosions, a result of which is depicted on the volume cover at Krafla Caldera, Iceland. In 1976, phreatic activity and a seismic crisis at Soufriere de Guadeloupe, French West Indies attracted world attention when a volcanologist said that the volcano would explode like an atomic bomb. About 75,000 people were evacuated from the volcano's vicinity, an especially great misfortune for those who were farmers. No such magmatic eruption followed, and this experience demonstrated two things: (1) disaster mitigation efforts can do more harm than good; and (2) careful monitoring and understanding of such restless systems are required. Tamburello et al. [3] report a state-of-the-art analysis of current increased restlessness at Soufriere. They provide data and interpretation that was not possible in 1976 and argue that the current activity is due to magma at 6–7 km depth rather than rise of new magma into the summit dome. Probably the failed eruption of Soufriere in 1976 influenced many volcanologists to think that > 2 months of phreatic activity at Mount St Helens, USA, in 1980 would fizzle out as well. Instead, a catastrophic magmatic eruption followed.

However, even phreatic eruptions, though generally smaller than magmatic ones, can be deadly. Fumarolic manifestations attract tourists and explosions are notoriously hard the forecast. In Japan, the eruption of Mount Ontake on 27 September 2014 killed 63 climbers. Tajima et al. [4] were fortunate to be able to obtain a time-series of fluid chemistry and temperature data leading up to a magmatic hydrothermal eruption of Ebinokogen Ioyama on 18 April 2018. There were no injuries, but as with Mount Ontake the summit fumaroles of Ebinokogen Ioyama attract hikers. They found that the hydrothermal fluids prior to eruption contained both magmatic and meteoric components, as well as evidence that the magmatic input required about 2 months to arrive in the shallow system. It appears that an increase in gas release from the magma at depth provided a kick that induced top-down flashing to steam in multiple small shallow hydrothermal reservoirs, producing the explosion.

The most recent magma–hydrothermal event that made world news was the explosion at a popular tourist destination, Wakaari/White Island Volcano, New Zealand on 9 December 2019. Forty-seven tourists and their guides were in the crater at the time of the eruption, of whom 21 died and 26 were injured, most of the latter seriously. Again, there is the deadly combination of a tourist attraction with a high level of baseline fumarolic and seismic activity (Figure 1), so it is difficult to distinguish definite precursory signals. A heightened alert level had been announced.

Figure 1. A tour group explores the fumarole field in the crater of Whakaari/White Island Volcano, New Zealand. Photo credit: Patjo/Shutterstock.com.

White Island Volcano (Whakaari in Te Reo Māori) is a composite, mostly submerged basaltic andesite to dacite volcano characterized by frequent small eruptions and extensive fumarolic activity. Jolly et al. [5] describe how magma rose intact through the hydrothermal system and reached the surface in 2012. The ascent is surmised to have begun after ash venting on 2 September and lava was first observed at the surface in late November, all rather quietly. This is contrary to expectations and to the premonitory seismic events of late 2019. Also remarkable is the gradual increase in frequency of tremor spectral lines during the rise of magma, perhaps due to shortening of the empty portion of the resonating conduit ahead of the magma.

Kennedy et al. [6] report on rock mechanical and geochemical investigations of ballistics from White Island explosions, which they infer to represent the material that normally occupies the upper conduit between eruptions. Cracks and clasts are sealed with alunite and anhydrite. This hydrothermally altered material exhibits highly pressure-dependent permeability in contrast to fresh rocks of similar porosity. They conclude that this interplay between pressure and permeability results in rapidly varying hydrothermal fluid and magmatic gas flow up the conduit.

Not all the surface hydrothermal manifestations of magma at depth are driven by water. Uzon Caldera in Kamchatka, Russia formed at about 40 ka. Within it is the Valley of Geysers, a famous tourist attraction. The most recent evidence of magmatic activity was detected by InSAR as 15 cm of uplift during 1999–2003. This deformation may have caused a debris flow in 2007 that dammed a river and drowned many of the geysers. The new magma would have also introduced new CO_2

into the hydrothermal system, with the result that some geyser activity switched to sub-boiling CO_2 bubble-lift (what water well drillers would call airlift) activity, rather than geysering due to boiling of water. Kiryukhin and Karpov [7] speculate that an abundance of CO_2 in the hydrothermal system could have unexpected consequences, such as diatreme formation.

3. Relationships between Magmatic Processes and Hydrothermal Evolution

Moving discussion deeper into magma–hydrothermal systems, Kiryukhin et al. [8] used a computer algorithm, Frac Digger, to pick seismic events that are coincident in time and align in a single plane. They interpret these as signaling "magma fracking". They occur in the upper 10 km of crust beneath the volcano and enhance hydrothermal activity as expressed in temperature and CO_2 emission.

Agroli et al. [9] investigated the Erdenet Cu–Mo Deposit, in Mongolia. The ore deposits formed in association with emplacement and cooling of pulses of granodioritic magma. Their emphasis is on the role of the supercritical fluids that were responsible for transport of copper and molybdenum and are also of much current interest as a greatly enhanced geothermal power source. They use a variety of geobarometric and geothermometric techniques, especially Ti-in-quartz geothermometry, coupled with fluid inclusion studies to track the evolution of the system as it cooled. They infer the system began with magma emplacement at about 200 MPa and 750 C. Veins with euhedral quartz started to develop with cooling to near the granodiorite solidus and this sealed fractures, resulting in a rise in fluid pressure and precipitation of ores. They make the interesting point that fluid pressure likely switches from lithostatic to hydrostatic as temperature falls below the brittle-ductile transition. The fluid is then sub-critical and only quartz is deposited.

Stearns et al. [10] provide a wealth of geochronological, apparent temperature, and geochemical data from a magma–hydrothermal system in Utah, USA that is exposed through a reconstructed depth range of >11 km. What volcanologists especially will find notable is that activity spanned more than 10 Ma, with volcanism ending midway through this time. Nevertheless, plutons continued to grow incrementally, accompanied by pulses of hydrothermal activity. Although at the conclusion of magmatic activity the intrusive and extrusive portions were comparable in volume, the intrusive volume accumulated much more slowly.

Bartley et al. [11] present both geologic and petrologic evidence that the rise of granitic magmas is an important means of water and heat transport in the crust, an upward flowing supercritical aqueduct. They envision that in general granitic plutons grow incrementally from the top down. Because water is contained primarily in solution in the melt phase, virtually all is released during crystallization, carrying a substantial portion of the magma's thermal energy upward through its older but cogenetic roof and into the suprajacent hydrothermal system, modifying both on the way. Using a novel diagram displaying enthalpy of the system vs. water content of magma, they show that the rise of supercritical magmatic water can remelt previously emplaced granite even though the latter is well below the solidus.

Working in the opposite, descending direction, Glazner [12] presents the provocative idea that climate exerts a strong control on the development of magma bodies. He derives a dimensionless number, M, that shows as a function of volume rate of accretion of incremental intrusions and the difference between heat flow into the base of the plutonic box vs. out through the top, whether a magma body will grow or the plutonic accumulation will remain as separate, mostly solid units. Because heat output is largely controlled by the vigor of hydrothermal circulation and this in turn by water supply, the growth of very large magma chambers may be favored by an arid climate.

4. Geophysical and Drilling Insights from the Krafla Magma–Hydrothermal System, Iceland

Lastly, we come to perhaps the most drilled and geophysically, geochemically, and volcanologically probed magma–hydrothermal system, Krafla Caldera, Iceland (Figure 2a). Arnason [13] provides an up-to-date synthesis of our understanding of the Krafla system. This is a must-read for anyone interested in this remarkable volcano growing astride the American–European Plate boundary, by an author who has spent much of his career studying Krafla.

Eichelberger [14] takes lessons from lava lake drilling and new insights, centered on Krafla, from accidental geothermal drilling encounters with silicic magma to further develop the concept of a conductive, magma–hydrothermal boundary (MHB) between magma and the hydrothermal system. MHB can move downward through thermal cracking at the top of MHB or upward by melting at its base. A conclusion, based on drilling-produced evidence of near-liquidus magma in contact with partial melting at the base of MHB, suggests that Krafla's rhyolite body is substantial and convecting. This conflicts with some concepts of magma chambers and much conventional wisdom about Krafla. This and other concepts of magma–hydrothermal systems can be rigorously tested by intentional scientific drilling through hydrothermal systems to magma. Such drilling could also lead to a quantum leap in production geothermal energy (Figure 2b) and reliable eruption forecasts.

Figure 2. (**a**) Viti Crater, Krafla, Iceland, looking south across a portion of the caldera. The crater formed in 1724 during the Myvatn Fires, which was a rifting event with basalt fissure eruptions. Most of the Viti ejecta are from the hydrothermal system, but minor components from rhyolitic and basaltic magma are present. The rhyolite matches magma encountered by geothermal drilling 500 m to the west (S.M. Rooyakkers, personal communication, 2020). Apparently basaltic magma intruded a quiescent rhyolitic body beneath the central caldera, and then together interacted with and erupted through the hydrothermal system. Photo credit: Yevhenii Chulovskyi/Shutterstock.com. (**b**) Flow test of Well KJ-36, Krafla Caldera, Iceland on 12 December 2007. Just two super-hot wells such as this near-magma borehole and IDDP-1, which penetrated rhyolite magma [14], could power Landsvirkjun's 60 MWe power plant at Krafla, if the engineering challenges can be overcome. The well slants under Viti Crater and reaches a vertical depth of about 2500 m approximately 500 m north of the crater. Photo credit: Ásgrímur Gudmundsson.

5. Conclusions

The editors are grateful to the many authors, from ten countries, who contributed to this volume. We hope this will stimulate more investigation, including scientific drilling, and discussion of the magma–hydrothermal connection. Much remains to be done to combine and reconcile the insights of volcanologists, igneous petrologists who study plutons, and geothermalists, to understand the magma–hydrothermal regime as a single system.

Funding: This research received no external funding.

Acknowledgments: We thank Richard Li of MDPI/*Geosciences* for suggesting the magma–hydrothermal theme for a special issue.

Conflicts of Interest: The authors declare no conflict of interest.

References

1. Carrigan, C.R. Biot number and thermos bottle effect: Implications for magma-chamber convection. *Geology* **1988**, *16*, 771–774. [CrossRef]
2. Hawkesworth, C.J.; Blake, S.; Evans, P.; Hughes, R.; Macdonald, R.; Thomas, L.E.; Turner, S.P.; Zellmer, G. Time scales of crystal fractionation in magma chambers—Integrating physical, isotopic and geochemical perspectives. *J. Petrol.* **2000**, *41*, 991–1006. [CrossRef]
3. Tamburello, G.; Moune, S.; Allard, P.; Venugopal, S.; Robert, V.; Rosas-Carbajal, M.; Deroussi, S.; Kitou, G.-T.; Didier, T.; Komorowski, J.-C.; et al. Spatio-temporal relationships between fumarolic activity, hydrothermal fluid circulation and geophysical signals at an arc volcano in degassing unrest: La Soufrière of Guadeloupe (French West Indies). *Geosciences* **2019**, *9*, 480. [CrossRef]
4. Tajima, Y.; Nakada, S.; Maeno, F.; Huruzono, T.; Takahashi, M.; Inamura, A.; Matsushima, T.; Nagai, M.; Funasaki, J. Shallow magmatic hydrothermal eruption in April 2018 on Ebinokogen Ioyama Volcano in Kirishima Volcano Group, Kyushu, Japan. *Geosciences* **2020**, *10*, 183. [CrossRef]
5. Jolly, A.; Caudron, C.; Girona, T.; Christenson, B.; Carniel, R. "Silent" dome emplacement into a wet volcano: Observations from an effusive eruption at White Island (Whakaari), New Zealand in late 2012. *Geosciences* **2020**, *10*, 142. [CrossRef]
6. Kennedy, B.M.; Farquhar, A.; Hilderman, R.; Villeneuve, M.C.; Heap, M.J.; Mordensky, S.; Kilgour, G.; Jolly, A.; Christenson, B.; Reuschlé, T. Pressure controlled permeability in a conduit filled with fractured hydrothermal breccia reconstructed from ballistics from Whakaari (White Island), New Zealand. *Geosciences* **2020**, *10*, 138. [CrossRef]
7. Kiryukhin, A.V.; Karpov, G. A CO_2-driven gas lift mechanism in Geyser Cycling (Uzon Caldera, Kamchatka). *Geosciences* **2020**, *10*, 180. [CrossRef]
8. Kiryukhin, A.; Chernykh, E.; Polyakov, A.; Solomatin, A. magma fracking beneath active volcanoes based on seismic data and hydrothermal activity observations. *Geosciences* **2020**, *10*, 52. [CrossRef]
9. Agroli, G.; Okamoto, A.; Uno, M.; Tsuchiya, N. Transport and evolution of supercritical fluids during the formation of the erdenet cu–mo deposit, Mongolia. *Geosciences* **2020**, *10*, 201. [CrossRef]
10. Stearns, M.A.; Bartley, J.M.; Bowman, J.R.; Forster, C.W.; Beno, C.J.; Riddle, D.D.; Callis, S.J.; Udy, N.D. Simultaneous magmatic and hydrothermal regimes in alta–little cottonwood stocks, Utah, USA, Recorded Using Multiphase U-Pb Petrochronology. *Geosciences* **2020**, *10*, 129. [CrossRef]
11. Bartley, J.M.; Glazner, A.F.; Stearns, M.A.; Coleman, D.S. The granite aqueduct and autometamorphism of plutons. *Geosciences* **2020**, *10*, 136. [CrossRef]
12. Glazner, A.F. Climate and the development of magma chambers. *Geosciences* **2020**, *10*, 93. [CrossRef]
13. Árnason, K. New conceptual model for the magma-hydrothermal-tectonic system of krafla, NE Iceland. *Geosciences* **2020**, *10*, 34. [CrossRef]
14. Eichelberger, J. Distribution and transport of thermal energy within magma–hydrothermal systems. *Geosciences* **2020**, *10*, 212. [CrossRef]

© 2020 by the authors. Licensee MDPI, Basel, Switzerland. This article is an open access article distributed under the terms and conditions of the Creative Commons Attribution (CC BY) license (http://creativecommons.org/licenses/by/4.0/).

Article

Spatio-Temporal Relationships between Fumarolic Activity, Hydrothermal Fluid Circulation and Geophysical Signals at an Arc Volcano in Degassing Unrest: La Soufrière of Guadeloupe (French West Indies)

Giancarlo Tamburello [1,2,*], Séverine Moune [2,3,4], Patrick Allard [2,5], Swetha Venugopal [3,6], Vincent Robert [2,4], Marina Rosas-Carbajal [2], Sébastien Deroussi [2,4], Gaëtan-Thierry Kitou [2,4], Tristan Didier [2,4], Jean-Christophe Komorowski [2], François Beauducel [2,7], Jean-Bernard De Chabalier [2], Arnaud Le Marchand [2], Anne Le Friant [2], Magali Bonifacie [2,4], Céline Dessert [2,4] and Roberto Moretti [2,4]

1. Istituto Nazionale di Geofisica e Vulcanologia, sezione di Bologna, 40128 Bologna, Italy
2. Institut de Physique du Globe de Paris, Université de Paris, UMR7154 CNRS, 75005 Paris, France; moune@ipgp.fr (S.M.); pallard@ipgp.fr (P.A.); robert@ipgp.fr (V.R.); rosas@ipgp.fr (M.R.-C.); deroussi@ipgp.fr (S.D.); kitou@ipgp.fr (G.-T.K.); didier@ipgp.fr (T.D.); komorow@ipgp.fr (J.-C.K.); beauducel@ipgp.fr (F.B.); dechabal@ipgp.fr (J.-B.D.C.); arnaudl@ipgp.fr (A.L.M.); lefriant@ipgp.fr (A.L.F.); bonifaci@ipgp.fr (M.B.); dessert@ipgp.fr (C.D.); moretti@ipgp.fr (R.M.)
3. Laboratoire Magmas Volcans-OPGC, Université Clermont Auvergne, 63170 Clermont Ferrand, France; swethav@sfu.ca
4. Observatoire Volcanologique et Sismologique de Guadeloupe, Institut de Physique du Globe de Paris, 97113 Gourbeyre, France
5. Istituto Nazionale di Geofisica e Vulcanologia, sezione di Catania, 95125 Catania, Italy
6. Department of Earth Sciences, Simon Fraser University, 8888 University Drive, Vancouver, BC V5A 1S6, Canada
7. Institut des Sciences de la Terre—Institut de Recherche pour le Développement—BPPTKG—PVMBG Badan Geologi, Yogyakarta 55166, Indonesia
* Correspondence: giancarlo.tamburello@ingv.it

Received: 1 October 2019; Accepted: 9 November 2019; Published: 15 November 2019

Abstract: Over the past two decades, La Soufrière volcano in Guadeloupe has displayed a growing degassing unrest whose actual source mechanism still remains unclear. Based on new measurements of the chemistry and mass flux of fumarolic gas emissions from the volcano, here we reveal spatio-temporal variations in the degassing features that closely relate to the 3D underground circulation of fumarolic fluids, as imaged by electrical resistivity tomography, and to geodetic-seismic signals recorded over the past two decades. Discrete monthly surveys of gas plumes from the various vents on La Soufrière lava dome, performed with portable MultiGAS analyzers, reveal important differences in the chemical proportions and fluxes of H_2O, CO_2, H_2S, SO_2 and H_2, which depend on the vent location with respect to the underground circulation of fluids. In particular, the main central vents, though directly connected to the volcano conduit and preferentially surveyed in past decades, display much higher CO_2/SO_2 and H_2S/SO_2 ratios than peripheral gas emissions, reflecting greater SO_2 scrubbing in the boiling hydrothermal water at 80–100 m depth. Gas fluxes demonstrate an increased bulk degassing of the volcano over the past 10 years, but also a recent spatial shift in fumarolic degassing intensity from the center of the lava dome towards its SE–NE sector and the Breislack fracture. Such a spatial shift is in agreement with both extensometric and seismic evidence of fault widening in this sector due to slow gravitational sliding of the southern dome sector. Our study thus provides an improved framework to monitor and interpret the evolution of gas emissions from La Soufrière in the future and to better forecast hazards from this dangerous andesitic volcano.

Keywords: la soufrière; guadeloupe; volcanic gas; volcanic unrest; hydrothermal gas; multigas; extensometry

1. Introduction

Active arc volcanoes that erupt with evolved magmas commonly display long periods (decades to centuries) of dormancy between their magmatic eruptions. However, these quiescent periods are often characterized by hydrothermal manifestations (fumaroles, boiling pools, thermal springs, etc.) that attest to a persistent heat and gas supply from a magmatic source at depth. While interacting with shallow groundwater and host rocks, this continued magmatic supply sustains a hydrothermal system (e.g., [1])) and generates acid hydrothermal fluids that promote intense alteration of the host rocks. This could also lead to mechanical weakening of the volcanic edifices and their potential collapse. Moreover, an increasing gas supply from depth or/and gradual self-sealing of a constantly fed hydrothermal system are processes that can lead to shallow overpressurization followed by violent eruption of a closed-conduit volcano (e.g., [1–3]). As recorded by the cases of Ontake in Japan in 2014 [4] and Tongariro in New Zealand in 2012 [5], even purely phreatic eruptions at volcanoes displaying prolonged unrest can involve major hazards and risks owing to their sudden and often unpredictable onset [6]. Therefore, monitoring the spatial distribution and temporal evolution of hydrothermal manifestations at such volcanoes, in combination with geophysical surveys, is crucial to detect and interpret precursors of either non-magmatic explosive activity or, instead, a magmatic eruption.

Compositional changes in fumarolic exhalations have in fact been recognized as signals of unrest or even precursors of several eruptions at dormant volcanoes (e.g., [2,7–10]). However, unequivocal interpretation of these chemical changes is often challenging, owing to the complexity of chemical reactions and buffering effects involved in water-gas-rock interactions (e.g., [1]). Additional insight into the significance of chemical changes can be obtained by quantifying the emission rate of fumarolic gases (e.g., [11,12]). Nevertheless, flux measurements of fumarolic manifestations are not straightforward, for two main reasons: (i) hydrothermal gas emissions are often too weak to generate a sizable volcanic plume and, hence, to allow gas flux quantification with remote sensing or airborne measuring tools; and (ii) low-temperature (<100–300 °C) fumarolic gases generally contain little SO_2, which impedes the use of UV sensing tools commonly applied to quantify SO_2-rich gas emissions from hotter vents or erupting volcanoes (e.g., [13]). The prevalent sulfur species in hydrothermal gas emissions is usually H_2S (e.g., [8]) whose remote detection still remains challenging (e.g., [14]). Alternative approaches targeting the fumarolic fluxes of H_2O and CO_2 were tested with success on a few volcanic sites: these include ground-based eddy gas profiling [15], CO_2 plume imaging with tunable diode laser spectroscopy [16] or differential absorption lidar [17]. However, these methods are relatively difficult to carry out in the field, because they require gentle volcano topography, easy access to fumarolic fields and favorable weather conditions. Moreover, the high abundances of H_2O and CO_2 in the atmospheric background require substantial volcanic enrichments of these two components to allow reliable quantification.

Recently, Allard et al. [12] demonstrated that the flux of H_2S-bearing fumarolic gases at dormant volcanoes in hydrothermal activity can reliably be determined from in situ gas plume concentration profiles measured with a portable Multi-component Gas Analysing System (MultiGAS). MultiGAS is a light and compact device composed of an infrared spectrometer and electrochemical sensors (plus air temperature, atmospheric pressure, and relative humidity sensors) that allows simultaneous and analysis of H_2O, CO_2, SO_2, H_2S and H_2 mixing ratios in air-diluted volcanic plumes (e.g., [18–20]). The MultiGAS can be used for discrete measurements but also for permanent gas surveys such are currently operated on several volcanoes worldwide (e.g., [2,21]).

Using a MultiGAS instrument, Allard et al. [12] determined the mass output of fumarolic gas emissions from La Soufrière volcano in Guadeloupe, an andesitic volcanic dome of the Lesser Antilles arc that threatens several tens of thousands of people and has raised recent concern due to increasing degassing unrest over the past two decades ([22] and references therein). By coupling the measured gas composition to the horizontal and vertical distribution of H_2S in the plume cross-sections and then scaling to the wind speed measured at the vent, Allard et al. [12] found that the total gas flux from La Soufrière had increased by a factor of ~3 in 2012 compared to a first measurement made in 2006 (using one simple H_2S electrochemical sensor). Because isotopic tracers demonstrate a persistent supply of magma-derived volatiles and heat to La Soufrière hydrothermal system (e.g., [12,23] and references therein; [24–27]), such an increase in the gas discharge, together with other phenomena recorded by the local volcano Observatory [28], have raised concerns about the evolution of current unrest and, therefore, deserves further investigations.

Here we report on new and more extensive measurements of the gas compositions and fluxes of fumarolic emissions from the La Soufrière volcano in 2016–2017, while degassing unrest at its summit displayed continued expansion of thermal (>50 °C) ground areas, new fumaroles and the reactivation of several other vents since July 2014. For these measurements, we used a novel instrumental geometry consisting of an array of three MultiGAS devices operated simultaneously at various heights (1 to 3 m) above the ground. This instrumental array allowed us to accurately determine the fumarolic gases emitted from the different active vents of La Soufrière lava dome, as well as gas emission rates from the three major degassing vents. Moreover, in addition to be compared with previous gas data, our results are interpreted in light of three complementary information: (i) recent electrical conductivity imaging of the underground circulation of hydrothermal fluids inside the lava dome [29], (ii) ground deformation of the lava dome as revealed by extensometric survey of its main fractures since 1995 [22,30], and (iii) seismic data recorded by the Observatoire Volcanologique et Sismologique de Guadeloupe (OSVG-IPGP). With such an approach, our study provides unprecedented insight into the spatio-temporal relationships between the evolution of surface activity (fumarolic degassing and the propagation of ground thermal anomalies) and underground phenomena (hydrothermal circulation, near-field ground deformation and seismicity) at La Soufrière. Thus, we define an improved framework to interpret temporal changes in gas emissions from the volcano during its present unrest phase and in the future, which also bears broader implications for the monitoring of dormant active volcanoes in hydrothermal unrest elsewhere.

2. Volcanological Background and Recent Activity

La Soufrière of Guadeloupe is one of the most active volcanoes of the Lesser Antilles island arc (Figure 1a). It is the youngest eruptive centre of a larger composite volcano, the Grande Découverte massif, located in the southern part of Basse-Terre Island [6,31]. Its main feature is a ~0.05 km^3 andesitic lava dome (1427 m a.s.l.; Figure 1b,d), cut by numerous fractures (Figure 1b,c), that was emplaced during the last major magmatic eruption in 1530 AD [32].

Since then, intense hydrothermal activity has persisted on and around this lava dome (fumaroles, solfataras, hot grounds, thermal springs), culminating in six phreatic eruptions of varying intensity in 1690, 1797–1798, 1812, 1836–1837, 1956, and 1976–1977 [6,33]. The last phreatic events in 1976–1977 were accompanied by an intense seismic crisis (Figure 2) and resulted in a four-month evacuation of 75,000 people from the surroundings [33,34]. Exegesis and re-analysis of historical chronicles have shown that the three most violent phreatic eruptions in 1797–1798, 1836–1837 and 1976–1977 generated small-volume but hazardous manifestations, such as laterally-directed explosions, cold dilute turbulent pyroclastic density currents (PDCs) with runouts ≤1.5–2 km, and rockslides and/or debris avalanches resulting from partial collapses of the dome [6,29,35].

After the 1976–1977 eruption, La Soufrière has become increasingly studied and monitored with multi-parameter networks managed by the Observatoire Volcanologique et Sismologique de Guadeloupe (OVSG-IPGP). Monitoring data are provided by continuous seismic, global navigation satellite system (GNSS), and meteorological networks, as well as periodic extensometric surveys of the evolution of the lava dome fractures [30] and routine sampling/analysis of the fumarolic gases and thermal springs (OVSG-IPGP, 1999–2017; [10,12,36]. Recorded data are processed and available online on the WebObs internal server [37,38]. Seismic and GNSS data are distributed on the public access Volobsis server of the IPGP Dater Center (http://volobsis.ipgp.fr/). Seismic data and petro-geochemical investigations indicate that the volcano is fed by an andesitic magma reservoir located at about 6–7 km depth beneath the summit [32,33,39–41]. According to C, He and Cl isotopic ratios of the hydrothermal fluids, persistent degassing of this magma reservoir continuously supplies fluids and heat to a shallower hydrothermal system [10,12,24–28,42,43].

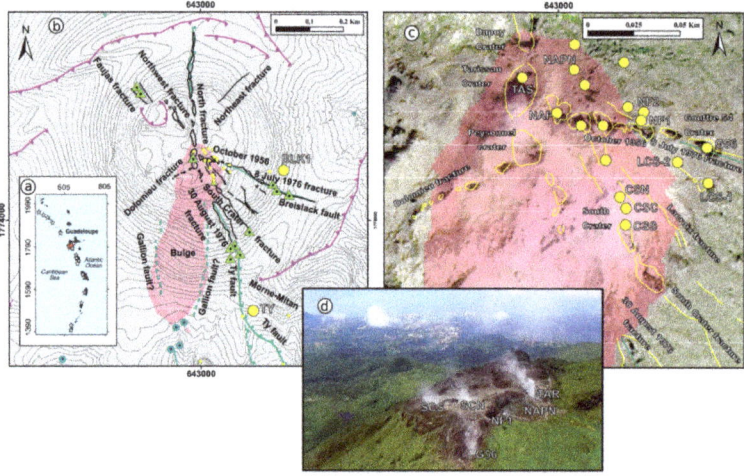

Figure 1. (**a**) Map of the Lesser Antilles volcanic arc; (**b**) main structures and manifestations of the hydrothermal system at La Soufrière volcano showing the locations of thermal springs (blue circles), sites of exurgence of pressurized hydrothermal fluid (green triangles) during 1797–98, 1836 and 1976–77 phreatic eruptions [6,29], the main faults (blue-green), historical eruptive fractures and craters (black) and collapse structures (purple triangle on trace), the region of highest electrical conductivity (>1 S/m, light purple area) determined by Rosas-Carbajal et al. [29], active fumaroles (small and big yellow circles), 10 m DEM from GeoEye image, Latitude Geosystems; (**c**) 1 m resolution orthophoto (GeoEye) of the lava dome showing the main active fumaroles of the summit area (yellow circles) TAS: Tarissan crater; NAPN: Napoléon Nord; NAP: Napoléon 1 NPE1: Napoleon Est 1; NPE2: Napoléon Est 2; CS: Cratère Sud, which is divided into northern (CSN), central (CSC) and southern (CSS) vents; G56: Gouffre-56; LCS: Lacroix Supérieur, that is divided into LCS-1 and LCS-2; BLK1: Breislack fumarole; TY: Morne-Mitan fumarole along the Ty fault; (**d**) aerial photo of La Soufrière lava dome (October 2016) showing vegetation impacted by prolonged H_2S- and HCl-rich acid gas emissions, photo taken by A. Anglade, OVSG-IPGP, with a drone from OBSERA and with permission by the Parc National de Guadeloupe).

After the 1976–1977 events, fumarolic degassing around and on top of the lava dome strongly diminished to ultimately disappear by 1984, synchronously with a gradual decline of seismic activity (Figure 2a). A period of deep rest extended up to 1992, with only minimal fumarolic activity persisting along the volcano-tectonic Ty fault at the SW base of the dome [6,36,44,45]. However, towards the end of 1991, the Tarade (TA) thermal spring that was dry since 1977 reactivated and a new thermal spring (Pas du Roy, PR) appeared at the southern base of the dome (Figure 1b). Then, in May 1992 a new phase of degassing unrest began on top of the lava dome in concomitance with a renewed increase of shallow seismicity (Figure 2). For about five years, increasing fumarolic degassing remained focused at the Cratère Sud (Figure 1b–d), but subsequently extended along the Napoleon fracture (March 1997) then at Gouffre Tarissan pit crater (late 1998). Gas emissions from both Cratère Sud and Tarissan gradually became intense enough to generate a permanent volcanic plume, visible from several kilometers distance always on clear days. Moreover, in 1998 fumarolic exhalations from both craters started to become extremely acidic (mean pH of 0.95 ± 0.64 at CSN between 1998–2001) due to their marked enrichment in chlorine. A surficial boiling acid lake (pH= −0.8 to 1.6 and T °C = 88.8 ± 8.6) formed and persisted from April 1997 to about December 2004 in Cratère Sud (see Figure S4 in Rosas-Carbajal et al. [29]), while another boiling acid lake (pH= −1.3 to 0.8 and T °C = 78.3 to 100.3) developed since late 2001 at the bottom of the 30 m wide and 80–100 m deep Gouffre Tarissan (OVSG-IPGP, 1999–2018). Up to now, this acid lake in Tarissan has remained active and been regularly sampled by OVSG [6,10,28]. Acid gas emissions from both vents over the past two decades have considerably impacted the vegetation growing on the downwind summit and W-SW flanks of the dome (Figure 1b).

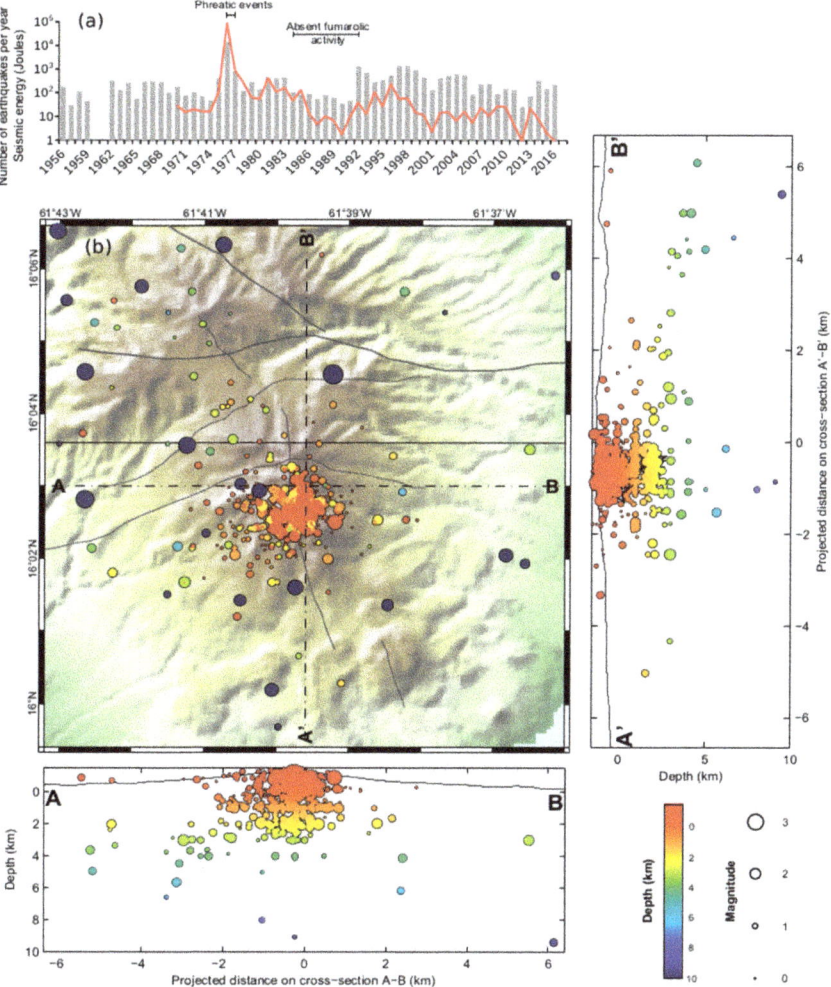

Figure 2. (**a**) Number of earthquakes per year (grey bars) and associated released seismic energy (red line). (**b**) Spatial distribution, magnitude (circles size) and depth (false color scale) of 1799 seismic events recorded in 2007–2017 with longitudinal and latitudinal projections of their hypocenters (OVSG-IPGP, 1999–2018).

Since 1998 volcanic seismicity at La Soufrière has fluctuated in terms of number of events and released energy (Figure 2a). The prevalent seismicity was characterized by numerous swarms of volcanic earthquakes of very low magnitude (dominantly Md ≤ 1), lasting over periods of a few days to a few weeks, most of which originated within 1–4 km depth right below the summit lava dome (Figure 2b). Since 1992, however, 20 felt volcanic earthquakes were recorded, among which 5 in 2013, 1 in 2014, 1 February (1st) 2018, 1 in April (16) 2018, and 2 on 27 April 2018 when the strongest (M = 4.1) seismic event occurred in 42 years [22,28]. Concurrently, the degassing unrest phase has continued to evolve, with gradual reactivation of other vents and the recent opening of new vents. While fumarolic activity remained dominantly concentrated at Cratère Sud and Gouffre Tarissan until 2007, more recently fumarolic degassing resumed or initiated at several other vents, gradually migrating from the lower eastern flanks of the volcano, along the main October 1956–8 July 1976 fracture and the Breislack fault system (Figure 1b,c), up to the summit area. In particular, fumarolic degassing progressively renewed and increased at y Gouffre-56 (G56, the site of 1956 phreatic eruption [46]), then propagated to the nearby Lacroix Supérieure fumarole (LCS- and LCS-2) in December 2011 and, in October 2013, further east in the vicinity of the Breislack crater (Figure 1b, site of the 1797, 1812 and 1836 phreatic eruptions [6]). On top of the lava dome, a new fumarole (NAPN) appeared in July 2014 north of the Napoleon fracture that was reactivated in numerous sites along its 200 m stretch. Between 8 and 10 February 2016, two new vents (NPE1 and NPE2) opened further northeast (Figure 1c). Increasing fumarolic activity in that sector of the lava dome and along the Breislack fault system over the past decade also coincides with enhanced shallow seismicity beneath that part of the volcano (Figure 2b). The Breislack fracture system was involved in all of the phreatic eruptions of La Soufrière since 1797 [6,29,35].

A regular survey of La Soufrière fumarolic gases has long been conducted at the Cratère Sud Central (CSC), the only vent accessible for gas sampling [10,12,43]. The new fumarole opened in 2014 north of Napoleon crater (NAPN; Figure 1c) has been sampled occasionally [28]. Otherwise, chemical data available for the other fumarolic vents unaccessible to gas sampling were obtained recently from in-situ MultiGAS survey (March 2006 and March 2012 campaigns [12]). In this study we report on gas compositions for all the fumarolic vents (see Figure 1c) that were active on top of the lava dome in 2016–2017, as well as gas fluxes from the three major emitting vents (Tarissan, Cratère Sud and Gouffre-56). We also report data on fumarolic degassing at the base of the lava dome, Morne Mitam site, along the Ty fault. Our chemical and flux results are then compared to previously obtained data [10,12,43]. From here in, we use the acronyms listed in the caption of Figure 1 for the fumarolic vents, whereas full names will be maintained for major structural elements such as faults, fractures and craters.

3. Methodologies for Gas Measurements and Extensometric Survey

In-situ field determination of fumarolic gas compositions at La Soufrière was performed on several occasions in 2016 (May 10–12, June 18, September 6, and December 9–10) and 2017 (March 26 and October 31) by using MultiGAS. Two types of devices were operated. The first one, built in Palermo University, consists of a Gascard IR spectrometer for CO_2 determination (calibration range: 0–3000 ppmv; accuracy: ± 2%; resolution: 0.8 ppmv) and of City Technology electrochemical sensors for SO_2 (sensor type 3ST/F; calibration range: 0–200 ppm, accuracy: ± 2%, resolution: 0.1 ppmv), H_2S (sensor type 2E; range: 0–100 ppm, accuracy: ±5%, resolution: 0.7 ppmv) and H_2 (sensor type EZT3HYT; range: 0–200 ppm, accuracy: ± 2%, resolution: 0.5 ppmv), all connected at a Campbell Scientific CR6 datalogger. The second one, built at Simon Fraser University (SFU, Canada), consists of two Alphasense non-dispersive IR (NDIR) solid-state detectors for CO_2 (range: 0–5000 ppmv and 0–5%; accuracy: ± 1 and ± 1.5%; resolution: 0.1 and 1 ppm, respectively) and Alphasense electrochemical sensors for SO_2 (sensor type EZT3ST/F; calibration range: 0–2000 ppm; accuracy: 1%; resolution: 0.5 ppm) and H_2S (sensor type EZT3H; range: 0–2000 ppm; accuracy: 1%; resolution: 0.25 ppm). Each instrument also includes a relative humidity sensor (Galltec, range: 0–100% Rh, accuracy: ± 2%), coupled with a temperature sensor (range: −30–70 °C, resolution: 0.01 °C) and atmospheric pressure (P_{atm}) sensor, all fixed externally, that permit to determine the concentration of water vapor (ppmv). The latter was obtained by combining the sensor readings of Rh% and P_{atm} following the procedure described in Moussallam et al. [47]. H_2O determination with these external sensors allowed us to circumvent the potential influence of steam condensation in the MultiGAS inlet tubing and, therefore, to avoid underestimating the measured water/gas ratios. Unfortunately, this setting has been used only during the May 2016 campaign when the SFU-type MultiGAS was available.

Prior to field measurements, all sensors were calibrated in laboratory using target gases of known concentration. The different MultiGAS instruments have been tested in the field by measuring the same gas (the inlets were close together), and the results show a good comparison between the measured concentrations. Time-averaged gas compositions (H_2O, CO_2, SO_2, H_2S and H_2) were determined at all fumarolic vents during stationary measurements lasting a few minutes in the downwind air-diluted plumes. Post-processing of data was performed using the RatioCalc software [48]. Differences in the response time of the sensors were taken into account from lag times in correlation analysis of the various time series, and potential interference between SO_2 and H_2S sensors was calibrated and corrected. CO_2 and H_2O contents were corrected for the ambient air composition, measured in the clean atmosphere outside the volcanic plumes. Figure 3 shows an illustration of MultiGAS recordings at the three main craters.

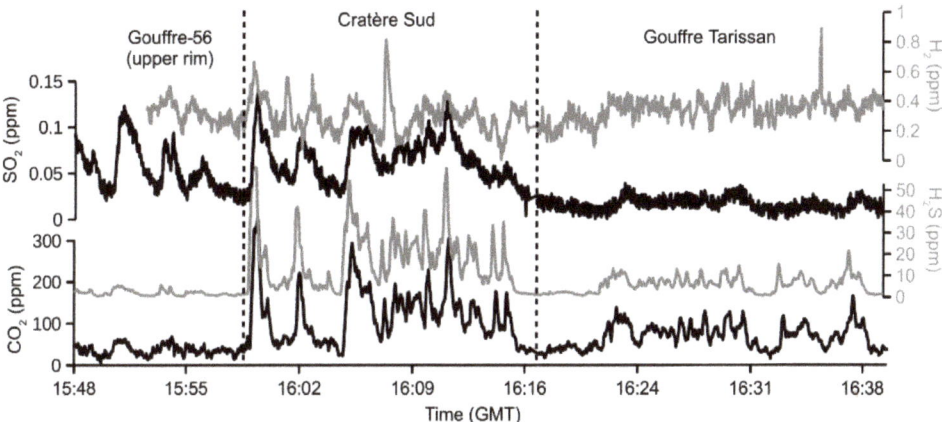

Figure 3. Example of MultiGAS recording of CO_2, SO_2, H_2S and H_2 co-variations in gas emissions from Gouffre-56 (G56), Cratère Sud (CS) and Gouffre Tarissan (TAS) fumaroles during walking profiles across the plumes in May 2016.

Fumarolic gas fluxes were determined on six occasions in 2016 (March, May and December) and 2017 (March and October) at the three main degassing craters: CS, TAS and G56 (Figure 1c). As detailed below, gas fluxes were derived by scaling the integrated amount of each gas species in the air-diluted plume cross-sections to the wind speed measured during the gas survey with a hand-held anemometer. During most of our measurements, the volcanic gas plumes were flattened to the ground by relatively strong trade winds (4–16 m·s^{-1}) and, according to both field observations and video-camera footage, had a maximum height of ca. 3 m above the ground at each of our measuring sites. The horizontal and vertical distributions of gas species in the plume cross-sections were measured during walking traverses orthogonal to the plume transport direction, a few meters downwind to the vents, with hand-held GPS in one-second track mode. One key improvement in our measurements, with respect to previous studies, has been the simultaneous use of multiple MultiGAS devices allowing to record gas concentrations in real time at different heights within the plume cross-sections. On 12 May 2016 we simultaneously operated three MultiGAS (two UniPA-type and one SFU-type) instruments whose gas inlets were fixed at 1, 2 and 3 m height above the ground along one single (4-m long) vertical pole held vertically (Figure 4a).

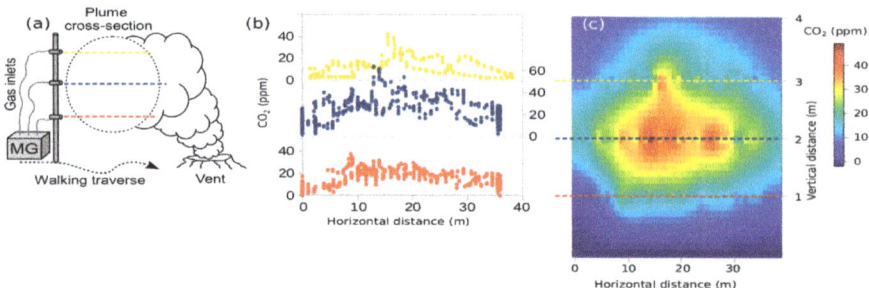

Figure 4. (a) MultiGAS setting for measuring gas concentrations between 1 to 3 m above the ground along a traverse perpendicular to the plume's main axis and resulting (b) CO_2 concentration profiles crossing G56 volcanic plume acquired on 10 May 2016. (c) Calculated fitting surface interpolating CO_2 concentrations.

On 9 December 2016, we used the same configuration but with only two UniPA-type MultiGAS instruments whose inlets were positioned at ~1.7 and ~3 m heights. On 26 March 2017, multiple (2–3) plume transects at the three main vents were made using one single UniPA-type MultiGAS but sequentially switching the gas inlet position from ~0.8 to ~2.7 m height above the ground. The maximum height of ~3 m for our measurements corresponds to the observed maximum elevation of the volcanic plumes at each measuring site, recurrently grounded by strong winds, and allowed us to accurately quantify gas distributions within the entire, or almost entire, plume cross-sections.

While H_2S, SO_2 and most of the measured H_2 are purely volcanic-hydrothermal in origin, H_2O and CO_2 concentrations need to be corrected for the atmospheric background in which these two species are abundant. Accordingly, each of our walking profiles (Figure 4b) was initiated in pure atmosphere, upwind of the volcanic plumes, in order to characterize and then subtract the ambient air composition from our recorded data. The concentrations of purely volcanic CO_2 and H_2S retrieved in each plume cross-section were interpolated with a smoothing interpolant function with an $R^2 > 0.95$. The interpolating surfaces (Figure 4c) were integrated to obtain integrated concentration amounts (ICAs, in ppm·m^2) in each plume cross-section (Table 2). Our series of traverses for each vent reveal a similar CO_2 and H_2S distribution pattern at the different heights, with maximum concentrations at half plume height (see Figure 4b), indicating steady plume structures with maximum gas density centered at between ~1.5 and 2 m above the ground (Figure 4b). In agreement with both field visual observations and video-camera footage, volcanic gas concentrations were considered below the detection limit at above the maximum plume elevation of 3–3.5 m. Both our MultiGAS procedure and the field conditions thus provided us with a good coverage of the plume structures and gas emissions from the three main vents on top of La Soufrière lava dome.

Wind speed is one key parameter and a main source of error in quantifying volcanic gas fluxes. On La Soufrière, we repeatedly measured the wind speed, as well as atmospheric pressure, temperature and relative humidity, with a hand-held weather sensor device. The measurements were performed inside the gas plume in order to achieve a fair match between plume and wind speeds. In fact, these two velocities can significantly differ from one another if measured at different sites, contributing to high and usually unquantified errors [49]. Hence, the measured wind/plume speed was used (error 1σ) in our calculations of the volcanic gas fluxes (Table 2). Weather conditions varied rapidly during our measurements: sunny intervals alternated with episodes of fog and occasional rain, and trade winds blowing from the northeast varied in speed from moderate (3–5 m·s^{-1}) to strong (12–18 m·s^{-1}). Strong winds were less variable (relative standard deviation of ~30%) than moderate winds (RSD of

~60%). Despite the fact that the derived gas fluxes are affected by these changing conditions in terms of variability, our measured chemical compositions show, instead, constant values at single vents on time scales of a few days, months and even one year.

The flux of H_2S and CO_2 are thus obtained from the integrated column amounts (ICAs) of H_2S and CO_2 directly measured in each plume cross-section and then by multiplying with the average wind speed (Table 2). The fluxes of other gas species were derived by multiplying the H_2S or CO_2 fluxes by the average X_i/H_2S or X_i/CO_2 weight ratio of each gas emission (calculated from the molar ratios shown in Table 1).

We integrate here our volcanic gas observations with data from extensometric survey of the fractures of La Soufrière lava dome (Figure 1b–c) monitored since 1995 by OSVG manual recording of the width of 15 fractures performed every three months [22,30]. Each of the monitored sites is equipped with two stainless steel hooks anchored to rock on both side of the fracture. A tape-type extensometer allows operators to measure the distance in between the two hooks. The instrument consists of a steel tape, a tape tensioning apparatus and an embedded caliper. An indexing mark is used to apply tension on the tape at constant values for each measurement. The standard error ranges between 0.1 and 0.5 mm, mainly depending on contemporaneous wind conditions.

4. Results

4.1. Fumarolic Gas Compositions

Table 1 reports the H_2S-normalized molar ratios as well as the overall molar compositions of fumarolic gases from the different active vents of La Soufrière measured in 2016 and 2017. The bulk molar percentages for H_2O, CO_2, H_2S, SO_2 and H_2 are reported only when water was successfully determined using an external Rh sensor. We observe that gas compositions vary significantly as a function of the gas exit temperature but also the spatial location of the vents. Water vapour greatly prevails (~86–97%) in all gas mixtures emitted at above or near the water boiling temperature (96 °C) at ambient elevation (CS, NAPN), but is typically depleted in the 'colder' (40–60 °C) emissions (NAP1, NPE1, NPE2 and the peripheral TY and BLK1) due to shallow steam condensation in the ground. Carbon dioxide is the second most abundant component, followed by H_2S. Figure 3 illustrates clear co-variations of CO_2, H_2S, SO_2 and H_2 in cross-sections of the volcanic plumes, as recorded with MultiGAS. CO_2/H_2S ratios display a relatively restricted range (2.9–6.5) in the hottest fluids but also in cooled emissions from Gouffre-56, except for NaPN (Figure 5b), while more variable and higher values (up to 190) in the 'coldest' gas emissions (NA1, NF2, BLK1 and TY). SO_2/H_2S ratios vary by three orders of magnitude among the different fumaroles and are generally higher in emissions from the east-southeast sector of the dome (G56, NAP1, NAPN and NPE2). Finally, both H_2/H_2S and H_2/H_2O ratios tend to be about 10 times higher in cooler than in hotter gas emissions (Table 1), supporting the idea of simultaneous fractionations due to partial water condensation and sulfur loss prior to gas exit. Figure 5 provides further insight into the compositional differences and temporal evolution of La Soufrière fumaroles in 2016 and 2017 compared to previous periods.

Figure 5a shows the fumarole compositions in an H_2O-CO_2-S_{tot} ternary diagram. Displayed here are only our May 2016 gas samples in which the H_2O molar proportion was accurately determined with an external Rh sensor. The data are compared with the 2012 MultiGAS dataset [12], unpublished data for Col de l'Echelle fumarolic emissions during the 1976 eruptive crisis (P. Allard, in prep.) and high-temperature (720 °C) SO_2-rich gas collected in 1996 from extruding andesite in nearby Montserrat island [9]; the latter is taken as a reliable proxy for the andesitic magmatic end-member at La Soufrière. The diagram reveals a relatively restricted compositional domain for gas emissions from the major fumarolic vents (TAS, SCS and G56) in 2012–2017, at least in terms of H_2O/S_{tot} ratios. Instead, 'colder' gas emissions from the TY, BLK1 and NA1, as well as NPE1 and NPE2 vents, display widely different and variable H_2O/S_{tot} and CO_2/S_{tot} ratios.

Table 1. Molar gas ratios in fumarolic emissions from La Soufrière volcano measured with MultiGAS in 2016-2017. Coordinates and gas temperature ranges are given for each fumarole. Overall molar compositions are reported only when H_2O was determined.

Vent Latitude Longitude Temperature	Time (GMT)	H_2O/H_2S	CO_2/H_2S	SO_2/H_2S	H_2/H_2S	H_2/H_2O	$H_2O\%$	$CO_2\%$	$H_2S\%$	$SO_2\%$	$S_{tot}\%$	$H_2\%$
Gouffre 56 −61.66238 16.04347 nd	10/05/16 15:20	157	4.7	3.8×10^{-2}	0.04	2.4×10^{-4}	96.4	2.9	0.6	0.02	0.6	0.02
	10/05/16 15:15	157	3.7	2.6×10^{-2}	0.03	1.7×10^{-4}	97.1	2.3	0.6	0.02	0.6	0.02
	12/05/16 15:29	nd	4.5	4.0×10^{-2}	0.04	nd	nd	nd	nd	nd	nd	nd
	12/05/16 16:04	nd	3.9	nd	0.03	nd	nd	nd	nd	nd	nd	nd
	12/05/16 15:30	161	3.6	2.0×10^{-2}	0.03	1.7×10^{-4}	97.2	2.2	0.6	0.0	0.6	0.02
	18/06/16 17:22	nd	3.8	2.0×10^{-2}	0.03	nd	nd	nd	nd	nd	nd	nd
	06/09/16 16:45	nd	3.7	1.3×10^{-2}	0.7	nd	nd	nd	nd	nd	nd	nd
	09/12/16 18:36	nd	3.9	2.8×10^{-3}	0.04	nd	nd	nd	nd	nd	nd	nd
	24/03/17 19:32	nd	3.2	2.7×10^{-2}	nd	nd	nd	nd	nd	nd	nd	nd
	31/10/17 16:15	nd	1.3	nd	nd	nd	nd	nd	nd	nd	nd	nd
South Crater N −61.66286 16.04305 93.8–100.4 °C	10/05/16 16:00	106	4.1	2.1×10^{-3}	0.01	4.7×10^{-5}	95.4	3.7	0.9	0.002	0.9	0.004
	10/05/16 16:35	106	3.4	8.1×10^{-4}	0.02	1.9×10^{-4}	96.0	3.1	0.9	0.001	0.9	0.02
	12/05/16 16:20	106	3.9	3.4×10^{-3}	0.012	1.1×10^{-4}	95.6	3.5	0.9	0.003	0.9	0.01
	12/05/16 16:42	106	3.2	3.4×10^{-3}	0.012	1.1×10^{-4}	96.1	2.9	0.9	0.003	0.9	0.01
	18/06/16 16:22	nd	3.7	3.0×10^{-3}	nd	nd	nd	nd	nd	nd	nd	nd
	06/09/16 16:45	nd	2.9	nd	nd	nd	nd	nd	nd	nd	nd	nd
	09/12/16 19:30	nd	3.1	1.0×10^{-3}	0.012	nd	nd	nd	nd	nd	nd	nd
	24/03/17 20:00	nd	2.9	1.5×10^{-3}	nd	nd	nd	nd	nd	nd	nd	nd
	31/10/17 19:25	nd	0.9	nd	nd	nd	nd	nd	nd	nd	nd	nd
South Crater S −61.66281 16.04286	12/05/16 16:58	47	3.2	9.8×10^{-3}	0.007	1.6×10^{-4}	91.7	6.3	2.0	0.019	2.0	0.01
	12/05/16 16:30	47	3.3	$9.8 \cdot 10^{-3}$	0.007	$1.6 \cdot 10^{-4}$	91.6	6.4	1.9	0.019	2.0	0.01
	06/09/16 16:58	nd	2.9	$2.0 \cdot 10^{-3}$	nd	nd	nd	nd	0.0	0.000	nd	nd
	09/12/16 19:38	nd	4.0	nd	0.07	nd	nd	nd	nd	nd	nd	nd
	24/03/17 20:16	nd	2.9	$1.5 \cdot 10^{-3}$	nd	nd	nd	nd	nd	nd	nd	nd
Tarissan −61.66361 16.04373	10/05/16 16:25	70	6.5	$2.3 \cdot 10^{-3}$	0.01	$1.4 \cdot 10^{-4}$	90.3	8.4	1.3	0.003	1.3	0.013
	10/05/16 18:25	210	5.8	$3.3 \cdot 10^{-3}$	0.01	$4.8 \cdot 10^{-5}$	96.8	2.7	0.5	0.002	0.5	0.005
	10/05/16 18:30	177	5.6	nd	nd	nd	nd	nd	nd	nd	nd	nd
	12/05/16 17:31	130	5.0	nd	nd	nd	nd	nd	nd	nd	nd	nd
	12/05/16 17:34	218	5.8	2.0×10^{-3}	nd	nd	nd	nd	nd	nd	nd	nd
	09/12/16 20:03	nd	5.5	nd	nd	nd	nd	nd	nd	nd	nd	nd
	24/03/17 20:30	nd	4.7	nd	nd	nd	nd	nd	nd	nd	nd	nd
	31/10/17 20:20	nd	5.6	nd	nd	nd	nd	nd	nd	nd	nd	nd
		nd	2.4	nd	nd	nd	nd	nd	nd	nd	nd	nd
Napoleon 16.04369 −61.66327 40–60 °C	10/05/16 18:02	157	37.8	8.6×10^{-3}	0.04	2.6×10^{-4}	80.2	19.3	0.5	0.004	0.5	0.021

Table 1. Cont.

Vent Latitude Longitude Temperature	Time (GMT)	H_2O/H_2S	CO_2/H_2S	SO_2/H_2S	H_2/H_2S	H_2/H_2O	$H_2O\%$	$CO_2\%$	$H_2S\%$	$SO_2\%$	$S_{tot}\%$	$H_2\%$
Napoleon N 16.04389 −61.66316 93.5–95.5 °C	10/05/16 17:28	23	6.3	2.5×10^{-2}	0.03	1.3×10^{-3}	75.8	20.8	3.3	0.1	3.4	0.1
	18/06/16 17:44	nd	4.3	2.0×10^{-2}	nd	nd	nd	nd	nd	nd	nd	nd
	06/09/16 16:58	nd	4.5	2.8×10^{-2}	nd	nd	nd	nd	nd	nd	nd	nd
	10/12/16 20:02	nd	4.0	9.5×10^{-2}	nd	nd	nd	nd	nd	nd	nd	nd
NF1 −61.662671 16.043580 40–60 °C	10/05/16 17:44	3	5.6	8.3×10^{-4}	0.02	5.4×10^{-3}	33.6	56.2	10.1	0.008	10.1	0.2
	10/05/16 17:37	5.1	4.1	nd	0.02	3.1×10^{-3}	49.8	40.3	9.8	nd	9.8	0.2
	10/12/16 20:25	nd	4.8	nd	nd	nd	nd	nd	nd	nd	nd	nd
NF2 −61.662626 16.043669 40–60 °C	10/05/16 17:37	175	192.4	6.7×10^{-1}	0.25	1.4×10^{-3}	47.4	52.1	0.3	0.2	0.5	0.1
	10/05/16 17:43	175	123.7	2.6×10^{-1}	0.17	9.7×10^{-4}	58.3	41.2	0.3	0.1	0.4	0.1
	10/12/16 20:15	nd	123	3.3×10^{-2}	nd	nd	nd	nd	nd	nd	nd	nd
Breislack −61.66111 16.04330 40–60°C	10/05/16 14:25	458	43.8	nd	0.22	4.7×10^{-4}	91.1	8.7	0.2	nd	0.2	0.04
Ty fault −61.66188 16.03871 40–60 °C	12/05/16 20:22	147	10.7	5.0×10^{-4}	0.46	3.1×10^{-3}	92.4	6.7	0.6	3×10^{-4}	0.6	0.3
	12/05/16 20:27	147	23.7	5.0×10^{-4}	0.30	2.0×10^{-3}	85.5	13.8	0.6	3×10^{-4}	0.6	0.2
	12/05/16 20:31	147	10.7	5.0×10^{-4}	0.43	3.0×10^{-3}	92.4	6.7	0.6	3×10^{-4}	0.6	0.3

nd: not determined.

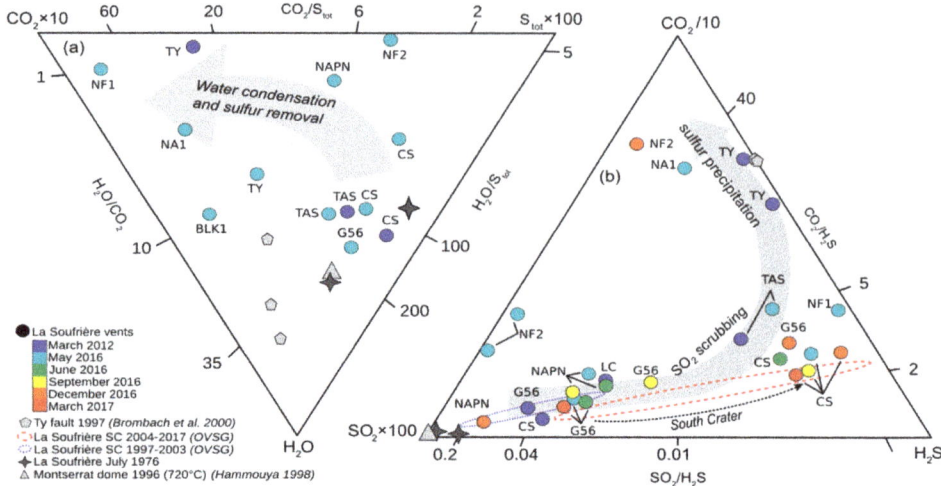

Figure 5. Gas compositions and trends in ternary diagrams (**a**) H_2O-CO_2-S_{tot} and (**b**) CO_2-SO_2-H_2S. Coloured circles represent MultiGAS-derived molar ratios of La Soufrière fumaroles in the period 2012–2017. Grey triangle: Montserrat-type magmatic end-member. Grey crosses: La Soufrère fumarolic gases during the 1976 eruptive crisis (see text). Dashed blue and red lines delineate the compositional ranges of CS fumarole obtained from direct sampling during the period 1997–2003 and 2004–2017, respectively.

Figure 5b provides further insight into the spatial and temporal variations of sulfur species in a ternary diagram CO_2-SO_2-H_2S. From 2012 to 2016, one observes that fumarolic gases from SC and to a lesser extent NAPN and G56 display large temporal variations in SO_2/H_2S ratio at relatively steady CO_2/H_2S ratio, along a trend that extends from the SO_2-rich magmatic end-member (left corner) towards strongly SO_2-depleted samples with very low SO_2/H_2S ratio. Such a trend is best explained by variable SO_2 scrubbing in the hydrothermal liquid water. Gas emissions from Tarissan are systematically impoverished in SO_2 through this process. Colder gas emissions from TY, BLK1, NA1 and generally NF1 and NF2 vents are not only strongly depleted in SO_2 but also variably impoverished in H_2S, as shown by their plot on a second trend of increasing CO_2/H_2S ratios. Such a trend strongly suggests a variable but extensive loss of H_2S in the volcanic ground prior to gas exit. Only fumarole NPE1 in May 2016 deviates from this trend, but we cannot exclude an influence of the measuring conditions.

4.2. Fumarolic Gas Fluxes

CO_2 and H_2S concentrations typically exceed tens of ppmv in the core of the volcanic plumes and progressively decrease toward the plume margins (Figure 4b–c). Single fluxes of H_2S and CO_2 at each vent were determined from the respective ICAs values and the wind speed. Total gas fluxes were then computed by scaling the overall gas composition to either the H_2S flux or the CO_2 flux. We found that total gas fluxes derived from the H_2S flux tend to be lower (by up to ~70%) than those derived from the CO_2 flux. In addition to a more conservative behavior of CO_2, compared to more reactive H_2S, such a discrepancy most likely results from the slower response of the electrochemical H_2S sensor compared to the infrared CO_2 sensor [50]; while the latter is able to detect rapid concentration changes during a plume transect, the H_2S sensor needs comparatively more time to reach a full read at each position and, therefore, tends to provide smoothed concentration profiles. Therefore, our flux calculations were safely based on single determinations of the CO_2 flux at each vent. The H_2S flux was inferred by multiplying the CO_2 flux by the H_2S/CO_2 weight ratio.

Table 2 reports the computed CO_2 and H_2S gas fluxes from La Soufrière in 2016–2017 and compares them to previous data. Note that flux data in 2006 and 2012 [12] were based on the H_2S flux measured with either MultiGAS (2012) or a specific H_2S sensor (2006) and, therefore, represent minimum figures. Fumarolic steam, calculated for all the three vents only on May 2016 (and thus not displayed in Table 2), contributes a predominant fraction (75–203 $t \cdot d^{-1}$) of the total gas flux emitted by La Soufrière volcano over the past decade, followed by CO_2 (2–18 $t \cdot d^{-1}$) and H_2S (1–4 $t \cdot d^{-1}$). The greater variability of H_2O fluxes, compared to other gas fluxes, simply reflects the larger uncertainty in H_2O determination due to both high ambient humidity on top of La Soufrière (Rh close to 100% [12]) and occasional partial steam condensation on cold surfaces at the inlet of analytical instruments. The lack of an external water sensor in our MultiGAS setting after the May 2016 campaign prevented reliable calculation of H_2O/H_2S ratio in the fumarolic emissions (Table 2). SO_2 and H_2 gas fluxes are in the order of 10^{-2} and 10^{-4} $t \cdot d^{-1}$ and contribute negligibly to the total gas output. At the TAR crater, fumarolic degassing appears relatively steady over time. Instead, since 2012, the total gas flux from SC, the historically most active vent, tended to decrease, whereas in the same time G56 displayed a noticeable flux increase accompanying its progressive reactivation since 2007 [10]. From March 2012 to October 2017, gas fluxes from G56 have varied from below detection limit to values that are comparable to those at SC and TAR. As a whole, we find that the total gas discharge from La Soufrière measured after 2016 was approximately equivalent to that determined in March 2012 (see Section 5.3).

4.3. Patterns of Fracturing and Shallow Ground Deformation

Among the numerous fractures dissecting La Soufrière lava dome (Figure 1b–c), 12 have been monitored since 1995 by OSVG using extensometers [30]. The dataset collected until 2017 reveals that some fractures displayed systematic trends in extension over the last 20 years, whereas others displayed no extension or even a contraction. Figure 6f shows the displacement vectors computed from the extensometric data set since 2007, with particular focus on deformation that affected the fractures hosting most of the fumarolic activity. The main fractures of Napoléon (NAP1, in Figure 6), Faille du 30 Août (F303, in Figure 6), Breislack (BLK1, in Figure 6), 8 July 1976 (F8J1, in Figure 6), all show an extension pattern since 1995. For these fractures, three main temporal phases of extension can be identified (Figure 6d,e): (i) a period of moderate extension from 1995 to 1999, coinciding with the initial seismogenic and degassing phase of the ongoing unrest; (ii) a period of no extension or minor contraction between 1999 and 2004; then (iii) a new period of extension, at a much higher rate, from 2004 to 2017. Phases of increased extension were previously attributed to a pressure increase upon the solid dome rocks that host the upper hydrothermal reservoir of La Soufrière [30].

Here we find that, since 2004, the largest displacement change has affected the WNW-ESE oriented Breislack fracture (Figure 1b). The maximum opening rate averages 2.49 ± 0.18 $mm \cdot yr^{-1}$ in its upper part at the Napoleon Crater (blue arrow in Figure 6f), with a total opening of 40.54 mm over the past 10 years, and 67.25 mm since June 1995. This extensional trend along the Breislack fracture coincides with the renewal (since 2007) of fumarolic degassing along this structure cutting the lava dome, especially marked at the G56 vent. Increased extension along this fault zone propagated more recently to other summit structures such as Peyssonnel Crater and Dupuy Crater, where thermal anomalies and diffuse degassing started to become observed since 2016 (OVSG-IPGP Report). Jacob et al. [30] modelled the displacements of four of the most important fractures and suggested that the main source of displacement can be accounted by a hydrothermal reservoir of ellipsoidal shape centered within the lava dome, at ~100 m depth, undertaking pressure changes.

Table 2. CO_2 and H_2S gas fluxes at La Soufrière in 2016–2017 computed from both CO_2 and H_2S ICAs and CO_2/H_2S average molar ratios (see Section 4.1). Dry gas flux is calculated as the sum of CO_2 and H_2S. The sensor heights above the ground during MultiGAS traverses and the wind speed used for gas flux calculation are indicated.

Date	Sensors Heights (m)	ICA H_2S (ppm·m^{-2})	ICA CO_2 (ppm·m^{-2})	Molar CO_2/H_2S	Plume Speed (m·s^{-1})	H_2S Flux (ton·d^{-1}) ICA H_2S	CO_2 Flux (ton·d^{-1}) ICA H_2S CO_2/H_2S	H_2S Flux (ton·d^{-1}) ICA CO_2 CO_2/H_2S	CO_2 Flux (ton·d^{-1}) ICA CO_2	Dry Flux (ton·d^{-1}) ICA H_2S	Dry Flux (ton·d^{-1}) ICA CO_2
South Crater (SCN+SCC+SCS)				daily average							
27/03/2006 Allard et al. 2014	nd	2610	nd	2.2	3.6 ± 1.8	1 ± 0.5	2.8 ± 1.4	nd	nd	3.8 ± 1.9	nd
07/03/2012 Allard et al. 2014	nd	3940	nd	2.5	7 ± 2.1	2.8 ± 0.8	9 ± 2.7	nd	nd	11.8 ± 3.6	nd
10–12/05/2016	1, 2, 3	1447	5798	3.5	6 ± 2.4	1 ± 0.4	4.7 ± 1.9	1.2 ± 0.5	5.4 ± 2.2	5.8 ± 2.4	6.6 ± 2.7
09/12/2016	1.7, 3	1047	6861	3.6	4.4 ± 2.8	0.6 ± 0.4	2.6 ± 1.7	1 ± 0.7	4.7 ± 3	3.1 ± 2	5.7 ± 3.7
24/03/2017	0.8, 2.7	2000	8139	2.9	5.4 ± 2.8	1.3 ± 0.7	4.9 ± 2.6	1.8 ± 1	6.8 ± 3.6	6.2 ± 3.3	8.6 ± 4.6
31/10/2017	0.8, 1.8	4749	7774	0.9	3.3 ± 1.5	1.9 ± 0.9	2.2 ± 1	3.5 ± 1.6	4 ± 1.9	4.1 ± 1.9	7.5 ± 3.5
Gouffre56											
10–12/05/2016	1, 2, 3	470	2930	4.1	8.2 ± 2.9	0.5 ± 0.2	2.4 ± 0.9	0.7 ± 0.3	3.7 ± 1.3	2.9 ± 1	4.4 ± 1.6
09/12/2016	1.7, 3	300	3416	3.9	14.1 ± 3.7	0.5 ± 0.2	2.6 ± 0.7	1.5 ± 0.4	7.5 ± 2	3.1 ± 0.9	9 ± 2.4
24/03/2017	0.8, 2.7	660	3300	3.2	6.6 ± 2.6	0.5 ± 0.2	2.2 ± 0.9	0.8 ± 0.4	3.4 ± 1.3	2.7 ± 1	4.2 ± 1.7
31/10/2017	0.8, 1.8	3207	7015	1.3	3.2 ± 1.4	1.2 ± 0.6	2 ± 0.9	2 ± 0.9	3.4 ± 1.5	3.3 ± 1.5	5.5 ± 2.5
Tarissan											
07/03/2012 Allard et al. 2014	nd	1360	nd	4.5	7.0 ± 2.1	1 ± 0.3	5.8 ± 1.8	nd	nd	6.8 ± 2	nd
10–12/05/2016	1, 2, 3	820	5160	5.6	9.1 ± 2.8	0.9 ± 0.3	6.5 ± 2	1 ± 0.3	7.3 ± 2.3	7.4 ± 2.3	8.3 ± 2.6
09/12/2016	1.7, 3	761	7336	4.7	4.9 ± 3	0.5 ± 0.3	2.7 ± 1.7	0.9 ± 0.6	5.6 ± 3.5	3.2 ± 2	6.5 ± 4
24/03/2017	0.8, 2.7	806	6000	5.6	4.1 ± 2.6	0.4 ± 0.3	2.9 ± 1.9	0.5 ± 0.3	3.8 ± 2.5	3.3 ± 2	4.3 ± 2.8
31/10/2017	0.8, 1.8	2468	7381	2.4	3.2 ± 2.2	1 ± 0.7	3 ± 2	1.2 ± 0.9	3.7 ± 2.6	3.9 ± 2.7	4.9 ± 3.4
Total											
27/03/2006 Allard et al. 2014		7997	nd			1 ± 0.5	2.8 ± 1.4	nd	nd	3.8 ± 1.9	nd
07/03/2012 Allard et al. 2014		7796	nd			3.8 ± 1.2	14.8 ± 4.5	nd	nd	18.6 ± 5.8	nd
10–12/05/2016		470	2930			2.4 ± 0.9	13.6 ± 4.8	2.9 ± 1.1	16.4 ± 5.7	16.1 ± 5.7	19.3 ± 6.75
09/12/2016		4167	13731			1.6 ± 0.9	7.9 ± 4	3.4 ± 1.6	17.8 ± 8.4	9.4 ± 4.8	21.2 ± 10
24/03/2017		2180	5160			2.2 ± 1.2	10 ± 5.5	3.2 ± 1.7	14 ± 7.3	12.2 ± 6.7	17.2 ± 9
31/10/2017		4035	20717			4.1 ± 2.1	7.2 ± 4	6.7 ± 3.3	11.1 ± 5.9	11.3 ± 6	17.9 ± 9.2

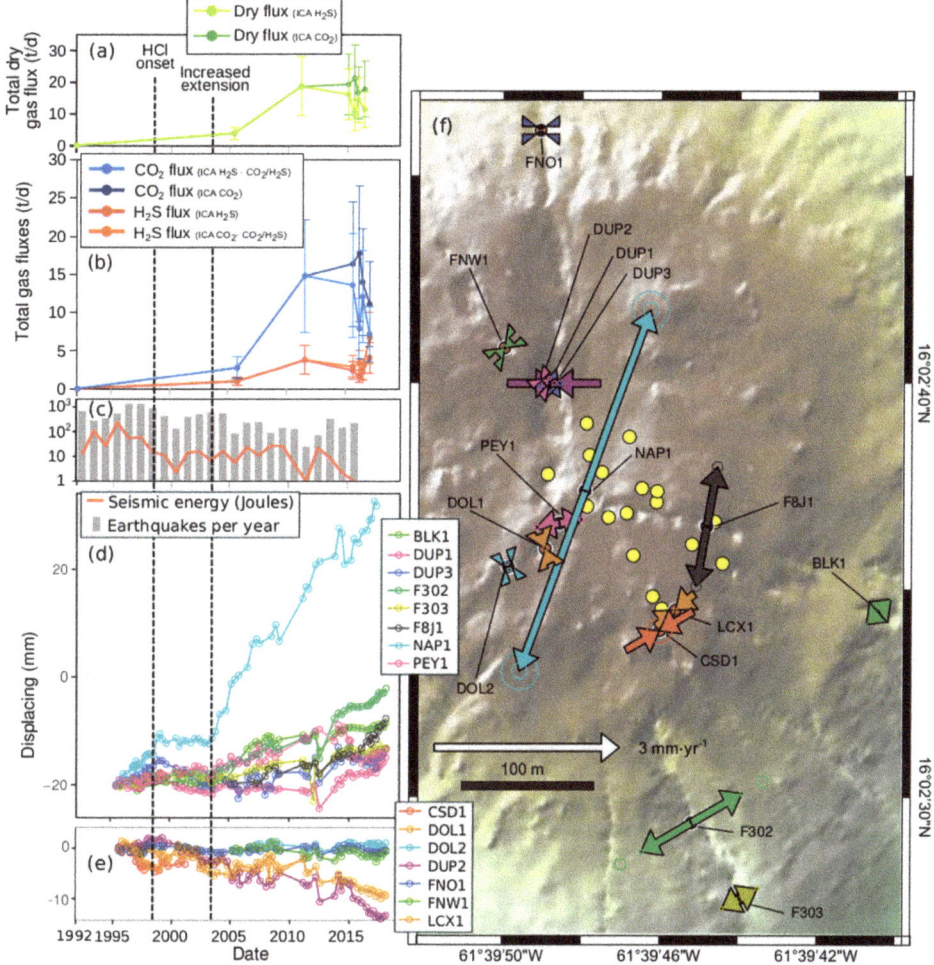

Figure 6. Compared temporal variations of fumarolic gas fluxes, seismicity and dome's fault extensometry at La Soufrière between 1992 and 2017. (**a**) Total dry gas flux and (**b**) and H_2S and CO_2 fluxes calculated from both H_2S and CO_2 ICAs (for 2006 and 2012 we used only H_2S ICA-derived values). (**c**) Seismicity and (**d**,**e**) extent of opening or closing of fractures cross-cutting La soufrière summit dome revealed by extrensometric survey. (**f**) Vectors and amplitudes of fracture width variations and location of active fumaroles (yellow circles) shown in Figure 1. BLK1: Gouffre Breislack, DOL1: Fracture Dolomieu Est, DOL2: Fracture Dolomieu Ouest, DUP1: Gouffre Dupuy Ouest, DUP2: Gouffre Dupuy Est, F302: Faille du Nord-Ouest, F303: Faille 30 Août Bas, F8J1: Faille 8 Juillet 1976, FNO1: Fente du Nord, FNW1: Faille du Nord-Ouest, LCX1: Fracture Lacroix, NAP1: Cratère Napoléon, PEY1: Gouffre Peyssonnel ([28,30], and this work).

5. Discussion

As previously mentioned, the actual source mechanism of the current degassing unrest and associated phenomena at La Soufrière still remains unclear. Hence it remains of key interest to decipher whether the current degassing unrest and associated phenomena may result from relatively shallow processes operating in the volcanic pile and the hydrothermal system or, instead, imply a deeper source mechanism involving the magma reservoir at about 5.6–8.5 km depth (e.g., [32,40,41]) or even a shallow magmatic intrusion [10,36,51]. Below we examine these possibilities in light of our results for La Soufrière gas emissions in 2016–2017 and their relations with both hydrothermal fluid circulation and geophysical signals.

5.1. 2016–2017 Gas Fluxes in the Context of La Soufrière Degassing Unrest

The magmatic origin of CO_2, S, HCl and He in La Soufrière fumarolic gases (e.g., [12] and references therein) and their increased emission from central vents of the summit lava dome since 1992 provides evidence of an enhanced release of magma-derived volatiles with respect to the 1984–1992 quiescent period (Section 2). As highlighted above and in Table 2, the first in situ measurements of fumarolic gas fluxes indicated a possible factor 3 increase in the emission rate of CO_2, H_2S and total dry gas from 2006 to 2012 [12], in broad agreement with an estimate based on thermal imaging [52]. If we consider gas fluxes derived from H_2S ICAs, our 2016–2017 data (Table 2) reveal a smaller but continuing increase of the overall gas discharge with respect to 2006. When compared to MultiGAS-based results in 2012, our data also reveal a spatial redistribution of the fumarolic gas pathways and emission rates in the shallow part of the edifice: the total gas flux is broadly modulated by increasing degassing at G56, whose magnitude has become of same order as that from TAS and CS. Moreover, because we have no flux data for the new fumaroles (e.g., NapN and NPE2) and new steaming ground that have extensively developed in the northern sector of the dome and along the Breislack fault, it is definitely possible that the overall gas flux from La Soufrière was increased in 2016–2017 compared to 2012. This would be fully coherent with the opening of new vents and the widening of fractures in the north-east sector of the lava dome since 2004, at an extension rate greater than for the 1995–2003 period. Given that fracture opening would have increased the permeability of that part of the volcanic edifice, a higher fumarolic gas flux in 2016-2017 would imply a higher fluid pressure in the hydrothermal system. This is supported by the increase in seismic energy release in 2017 then in early 2018 [22,28] and by the continued expansion of the thermal ground anomaly and degassing at the Napoleon fumarolic field.

5.2. Insight from the Fumarolic Gas Compositions

As shown by the ternary plots of Figure 5, the chemical composition of La Soufrière fumarolic gases in 2016–2017 and in previous years evidences two main trends with respect to a Montserrat-type magmatic compositional end-member and La Soufrière gases during the 1976 phreatic crisis. As discussed below, these two trends can be interpreted in terms of two main processes.

5.3. SO_2 Scrubbing in the Hydrothermal System

SO_2 scrubbing in liquid water is a common process at volcanoes displaying hydrothermal activity (e.g., [53,54]). During gas-water interactions magma-derived SO_2, which is much more soluble in liquid water than coexisting H_2S and CO_2, is efficiently removed from the gas phase through both hydrolysis and disproportionation reactions: $4SO_{2\,(g)} + 4H_2O_{(aq)} = H_2S_{(aq)} + 3H_2SO_{4\,(aq)}$ and $3SO_{2\,(g)} + 2H_2O_{(aq)} = S_{(s,l)} + 2H_2SO_{4\,(aq)}$. SO_2 scrubbing in two-phase hydrothermal systems with moderate temperature (100–300 °C), such as occurs beneath La Soufrière (e.g., [22,36,43]), thus leads to surface gas emissions essentially composed of CO_2 and H_2S besides water vapor. The observed variations of both CO_2/S_{TOT} and SO_2/H_2S ratios in La Soufrière fumaroles (Figure 5a,b) with respect to the hypothetical magmatic pole (Montserrat) typically demonstrate variable but extensive scrubbing of SO_2 in the liquid water phase of the hydrothermal system.

It is worth noting that the "hot" (~ 95 °C) NAPN fumarole keeps relatively high SO_2 proportions (Figure 4b) while being but is clearly affected by H_2S loss (Figure 4a). Around the fumarolic outlet, encrustations with multiple colored zoning can be observed, in addition to elemental sulphur. Although not analyzed yet, these encrustations are very likely determined by the precipitation of sulfide minerals, depleting the fumarolic fluid in H_2S. Reed and Palandri [55] showed that dilution and cooling of a hydrothermal fluid by cold water causes precipitation of metals into sulfide minerals, following the destabilization of chloride complexes (see run 9 by these authors). We thus suggest that NAPN fumarole is thermally and chemically buffered by near-surface interaction with very shallow cold groundwater, circulating in a very limited zone in the fumarole surroundings.

We emphasize that almost all gases were measured just at the vent exit, which excludes the possibility that higher SO_2/H_2S ratios in the peripheral emissions could result from an enhanced air oxidation of H_2S into SO_2 [56]. Such a possibility might only apply to G56 and TAS emissions that arise from a quite deep open vent and lake, respectively. Otherwise, the observed spatial contrast in SO_2/H_2S ratios between central and peripheral summit vents on La Soufrière dome strongly suggests greater gas interaction with liquid water and, thus, enhanced SO_2 scrubbing beneath the central summit vents, compared to the more peripheral vents. This chemical contrast was systematic in 2016–2017. However, we note that the extent of SO_2 scrubbing can vary over time: this is shown, for instance, by the evolution of G56 emission during 2016 or by the variations at both G56 and CSN from 2012 to 2016. On a longer-term basis, Figure 5b also reveals a potential trend of increasing SO_2 scrubbing over time at CSN when we compare the domains of variation (dotted areas) of fumarolic gases directly sampled at that vent in 2004–2017 with respect to 1997–2003.

5.4. Sulfur Precipitation in the Volcanic Ground

Gas emissions from the low-temperature and/or low-flux vents (TY, BLK1, and NPE2, NF1 and NA1) display a specific trend of H_2S depletion with respect to the hottest fluids, reflected in their much higher CO_2/H_2S ratios (Table 1 and Figure 5b). The degree of H_2S depletion depends on the exit gas temperature but can vary over time at a given site, as illustrated for instance by temporal CO_2/H_2S changes at NF2. We can discard that this pattern may result from partial H_2S oxidation in air-filled fractures, forming SO_2, since most of these low-T emissions are also depleted in total sulfur (higher CO_2/S_{TOT} ratios; Figure 5a). Metastable precipitation of elemental sulfur within the volcanic ground through the gaseous reactions $SO_{2\,(g)} + 2H_2S_{(g)} = 3S_{(native)} + 2H_2O_{(g)}$ and $2H_2S_{(g)} + O_{2\,(g)} = 2S_{(native)} + 2H_2O_{(g)}$ can deplete low-T volcanic gas in both H_2S and SO_2, but is likely to be of secondary importance at La Soufrière given the initial large predominance of H_2S over SO_2 in the hottest and SO_2-richest fumaroles (the least affected by SO_2 scrubbing). Field observations of abundant pyrite (FeS_2) in the shallow ground around the low-T vents rather suggest that H_2S depletion in these low-temperature and/or low-flux emissions mainly results from superficial gas-water-rock interactions involving ferrous iron in the wet volcanic ground. Reaction $2H_2S_{(g)} + FeO_{(rock)} + \frac{1}{2}O_2 = FeS_{2\,(rock)} + 2H_2O_{(g,l)}$ (e.g., [22,53,57,58]) well illustrates the buffering effect played by pyrite and the pyroclastic rock in the hydrothermal system. The presence of sulfur deposits at CS and other hot vents (e.g., NAPN) suggest a limited but additional effect of direct sulfur precipitation through rapid H_2S oxidation in air.

5.5. Compositional Gas Variations and Geophysical Signals

Since its marked reactivation at the onset of degassing unrest in 1992, volcanic seismicity at La Soufrière has remained essentially confined within a shallow depth interval (between 1 and 4 km below the summit; Figure 2b), with no peculiar temporal trend in the hypocenter distribution of seismic events. Together with a surface deformation field limited to the lava dome, as revealed by extensometric survey since 1995 (Figure 6d–f; see also ground deformation velocities determined by the Global Navigation Satellite System, GNSS, in [22]), this strongly suggests that relatively shallow processes, operating in a rather constant volume, have been responsible for both renewed seismicity and ground deformation over the past decades. Hydrothermal pressurization in this seismic volume could well account for the observed phenomena. In particular, the localized and minor deformations of the dome, with a fault opening rate of 1–2 mm·y^{-1}, are compatible with both hydrothermal pressure increase and local gravitational basal spreading of the southern sector of the lava dome ([30] and GNSS data in [22]). Preferential extension along the Breislack fault system in the past decade does coincide with shallow seismicity concentrating beneath this eastern sector of the lava dome (Figure 2c). Note, however, that these limited phenomena do not discard the possibility of a deeper triggering mechanism of the current unrest. In particular, it is much possible that increased degassing and geophysical signals recorded at La Soufrière since 1992 have resulted from an increased gas transfer from the crustal magma reservoir at 6–7 km depth, or even a new replenishment of that reservoir. Allard et al. [12] assessed that the fumarolic gas fluxes measured in 2012 could be accounted for by the bulk degassing of about 10^3 m$^3 \cdot$d^{-1} of parental basaltic magma feeding La Soufrière andesitic magma reservoir. Taken as representative for the degassing unrest phase from 1998 until present, such a rate would imply the cumulative degassing of ~6.5×10^6 m^3 of dense magma. Though an upper limit, this first-order estimation is intended to highlight that a magma intrusion of a few millions of cubic meters, sustaining the recent and current fumarolic gas fluxes, would have hardly escaped detection by the monitoring network of the OVSG-IPGP in terms of associated seismic and geodetic signals. Unless we imagine a series of small volume, sill-like intrusions emplaced at depth. Based on more recent gas data, Moretti et al. [22] estimated that a magmatic intrusion of 2.7×10^6 m^3 might have been responsible for the intense seismic crisis between February and late April 2018.

Above, we showed that two distinct periods can be recognized in La Soufrière degassing sequence over the past decades. We highlight here that the transition between these two periods coincides with the gradual but sustained widening of the dome fractures where new fumarolic activity has developed (NAP1, F8J1, BLK1, Figure 6d). As shown in Figures 5 and 6b, our MultiGAS measurements in 2016–2017 further verify a concomitant evolution of the fumarolic gas fluxes since 2004. However, they also reveal rapid (short-term) compositional gas changes (between March 2012 and May 2016; Figure 5), as well as at newly active vents (G56, NAP1, NAPN). We argue that such rapid gas changes do not result from deeply sourced mechanism but, instead, from fractionation processes affecting the fumarolic gases during their shallow paths. In particular, as outlined above, the quite short-lived variations of CO_2/H_2S and SO_2/H_2S ratios in various fumaroles between 2012 and 2017 but also during 2016–2017 (e.g., at G56; Figure 5) strongly support the idea of a key control of gas compositions at the different vents by the degree of SO_2 scrubbing in hydrothermal water and late-stage sulfur precipitation in the volcanic ground (Figure 5). Below we show how the variability in fumarolic gas compositions relates to the shallow circulation of hydrothermal fluids inside the lava dome.

5.6. Spatial Relationships of 2016—2017 Gas Compositions with Hydrothermal Fluid Circulation

The main fumarolic activity on La Soufrière currently extends along the upper sections of both the Breislack fault and the 30 August 1976 eruption fault (Figure 1b–c). The TAR pit crater, at the center of the dome, marks the intersection point of these two discontinuities. We present here two slices (Figure 7b,c) of the 3D electrical conductivity model obtained by Rosas-Carbajal et al. [29] for the period 2003–2011, in order to examine the relation between the pathways connecting the surface fumarolic vents to the hydrothermal reservoir and their influence upon the fumarolic gas compositions and fluxes. We have additionally superimposed the topographical survey of the cavities and pits on the summit plateau [59]; these reach depths of a few tens of meters before becoming narrow fissures or being obstructed by fallen blocks (Figure 7a). In particular, TAR crater is a 20–30 m wide and ≥100 m deep vertical pit that, since 2001, has been hosting a boiling acid pond whose level has varied between 60 and 100 m depth below the rim (average depth: 82 ± 9 m, n = 173 measurements from 26 November 2002 to 26 June 2017 [28]). In March 1993 a team of speleologists had also explored the CS pit crater [59] and found it to host an acid lake with bubbling gas at ~140 m depth (pH ~1.82, 74,060 ppm of SO_4^{2-} and 79.52 ppm of Cl$^-$; OVSG, G. Hammouya, personal communication). As previously mentioned, an acid boiling pond (pH ~−0.5, see Section 1), associated with geyser-like pulsating jets of boiling water, temporarily surfaced in CS vent as well between April 1997 and December 2004 (see Figure S4 in [29]).

One first important observation from the 3D electrical tomography of La Soufrière lava dome is that the upper boundary of the conductive hydrothermal region closely corresponds to the measured depth of the boiling acid water lake persisting in TAR pit crater and temporarily observed at the bottom of CS (Figure 7b,c). This means that the upper conductive boundary tracks the upper surface of the main body of thermal liquid water, saturated in dissolved sulfate and chlorine scrubbed from the magma-derived gas upflow. The TAR acid pond has an electrical conductivity of ~25 S/m at ambient temperature. Since its actual temperature is ~90 °C, it can be expected that the liquid saturating the extremely conductive region has an electrical conductivity of at least 50 S/m (e.g., [60]). This acidic fluid has infiltrated the pore space of the hydrothermally altered and fractured host-rock of the lava dome and likely accumulated in the resulting dome's cavities. Obviously, the potential water storage capacity of such cavities is larger at the centre of the dome than in its peripheral sectors. This inference is supported by the observations of recurrent exurgence (see Figure 1b) of hot acidic hydrothermal water, forming mud flows, during La Soufrière phreatic eruptions in 1797–1798, 1836–1837, 1956, and 1976–1977 [6,29,34].

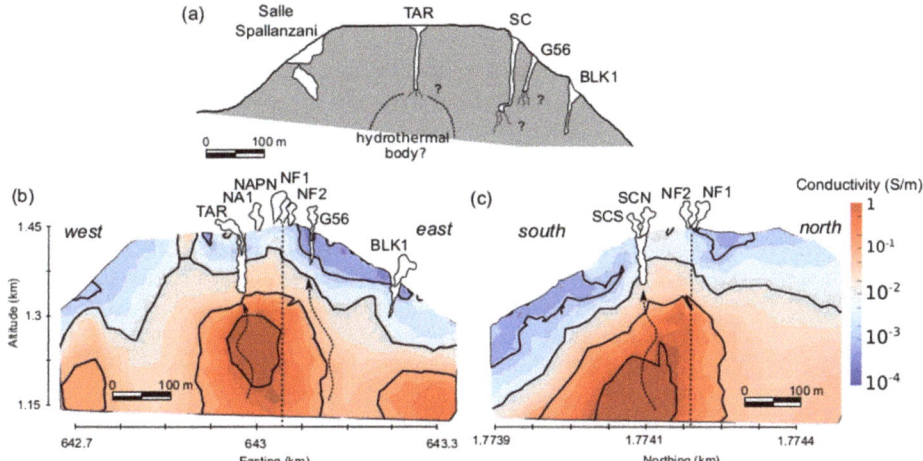

Figure 7. (**a**) Northwest-southeast cross-section of the La Soufrière lava dome with geometries of the main vents and cavities (from [59]). (**b**) West-East and (**c**) South-North slices of measured electrical conductivity with main isoconductivity contours shown for 1, 0.1, 0.001, and 0.001 S·m^{-1} from deep red to blue, respectively, as well as superimposed vents geometries of TAR, G56, BLK1 and CSS. Arrows with dashed lines represent hypothetical rising gas paths crossing the >0.1 S·m^{-1} hydrothermal water body and reaching TAR and CSS fumarolic vents. Instead, gases emitted from G56 and rising along the Breislack fault structure apparently bypass the main hydrothermal body (modified from [29]).

The highly conductive body detected below the southern part of the summit extends southwards and downwards to a few hundred meters depth below the base of the dome, but also rises vertically along a structural contact to reach the surface at the level of the Galion thermal springs (Figure 1b) [29]. This is in agreement with geochemical data that suggest a direct connection between the fluids reaching the dome summit and the fluids that ultimately feed these thermal springs [10,27]. It is noteworthy that the electrical conductivity is much smaller (<0.05 S·m^{-1}) in the upper region extending between the bottom of the central pits and the surface of the lava dome, where no stagnant liquid water occurs and where volcanic gas pathways are instead located (Figure 7). This feature, together with the high electrical conductivity values of the saturating liquid, suggests that the electrical conductivity of the host rock medium is mainly dependent on its degree of liquid saturation (e.g., [61] and references therein). This strong dependence seems to be particularly important in volcanic rocks, as shown by Ghorbani et al. (2018). The mid-range electrical conductivity values (>0.1 S·m^{-1}) found in this region most likely correspond to partially fluid-saturated rocks that have been altered by the intense hydrothermal activity and thus contain abundant clay-rich minerals which contribute to an increased bulk electrical conductivity. On the other hand, the region of the dome where G56 is located is characterized by a rather resistive massive dome rock (<0.1 S·m^{-1}). This indicates that host rocks in that sector are much less altered and not liquid-saturated (Figure 7b,c), as actually observed by speleologists who explored G56 [59].

Fumarolic gas from the CSC, the single vent regularly sampled and analyzed over the past 28 years, has long been considered to be the most representative of the mixed magmatic-hydrothermal end-member fluid at La Soufrière [10,12]. The boundaries of the liquid hydrothermal system, imaged through the 3D electrical tomography ([29] and our Figure 7b,c), as well as the depth of the explored pit craters in La Soufrière lava dome [59], offer new insight into the interplay between the shallow hydrothermal system and fumarolic degassing at the surface. Based on this imaging and our 2016–2017 gas results (Figure 5), here we demonstrate that fumarolic gases from both CS and TAR central pit craters strongly interact with the acid liquid water of the hydrothermal system, directly positioned at 80–100 m under these vents, and thereby undergo intense water/gas interactions. Gas scrubbing in fact accounts for their low SO_2 content (dissolved as SO_4^{2-} in the hydrothermal water) with respect to H_2S (much less soluble in acidic boiling water). Instead, higher SO_2/H_2S ratios in fumarolic gas emitted from the more peripheral G56 pit (Figure 1c), located on the NE border of the liquid hydrothermal system (Figure 7b,c), point to weaker (albeit variable) SO_2 scrubbing and thus more limited fractionation of sulfur species at that vent. As a matter of fact, the gas pathway beneath G56 is characterized by a weaker electrical conductivity anomaly, indicative of a small proportion of thermal liquid water in the rock column underneath. It is notable that the distinct chemical signature of fumarolic gas at that vent has persisted from 2012 through 2017, even though with temporal oscillations (Figure 5). Furthermore, our analyses of Napoleon vents (NAPN, NAP) and new peripheral fumaroles in 2016 (NPE1 and NPE2) reveal that other gas emissions markedly richer in CO_2 and SO_2 than CS fumarolic fluid are simultaneously active in the N-NE sector of the lava dome. Therefore, in contrast to previous exp1ectations, we argue that current gas emissions from peripheral vents on La Soufrière lava dome (G56 and other new smaller vents of the Napoleon fracture system) may be more closely representative of the unfractionated magmatic-hydrothermal gas end-member than emissions from the longer-lived, more central vents of the dome (SC and TAR). Based on that conclusion, we thus recommend that gas compositions and fluxes from peripheral vents of La Soufrière lava dome become carefully monitored in the future.

6. Conclusions

We present new results for the chemical composition and the mass flow rate of fumarolic gases emitted from different vents on top of La Soufrière volcano in 2016–2017, during an ongoing phase of degassing unrest that has developed since 1992. Our results reveal a wide range in gas compositions, reflecting the variable influence of shallow processes (SO_2 scrubbing in liquid water and near-surface sulfur precipitation in the volcanic ground), that closely relates to the evolution of the fumarolic activity with respect to the underground circulation of hydrothermal fluids inside the lava dome, as imaged from a recent electrical tomography [29]. Moreover, we find that the spatio-temporal evolution in degassing activity, gas compositions and gas emission rates coherently relate to the temporal deformation pattern (fracture widening/closing) of the lava dome since the onset of the current degassing unrest. When compared to previous data, fumarolic gas fluxes determined with the highest possible accuracy in 2016–2017 verify the persistency of an elevated degassing rate through the central conduits of La Soufrière. However, they also reveal a recent spatial shift in fumarolic degassing intensity towards the eastern and northern sectors of the lava dome where SW-NE fractures linked to the Breislack fault system, as well as the south-oriented Fracture du 30 Août 1976 (Figure 1c), are progressively widening, whereas other fractures tend to close or remain stable. These coupled geochemical and geophysical observations at La Soufrière provide a new framework to better elucidate the actual significance of fumarolic gas changes during the current unrest phase and in the future. At present, the available dataset for gas emissions and geophysical signals at La Soufrière do not support the hypothesis of a shallow magma intrusion as being responsible for the current unrest.

Instead, they suggest a possible increase of the magmatic gas and heat supply arising from the crustal magma reservoir emplaced at 6–7 km depth beneath the volcano. If correct, such a mechanism and its evolution over time must be carefully surveyed and quantified. The major phreatic eruptions at La Soufrière in past centuries (1797–1798, 1836–1837, 1976–1977) have often involved laterally directed explosions from pressurized regions of the shallow hydrothermal and associated hazards [6,29]. Our study illustrates the powerful potential of combining geochemical and geophysical investigations to better anticipate such events at La Soufrière, but also to interpret hydrothermal unrest at other active volcanoes in hydrothermal activity worldwide.

Author Contributions: Conceptualization, G.T., P.A., J.-C.K. and A.L.F.; Data curation, G.T., S.M., S.V., M.R.-C., S.D., F.B., J.-B.D.C., A.L.M. and M.B.; Formal analysis, G.T., S.M., S.V., M.R.-C., S.D., J.-C.K., F.B., J.-B.D.C. and A.L.M.; Funding acquisition, S.M. and J.-C.K.; Investigation, G.T., S.M., P.A., S.V., V.R., G.-T.K., T.D., J.-C.K., F.B., J.-B.D.C. and C.D.; Methodology, G.T., S.M., P.A., V.R., S.D. and J.-C.K.; Project administration, P.A., J.-C.K. and R.M.; Resources, S.M., P.A., S.D., G.-T.K., T.D., J.-C.K., F.B., A.L.M., C.D. and R.M.; Software, G.T.; Supervision, J.-C.K.; Visualization, G.T. and J.-C.K.; Writing—original draft, G.T.; Writing—review & editing, S.M., P.A., S.V., M.R.-C., J.-C.K., F.B., C.D. and R.M.

Funding: This research received no external funding.

Acknowledgments: Our study was supported by funding from a CNRS-INSU contract (Aleas-2015, S. Moune), from University of Palermo (ref. G. Tamburello) and from IPGP (equipment grant and postdoctoral contract to G. Tamburello). We thank G. Hammouya (OVSG) and O. Crispi (OVSG) for gas and acid lake sampling and measurements; D. Gibert, F. Nicollin, and participants of the ANR projects DOMOSCAN (2009-2013) and DIAPHANE (ANR-14-ce04-0001; 2014-2018) for their assistance with electrical resistivity measurements and TAR acid lake pH measurements; M. Bitetto and A. Aiuppa (University of Palermo) for their support to the development and testing of the UniPa-type MultiGAS; and Glyn Williams-Jones for help and calibration of the SFU MultiGAS. The staff of OSVG-IPGP provided us with invaluable assistance in the field, for maintenance of the stations, and with unpublished data. The Service National d'Observation en Volcanologie (SNOV) of INSU is acknowledged for recurrent funding, and the Swiss National Science Foundation and ANR Diaphane project for funding to M. Rosas-Carbajal. The authors wish to thank the three anonymous reviewers for their useful comments and suggestions upon reviewing the manuscript. Raw gas concentrations retrieved during the surveys are downloadable from the EarthChem Library (https://doi.org/10.1594/IEDA/111184).

Conflicts of Interest: The authors declare no conflict of interest.

References

1. Symonds, R.B.; Rose, W.I.; Bluth, G.J.S.; Gerlach, T.M. Volcanic-gas studies: Methods, results and applications. In *Volatiles in Magmas*; Carroll, M.R., Holloway, J.R., Eds.; Mineralogical Society of America: Chantilly, VA, USA, 1994; Volume 30, pp. 1–60.
2. De Moor, J.M.; Aiuppa, A.; Pacheco, J.; Avard, G.; Kern, C.; Liuzzo, M.; Martınez, M.; Giudice, G.; Fischer, T.P. Short-period volcanic gas precursors to phreatic eruptions: Insights from Poas Volcano, Costa Rica. *Earth Planet. Sci. Lett.* **2016**, *442*, 218–227. [CrossRef]
3. Fischer, T.P.; Sturchio, N.C.; Stix, J.; Arehart, G.B.; Counce, D.; Williams, S.N. The chemical and isotopic composition of fumarolic gases and spring discharges from Galeras Volcano, Colombia. *J. Volcanol. Geotherm. Res.* **1997**, *77*, 229–253. [CrossRef]
4. Maeno, F.; Nakada, S.; Oikawa, T.; Yoshimoto, M.; Komori, J.; Ishizuka, Y.; Takeshita, Y.; Shimano, T.; Kaneko, T.; Nagai, M. Reconstruction of a phreatic eruption on 27 September 2014 at Ontake volcano, central Japan, based on proximal pyroclastic density current and fallout deposits. *EarthPlanets Space* **2016**, *68*, 82. [CrossRef]
5. Procter, J.N.; Cronin, S.J.; Zernack, A.V.; Lube, G.; Stewart, R.B.; Nemeth, K.; Keys, H. Debris flow evolution and the activation of an explosive hydrothermal, system, Te Maari, Tongariro, New Zealand. *J. Volcanol. Geotherm. Res.* **2014**, *286*, 303–316. [CrossRef]
6. Komorowski, J.-C.; Boudon, G.; Semet, M.; Beaducel, F.; Antenor-Habazac, C.; Bazin, S.; Hammouya, G. Guadeloupe. In *Volcanic Atlas of the Lesser Antilles*; Lindsay, J.M., Robertson, R.E.A., Shepherd, J.B., Ali, S., Eds.; Seismic Research Unit, The University of the West Indies: Kingston, Jamaica, 2005; pp. 65–102.
7. Giggenbach, G.F.; Sheppard, D.S. Variations in the temperature and chemistry of White Island fumarole discharges 1972–85. *N. Z. Geol. Surv. Bull.* **1989**, *103*, 119–126.

8. Symonds, R.; Mizutani, Y.; Briggs, P. Long-term geochemical surveillance of fumaroles at Showa-Shinzan dome, Usu Volcano, Japan. *J. Volcanol. Geotherm. Res.* **1996**, *73*, 177–211. [CrossRef]
9. Hammouya, G.; Allard, P.; Jean-Baptiste, P.; Parello, F.; Semet, M.; Young, S. Pre- and syn-eruptive geochemistry of volcanic gases from Soufriere Hills of Montserrat, West Indies. *Geophys. Res. Lett.* **1998**, *25*, 3685–3688. [CrossRef]
10. Villemant, B.; Komorowski, J.C.; Dessert, C.; Michel, A.; Crispi, O.; Hammouya, G.; Beauducel, F.; De Chabalier, J.B. Evidence for a new shallow magma intrusion at La Soufrière of Guadeloupe (Lesser Antilles). Insights from long-term geochemical monitoring of halogen-rich hydrothermal fluids. *J. Volcanol. Geotherm. Res.* **2014**, *285*, 247–277. [CrossRef]
11. Aiuppa, A.; Tamburello, G.; Di Napoli, R.; Cardellini, C.; Chiodini, G.; Giudice, G.; Grassa, F.; Pedone, M. First observations of the fumarolic gas output from a restless caldera: Implications for the current period of unrest (2005–2013) at Campi Flegrei. *Geochem. Geophys. Geosyst.* **2013**, *14*, 4153–4169. [CrossRef]
12. Allard, P.; Aiuppa, A.; Beauducel, F.; Gaudin, D.; Di Napoli, R.; Calabrese, S.; Parello, F.; Crispi, O.; Hammouya, G.; Tamburello, G. Steam and gas emission rate from La Soufriere volcano, Guadeloupe (Lesser Antilles): Implications for the magmatic supply during degassing unrest. *Chem. Geol.* **2014**, *384*, 76–93. [CrossRef]
13. Oppenheimer, C. Ultraviolet sensing of volcanic sulfur emissions. *Elements* **2010**, *6*, 87–92. [CrossRef]
14. O'Dwyer, M.; Padgett, M.J.; McGonigle, A.J.S.; Oppenheimer, C.; Inguaggiato, S. Real time measurements of volcanic H_2S/SO_2 ratios by UV spectroscopy. *Geophys. Res. Lett.* **2003**, *30*. [CrossRef]
15. Todesco, M.; Chiodini, G.; Macedonio, G. Monitoring and modelling hydrothermal fluid emission at La Solfatara (Phlegrean Fiels, Italy): An interdisciplinary approach to the study of diffuse degassing. *J. Volcanol. Geotherm. Res.* **2003**, *125*, 57–79. [CrossRef]
16. Pedone, M.; Aiuppa, A.; Guidice, G.; Grassa, F.; Cardellini, C.; Chiodini, G.; Valenza, M. Volcanic CO_2 flux measurements at Campi Flegrei by tunable diode laser absorption spectroscopy. *Bull. Volcanol.* **2014**, *76*, 812. [CrossRef]
17. Aiuppa, A.; Fiorani, L.; Santoro, S.; Parracino, S.; Nuvoli, M.; Chiodini, G.; Minopoli, C.; Tamburello, G. New ground-based lidar enables volcanic CO_2 flux measurements. *Sci. Rep.* **2015**, *5*, 13614. [CrossRef]
18. Shinohara, H. A new technique to estimate volcanic gas composi- tion: Plume measurements with a portable multi-sensor system. *J. Volcanol. Geotherm. Res.* **2015**, *143*, 319–333. [CrossRef]
19. Aiuppa, A.; Federico, C.; Giudice, G.; Gurrieri, S. Chemical mapping of a fumarolic field: La Fossa Crater, Vulcano Island (Aeolian Islands, Italy). *Geophys. Res. Lett.* **2005**, *32*, 1–4. [CrossRef]
20. Aiuppa, A.; Shinohara, H.; Tamburello, G.; Giudice, G.; Liuzzo, M.; Moretti, R. Hydrogen in the gas plume of an open-vent volcano, Mount Etna, Italy. *J. Geophys. Res. Solid Earth* **2011**, *116*, 1–8. [CrossRef]
21. Aiuppa, A.; Fischer, T.P.; Plank, T.; Robidoux, P.; Di Napoli, R. Along-arc, inter-arc and arc-to-arc variations in volcanic gas CO_2/S_T ratios reveal dual source of carbon in arc volcanism. *Earth-Sci. Rev.* **2017**, *168*, 24–47. [CrossRef]
22. Moretti, R.; Komorowski, J.-C.; Ucciani, G.; Moune, S.; Jessop, D.; de Chabalier, J.-B.; Beauducel, F.; Bonifacie, M.; Burtin, A.; Vallee, M.; et al. The 2018 unrest phase at La Soufrière of Guadeloupe (French West Indies) andesitic volcano: Scrutiny of a failed but prodromal phreatic eruption. 2019; submitted.
23. Allard, P.; Dimon, B.; Morel, P. Mise en évidence d'hélium primordial dans les émissions gazeuses de la Soufrière de Guadeloupe, Petites Antilles. In Proceedings of the 9th Réunion des Sciences de la Terre, Paris, France, 17–19 March 1982; p. 454.
24. Ruzié, L.; Moreira, M.; Crispi, O. Noble gas isotopes in hydrothermal volcanic fluids of La Soufrière volcano, Guadeloupe, Lesser Antilles arc. *Chem. Geol.* **2012**, *304*, 158–165. [CrossRef]
25. Ruzié, L.; Aubaud, C.; Moreira, M.; Agrinier, P.; Dessert, C.; Gréau, C.; Crispi, O. Carbon and helium isotopes in thermal springs of La Soufrière volcano (Guadeloupe, Lesser Antilles): Implications for volcanological monitoring. *Chem. Geol.* **2013**, *359*, 70–80. [CrossRef]
26. Li, L.; Bonifacie, M.; Aubaud, C.; Crispi, O.; Dessert, C.; Agrinier, P. Chlorine isotopes of thermal springs in arc volcanoes for tracing shallow magmatic activity. *Earth Planet. Sci. Lett.* **2015**, *413*, 101–110. [CrossRef]
27. Jean-Baptiste, P.; Allard, P.; Fourré, E.; Parello, F.; Aiuppa, A. Helium isotope systematics of volcanic gases and thermal waters of Guadeloupe Island, Lesser Antilles. *J. Volcanol. Geotherm. Res.* **2014**, *283*, 66–72. [CrossRef]

28. OVSG-IPGP (1991–2018). Bilan Mensuel de l'activité Volcanique de la Soufrière de Guadeloupe et de la Sismicité Régionale. Monthly Public Report of Guadeloupe's Volcanic and Seismic Activity (1999–2018). Observatoire Volcanologique et Sismologique de Guadeloupe, Institut de Physique du Globe de Paris. Available online: http://www.ipgp.fr/fr/ovsg/bulletins-mensuels-de-lovsg (accessed on 11 November 2019).
29. Rosas-Carbajal, M.; Komorowski, J.-C.; Nicollin, F.; Gibert, D. Volcano electrical tomography unveils edifice collapse hazard linked to hydrothermal system structure and dynamics. *Sci. Rep.* **2016**, *6*, 29899. [CrossRef]
30. Jacob, T.; Beauducel, F.; Hammouya, G.; David, J.G.; Komorowski, J.C. Ten years of extensometry at Soufrière of Guadeloupe: New constraints on the hydrothermal system. In Proceedings of the Soufriere Hills Volcano—Ten Years On international workshop, Seismic Research Unit, University of West Indies, Kingston, Jamaica, 24–30 July 2005.
31. Boudon, G.; Semet, M.P.; Vincent, P.M. The evolution of la Grande Découverte (La Soufrière) volcano, Guadeloupe, F.W.I. In *Volcano Hazards: Assessment and Monitoring*; Latter, J., Ed.; Springer: Berlin, Germany, 1989; pp. 86–109.
32. Boudon, G.; Komorowski, J.-C.; Villemant, B.; Semet, M.P. A new scenario for the last magmatic eruption of La Soufrière of Guadeloupe (Lesser Antilles) in 1530 A.D: Evidence from stratigraphy radiocarbon dating and magmatic evolution of erupted products. *J. Volcanol. Geotherm. Res.* **2008**, *178*, 474–490. [CrossRef]
33. Feuillard, M.; Allègre, C.J.; Brandeis, G.; Gaulon, R.; Le Mouel, J.L.; Mercier, J.C.; Pozzi, J.P.; Semet, M.P. The 1975–1977 crisis of la Soufrière de Guadeloupe (F.W.I): A still-born magmatic eruption. *J. Volcanol. Geotherm. Res.* **1983**, *16*, 317–334. [CrossRef]
34. Le Guern, F.; Bernard, A.; Chevrier, R.M. Soufriere of Guadeloupe 1976–1977 eruption: Mass and energy transfer and volcanic health hazards. *Bull. Volcanol.* **1980**, *43*, 578–592. [CrossRef]
35. Komorowski, J.-C.; Peruzzetto, M.; Rosas-Carbajal, M.; Le Friant, A.; Mangeney, A.; Legendre, Y. New insights on flank-collapse and directed explosions hazards from hydrothermal éruptions at La Soufrière de Guadeloupe (Lesser Antilles). In Proceedings of the IAVCEI General Assembly, Portland, OR, USA, 14–18 August 2017.
36. Villemant, B.; Hammouya, G.; Michel, A.; Semet, M.P.; Komorowski, J.C.; Boudon, G.; Cheminée, J.L. The memory of volcanic waters: Shallow magma degassing revealed by halogen monitoring in thermal springs of La Soufrière volcano (Guadeloupe, Lesser Antilles). *Earth Planet. Sci. Lett.* **2005**, *237*, 710–728. [CrossRef]
37. Beauducel, F. Operational monitoring of French volcanoes: Recent advances in Guadeloupe. *Géosciences* **2006**, *2*, 64–68.
38. Beauducel, F.; Bosson, A.; Randriamora, F.; Anténor-Habazac, C.; Lemarchand, A.; J-MSaurel, A.; Nercessian, A.; Bouin, M.-P.; de Chabalier, J.-B.; Clouard, V. Recent advances in the Lesser Antilles observatories—Part 2—WEBOBS: An integrated web-based system for monitoring and networks management. In Proceedings of the European Geosciences Union General Assembly, Vienna, Austria, 2–7 May 2010.
39. Hirn, A.; Michel, B. Evidence of migration of main shocks during major seismovolcanic crises of la Soufrière (Guadeloupe, Lesser Antilles) in 1976. *J. Volcanol. Geotherm. Res.* **1979**, *6*, 295–304. [CrossRef]
40. Poussineau, S. Dynamique des Magmas Andésitiques: Approche Expérimentale et Pétrostructurale, Application à la Soufrière de Guadeloupe et à la Montagne Pelée. Ph.D. Thesis, Université d'Orléans, Orléans France, 2005.
41. Pichavant, M.; Poussineau, S.; Lesne, P.; Solaro, C.; Bourdier, J.L. Experimental Parametrization of Magma Mixing: Application to the ad 1530 Eruption of La Soufrière, Guadeloupe (Lesser Antilles). *J. Petrol.* **2018**, *59*, 257–282. [CrossRef]
42. Bernard, M.L.; Molinié, J.; Petit, R.H.; Beauducel, F.; Hammouya, G.; Marion, G. Remote and in situ plume measurements of acid gas release from La Soufrière volcano, Guadeloupe. *J. Volcanol. Geotherm. Res.* **2006**, *150*, 395–409. [CrossRef]
43. Brombach, T.; Marini, L.; Hunziker, J.C. Geochemistry of the thermal springs and fumaroles of Basse-Terre Island, Guadeloupe, Lesser Antilles. *Bull. Volcanol.* **2000**, *61*, 477–490. [CrossRef]
44. Zlotnicki, J.; Feuillard, M.; Hammouya, G. Water circulations on La Soufrière volcano inferred by self-potential surveys (Guadeloupe, Lesser Antilles). Renew of volcanic activity? *J. Geomagn. Geoelectr.* **1994**, *46–49*, 797–813. [CrossRef]

45. Allard, P.; Hammouya, G.; Parello, F. Diffuse magmatic soil degassing at Soufrière of Guadeloupe, Antilles. *C. R. Acad. Sci. Paris Earth Planet. Sci.* **1998**, *327*, 315–318.
46. Jolivet, J. La crise volcanique de 1956 à la Soufrière de la Guadeloupe. *Ann. Geophys.* **1958**, *14*, 305. [CrossRef]
47. Moussallam, Y.; Peters, N.; Masias, P.; Apaza, F.; Barnie, T.; Ian Schipper, C.; Curtis, A.; Tamburello, G.; Aiuppa, A.; Bani, P.; et al. Magmatic gas percolation through the old lava dome of El Misti volcano. *Bull. Volcanol.* **2017**, *79*, 46. [CrossRef]
48. Tamburello, G. Ratiocalc: Software for processing data from multicomponent volcanic gas analyzers. *Comput. Geosci.* **2015**, *82*, 63–67. [CrossRef]
49. McGonigle, A.J.S.; Inguaggiato, S.; Aiuppa, A.; Hayes, A.R.; Oppenheimer, C. Accurate measurement of volcanic SO_2 flux: Determination of plume transport speed and integrated SO_2 concentration with a single device. *Geochem. Geophys. Geosyst.* **2005**, *6*, Q02003. [CrossRef]
50. Roberts, T.J.; Braban, C.F.; Oppenheimer, C.; Martin, R.S.; Freshwater, R.A.; Dawson, D.H.; Griffiths, P.T.; Cox, R.A.; Saffell, J.R.; Jones, R.L. Electrochemical sensing of volcanic gases. *Chem. Geol.* **2012**, *332*, 74–91. [CrossRef]
51. Boichu, M.; Villemant, B.; Boudon, G. Degassing at La Soufrière de Guadeloupe volcano (Lesser Antilles) since the last eruptive crisis in 1975–77: Result of a shallow magma intrusion? *J. Volcanol. Geotherm. Res.* **2011**, *203*, 102–112. [CrossRef]
52. Gaudin, D.; Beauducel, F.; Coutant, O.; Delacourt, C.; Richon, P.; de Chabalier, J.-B.; Hammouya, G. Mass and heat flux balance of La Soufrière volcano (Guadeloupe) from aerial infrared thermal imaging. *J. Volcanol. Geotherm. Res.* **2016**, *320*, 107–116. [CrossRef]
53. Giggenbach, W.F. Redox processes governing the chemistry of fumarolic gas discharges from White Island, New Zeland. *Appl. Geochem.* **1987**, *2*, 143–161. [CrossRef]
54. Symonds, R.B.; Gerlach, T.M.; Reed, M.H. Magmatic gas scrubbing: Implications for volcano monitoring. *J. Volcanol. Geotherm. Res.* **2001**, *108*, 303–341. [CrossRef]
55. Reed, M.H.; Palandri, J. Sulfide mineral precipitation from hydrothermal fluids. *Rev. Mineral. Geochem.* **2006**, *61*, 609–631. [CrossRef]
56. Venturi, S.; Tassi, F.; Cabassi, J.; Vaselli, O.; Minardi, I.; Neri, S.; Caponi, C.; Capasso, G.; Di Martino, R.M.; Ricci, A.; et al. A multi-instrumental geochemical approach to assess the environmental impact of CO_2-rich gas emissions in a densely populated area: The case of Cava dei Selci (Latium, Italy). *Appl. Geochem.* **2019**, *101*, 109–126. [CrossRef]
57. Moretti, R.; Arienzo, I.; Civetta, L.; Orsi, G.; Papale, P. Multiple magma degassing sources at an explosive volcano. *Earth Planet. Sci. Lett.* **2013**, *367*, 95–104. [CrossRef]
58. Moretti, R.; De Natale, G.; Troise, C. A geochemical and geophysical reappraisal to the significance of the recent unrest at Campi Flegrei caldera (Southern Italy). *Geochem. Geophys. Geosyst.* **2017**, *18*, 1244–1269. [CrossRef]
59. Kuster, D.; Silve, V. *Guadeloupe-Canyons, Gouffres, Découverte*; Editions GAP: Rhone-Alpes, France, 1997.
60. Sen, P.N.; Goode, P.A. Influence of temperature on electrical conductivity on shaly sands. *Geophysics* **1992**, *57*, 89–96. [CrossRef]
61. Revil, A.; Finizola, A.; Piscitelli, S.; Rizzo, E.; Ricci, T.; Crespy, A.; Angeletti, B.; Balasco, M.; Cabusson Barde, S.; Bennati, L.; et al. Inner structure of La Fossa di Vulcano (Vulcano Island, southern Tyrrhenian Sea, Italy) revealed by high-resolution electric resistivity tomography coupled with self-potential, temperature, and CO_2 diffuse degassing measurements. *J. Geophys. Res. Solid Earth* **2008**, *113*, 1–21. [CrossRef]

© 2019 by the authors. Licensee MDPI, Basel, Switzerland. This article is an open access article distributed under the terms and conditions of the Creative Commons Attribution (CC BY) license (http://creativecommons.org/licenses/by/4.0/).

Article

Shallow Magmatic Hydrothermal Eruption in April 2018 on Ebinokogen Ioyama Volcano in Kirishima Volcano Group, Kyushu, Japan

Yasuhisa Tajima [1,*], Setsuya Nakada [2,3], Fukashi Maeno [3], Toshio Huruzono [4], Masaaki Takahashi [5], Akihiko Inamura [5], Takeshi Matsushima [6], Masashi Nagai [2] and Jun Funasaki [7]

1. Research & Development Center, Nippon Koei Co., Ltd., Tsukuba 300-1259, Japan
2. National Research Institute for Earth Science and Disaster Resilience, Tsukuba 305-0006, Japan; nakada@bosai.go.jp (S.N.); mnagai@bosai.go.jp (M.N.)
3. Earthquake Research Institute, The University of Tokyo, Tokyo 113-0032, Japan; fmaeno@eri.u-tokyo.ac.jp
4. Kirishima Nature Guide Club, Kobayashi 886-0004, Japan; kyushiu-huruzono@kir.biglobe.ne.jp
5. Geological Survey of Japan, National Institute of Advanced Industrial Science and Technology, Tsukuba 305-8567, Japan; mmst-takahashi@aist.go.jp (M.T.); a-inamura@aist.go.jp (A.I.)
6. Institute of Seismology and Volcanology Faculty of Science, Kyushu University, Shimabara 855-0843, Japan; takeshi_matsushima@kyudai.jp
7. Miyazaki Local Meteorological Observatory (Former), Miyazaki 880-0032, Japan; sp934m69@etude.ocn.ne.jp
* Correspondence: yasuhisa79@gmail.com; Tel.: +81-29-871-2065

Received: 16 February 2020; Accepted: 11 May 2020; Published: 14 May 2020

Abstract: The Kirishima Volcano Group is a volcanic field ideal for studying the mechanism of steam-driven eruptions because many eruptions of this type occurred in the historical era and geophysical observation networks have been installed in this volcano. We made regular geothermal observations to understand the hydrothermal activity in Ebinokogen Ioyama Volcano. Geothermal activity resumed around the Ioyama from December 2015. A steam blowout occurred in April 2017, and a hydrothermal eruption occurred in April 2018. Geothermal activity had gradually increased before these events, suggesting intrusion of the magmatic component fluids in the hydrothermal system under the volcano. The April 2018 eruption was a magmatic hydrothermal eruption caused by the injection of magmatic fluids into a very-shallow hydrothermal system as a bottom–up fluid pressurization, although juvenile materials were not identifiable. Additionally, the upwelling of mixed magma–meteoric fluids to the surface as a kick was observed just before the eruption to cause the top–down flashing of April 2018. A series of events was generated in the shallower hydrothermal regime consisting of multiple systems divided by conductive caprock layers.

Keywords: Kirishima volcano group; Ebinokogen Ioyama volcano; geothermal activity; multiple hydrothermal system; magmatic hydrothermal eruption; kick upwelling

1. Introduction

Explosive volcanic eruptions that do not contain juvenile materials have been called by several names, e.g., steam, phreatic or hydrothermal eruption. Although the terms phreatic eruption and hydrothermal eruption have similar meanings, the two are sometimes confused [1]. Hydrothermal eruption is often used in the geothermal fields in New Zealand and in the USA. The eruption termed here has both shallow and deep origins [2,3]. The general nomenclature for steam-driven eruptions is categorized into two types. First, phreatic eruption, which is an eruption that is caused by the heating and flashing of water produced when magma comes in contact with water. This results in the ejection of only country rock or overburden, i.e., no juvenile magmatic material. Second, hydrothermal eruption, which is an eruption that ejects at least some solid material and whose energy derives solely from heat loss and phase changes in a convecting hot water or steam-dominated hydrothermal system [3]. Steam-driven events are the most common phenomena in volcanic activity and are not easy to forecast, although there are many investigations on their precursor signals [1,4]. To explain steam-driven eruptions that contain non-juvenile materials, several mechanisms have been considered. The current perspective is that the eruption is triggered by the decompression driven boiling of ground hydrothermal fluids [2,3,5]. This type of mechanism is referred to as an abrupt departure from equilibrium conditions and has been experimentally produced by engineering researchers in Japan [6–8]. The 2014 eruption at Ontake Volcano in central Japan is considered one example of an abrupt departure from equilibrium conditions [9–11]. However, the true trigger of the eruption have not yet been made clear. One of the triggers of eruption in general may be pressurization of a hydrothermal system heated by input of magmatic fluids from the depths (bottom–up gas/fluid pressurization) [12]. In this case, an eruption is caused by the fluid pressure exceeding the lithostatic pressure [3,13]. Another trigger is decompression of the hydrothermal reservoir for external reasons. For example, a water level drop, drilling or a landslide can cause the flashing from the top to downward (top–down flashing) [12,14–16]. These processes are not simple, depending on the material heterogeneity under the ground, groundwater distribution and caprock development. In fact, the caprock structure is considered important for storing hydrothermal liquids. The caprock formation of silica or hydrothermal minerals is considered to be the layer that controls the sealing and release of hydrothermal fluids within different depths of a volcanic edifice e.g., [3,17–19]. In recent years, magnetotelluric (MT) surveys revealed the existence of highly conductive layers in the shallow parts of volcanic edifices, which are considered to be hydrothermally altered zones with an abundance of clay minerals such as smectite [20–24]. In Ebinokogen Ioyama Volcano, this conductive layer is located 200 to 700 m below the surface, with a pressure source below it [25]. It has been suggested that this layer is the low permeability.

Many casualties have been previously reported from steam-driven (phreatic) eruptions in Japan, including from events at Adatara Volcano in 1900, Kusatsu-Shirane Volcano in 1932, Aso Volcano in 1958 and Ontake Volcano in 2014 [10,26–28]. Meanwhile, geothermal areas—with the potential for steam-driven eruptions—often develop beneath a shallower hydrothermal system in a volcano and are beneficial as sightseeing spots. In the past, many laborers mined sulfur in these fumarolic areas in Japan. Expeditious information on the resumption of volcanic activity in such geothermal areas is crucial, and lack of timeliness has grave consequences. Sometimes, the occurrence of an explosive steam-driven eruption is preceded by a few signs. However, in other cases, an eruption may fail, even during an increase in geothermal activity. There are various precursory phenomena and event sequences that precede steam-driven eruptions: it is necessary to clarify what causes such a variety.

In Kirishima Volcano Group, steam-driven eruptions occurred at Shinmoedake in 1959, 1991–1992, 2008 and 2010 [29–31]. And they occurred at Ebinokogen Ioyama in 1768 [32]. These eruptions differed in impact and chronology. For example, at Shinmoedake Volcano, a phreatic eruption in 1959 was a strong explosive event with 8.6 million tons of ejecta in a short time and was accompanied by a blast [33]. However, the 1991–1992 eruption was extremely small, producing 360 tons of ejecta and leaving climbers unharmed on the summit crater rim [30]. Steam-driven eruptions have been dominant at Ebinokogen Ioyama Volcano over the long term, and there has been geothermal activity around there in recent times [32,34,35]. At Ebinokogen Ioyama located near tourist spots and mountain trails, volcanic activity has been observed since the 1910s. We recorded the activity just before the latest steam-driven eruption at Ebinokogen Ioyama Volcano in April 2018. Here, we discuss the sequence of geothermal activity and the possible phenomenology, especially focusing on the hydrothermal aquifer depth and the precursory signals of the eruption.

We use the above-mentioned terms of phreatic eruption and hydrothermal eruption, as suggested by Brown and Lawless [3]. Since juvenile materials were not found in the products from the 2018 Ebinokogen Ioyama eruption, hydromagmatic and phreatomagmatic eruptions are not considered here. In this study, a fumarolic jet is described as the phenomenon in which geothermal jetting occurs with jet-like roaring sounds [36], and the vent of such a fumarole is referred to as a jet fumarole vent. The term "fumarole area" refers to a place that contains many relatively weak to strong jet fumaroles. A small hole from which hot muddy hydrothermal fluids well up is called a mud pot [37], and a larger irregular-shaped pool with hydrothermal fluids is called a hydrothermal pond. The time description in this study is the Japan Standard Time (JST), UTC+9.

2. Geological Setting

The Kirishima Volcano Group is a compound group of volcanoes extending from the southeast to the northwest (Figure 1). This volcano has grown since the Kakuto ignimbrite eruption in 340 ka [38]. Shinmoedake, Nakadake, Ohachi and Takachihomine volcanoes produced frequent magmatic eruptions during the last 10 ky. Furthermore, Miike Volcano produced a Plinian eruption in 4.6 ka characterized a single massive eruption. Ebinokogen Ioyama and Ohatayama volcanoes produced dominant steam-driven eruptions during the last 10 ky e.g. [32]. The historical eruptions are recorded in Ebinokogen Ioyama, Shinmoedake and Ohachi volcanoes [29].

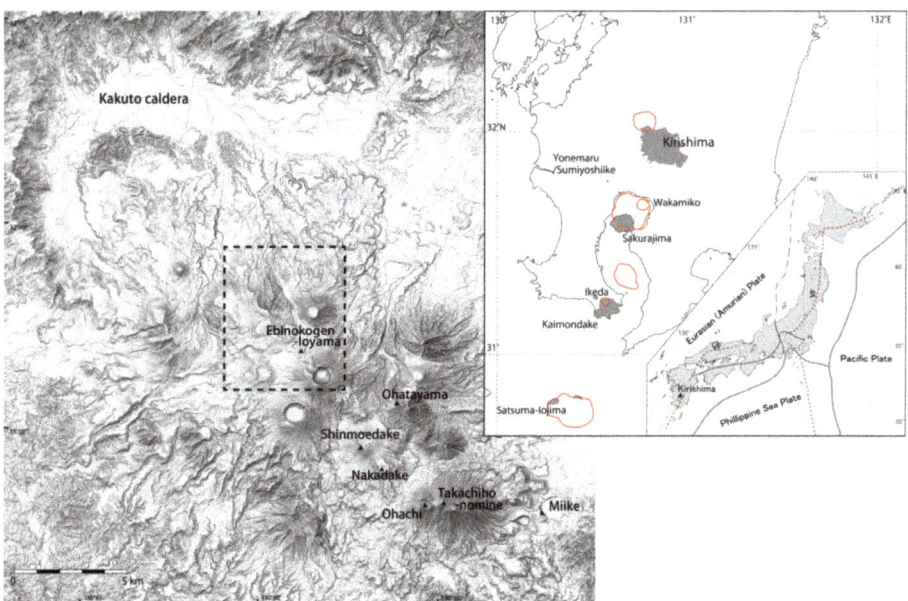

Figure 1. Geographic map of Kirishima Volcano Group. Dashed box shows the map of the study area in Figure 2. The base map with Kashimir3D [39]. In the upper right corner, the red circles denote the calderas while the gray areas indicate active volcanoes. VF: Volcanic front.

Around the Ebinokogen area, Karakunidake Volcano produced a Plinian eruption in 16.7 ka, forming the present Karakunidake cone [29,40]. Magmatic eruptions did not occur in this area between 16.7 and 9 ka. The small eruptions were repeated in the saddle place surrounded by older Karakunidake, Koshikidake and Ebinodake volcanoes (Figure 2a). The eruption of Fudoike lava flow with tephra occurred at the Fudoike Crater in 9 ka. The small amphitheater topography was formed by steam-driven eruptions with an avalanche (Karakunidake debris avalanche) in 4.3 ka. It was followed by a steam-driven eruption in 1.6 ka at Fudoike Crater, a lava flow between 0.3 and 0.4 ka at Ioyama and a steam-driven eruption in 1768 AD at Ioyama east crater. Ebinokogen Ioyama Volcano has formed from 9 ka [32].

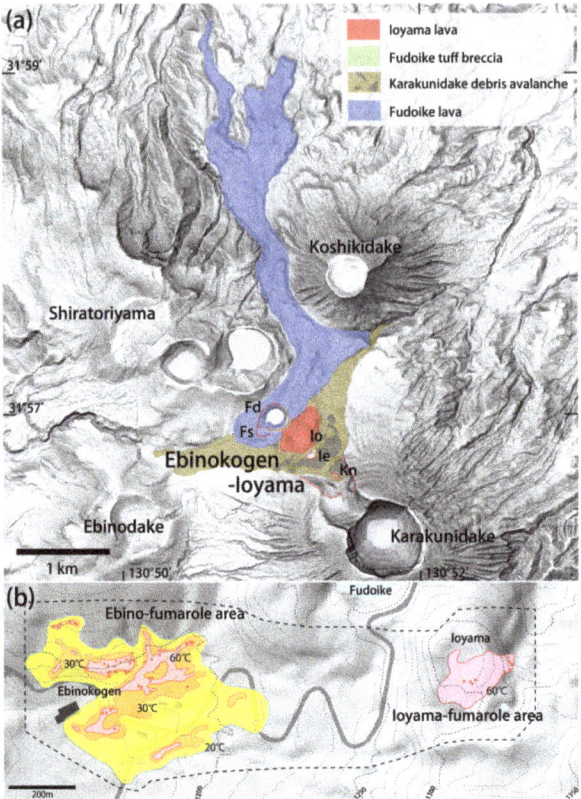

Figure 2. Shaded relief maps showing geological and geothermal features around Ebinokogen Ioyama Volcano. (**a**) Distribution of lavas and craters over 10,000 years [32]. Kn: Karakunidake north crater, Ie: Ioyama east crater, Io: Ioyama crater, Fd: Fudoike Crater, Fs: Fudoike south crater. Red circular lines denote the crater, red curves with ticks denote the amphitheater. The relief map using Kashimir3D [39]; (**b**) The distribution of Ebino-fumarole and Ioyama-fumarole areas in the 1950s [41]. Red points: active fumaroles. Numerals: the temperature of isothermals.

Geothermal activity is known to have continued at Ebinokogen Ioyama since around the 1900s (Meiji Era). The geothermal area was divided into two areas in the east as the Ioyama-fumarole area and in the west as the Ebino-fumarole area in the 1950s (Figure 2b). The center of the Ebino-fumarole area was about 1 km west from the center of the Ioyama-fumarole area [41]. Sulfur mining was worked in the Ioyama-fumarole area in the 20th Century. A fumarole temperature in the Ioyama lava was over 80 °C on 12 August 1916 [42]. Akiko Yosano, a famous poet in Japan, described fumaroles in this area in one of her poems in 1929 [43]. The temperatures of the Ioyama-fumarole area were between 96 and 120 °C in 1954 [41]. The highest temperature recorded 247 °C in the Ioyama-fumarole area in March 1975 [34]. It gradually declined in the 1990s, and its activity had stopped by around 2008 [43].

Alteration identified kaolinite, alunite and montmorillonite zones at the geothermal or alteration areas in the western part of Kirishima Volcano Group. The kaolinite zone was distributed within small area around the Ioyama, while the alunite zone surrounded the kaolinite zone. The presence of the following minerals reported alunite, kaolinite, quartz, cristobalite, halloysite, montmorillonite, tridymite and jarosite in the alteration area [44].

3. Methods

3.1. Geothermal Observations

Geothermal features were ordinarily mapped to monitor changes in geothermal activity and were used to understand precursory signals in Ebinokogen Ioyama Volcano. We monitored changes in the geothermal area from December 2015 by measuring the temperatures of the springs and fumaroles at fixed points. A thermal anomaly was observed in the area about 10 cm below the surface where the temperatures were higher than 50 °C. In the early stage of this observation, the area of geothermal anomaly was measured by trilateration using measuring tapes. The measurements were plotted on a plan map to calculate the area that was 50 °C and above in temperature. However, the anomaly area rapidly expanded, and we put colored markers at 50 °C points on the ground and took aerial photographs using drones (DJI Phantom 3 and 4). The markers, 18 cm in diameter, were visible from tens of meters above. The areas over 50 °C in temperature were plotted on the ortho-image to measure the area based on the drone observation made (Figure 3). A Google Earth [45] map was also used for the area uncovered by the drone observation. Simultaneously, we measured the temperatures and pH values of water at the spring points west of the Ioyama.

Figure 3. Drone image showing the distribution of geothermal features, ejecta and craters formed during the April 2018 eruption. Yellow areas: fumaroles, as observed on 3 February 2018 and jet fumarole vents (thin red circles). Orange circles: Eruption and hydrothermal vents appeared as Y1 (7 April 2018), S1 to S7 (19 to 20 April 2018) and W1 to W7 (20 to 21 April 2018). The distribution of ballistics is shown for over φ8 cm. Gray dash lined: ashy deposit with a thickness of 0.1 cm. Bold red lines and texts: eruption craters Y2a, Y2b, Y3, Y1, W3 and W4. Blue areas: hydrothermal ponds and streams. R is a mud pot that appeared after the eruption. The red line on the road is a crack.

3.2. Geochemical Analyses

Analyses of eruption products from the 2017 event and the 2018 eruption at Ebinokogen Ioyama were conducted using X-ray diffraction (XRD). Samples smaller than 0.063 mm were washed in extra pure water and a randomly oriented powder prepared and mounted on a glass slide according to the method of Itoh et al. [46]. Rigaku MiniFlex 600 and the PDXL2.6 software were used. Smectite and chlorite peaks at 15 Å were determined by using the peak shift of smectite to 16–18 Å after the samples were treated in ethylene glycol. Kaolinite and chlorite peaks at 7 Å were determined by utilizing the peak disappearance of chlorite after treatment in HCl solution for 2 h. Relative amount of minerals was analyzed by Reference Intensity Ratio (RIR) in the PDXL 2.6 software.

Oxygen isotope ratios of water samples were analyzed using a mass spectrometer from Thermo Fisher Scientific, Inc. (Waltham, MA, USA) (MS: Delta plus) with the automated CO_2–H_2O equilibrium method or using a Cavity Ringdown Spectrometer from Picarro, Inc.(Santa Clara, CA, USA) (CRD: L2130-i Isotopic H_2O). The precisions for both analyses are ±0.1‰.

Hydrogen isotope ratios of samples were analyzed using MS (Delta V) with automated H_2–H_2O reduction method (H-device) or by CRD. The precisions for both analyses are ±1‰.

4. Events at Ebinokogen Ioyama Volcano

4.1. Stage-1: December 2013 to December 2015

Seismic activity began in the northern area of Karakunidake in December 2013, after the sub-Plinian eruptions at Shinmoedake Volcano [47] in January 2011. A volcanic tremor event lasting 7 min was observed beneath Ebinokogen Ioyama Volcano on 20 August 2014 [48]. The seismic activity tended to gradually increase in 2015, swarmed in July, and tremor events were observed from July to October [49]. Very weak signs of unrest such as a hydrogen sulfide (H_2S) smell near the western foot of the Ioyama lava were observed from October 2015 (Figure 4). Stage 1 therefore, is the period from December 2013 to December 2015, during which there was seismic activity, but without any noticeable geothermal activity on the surface.

4.2. Stage-2: December 2015 to January 2017

On the afternoon (around 4 pm) of 14 December 2015, a hiker found a very small fumarole on the southwest rim of the Ioyama old crater at the A point in Figure 4. The location had previously been identified as the area of fumaroles [34]. The fumarole temperature was measured at 80 °C that day. On 25 and 27 December, temperatures of between 93 and 96 °C were recorded at the summit area of Ioyama. The water boiling temperature at an altitude of 1300 m is about 96 °C. Since then, the area with a thermal anomaly expanded north and east, mainly in the southern part of the Ioyama lava (Figure S1a–f). The water temperature at the spring K3 increased from about 21 °C at the end of December 2015 to about 27 °C in March 2016, synchronized with the expansion of the thermal anomaly area. There was a brief stability of the area of thermal anomaly from April to August 2016. The thermal anomaly area began to expand again from late 2016 (Figure S1g), and the expansion rate increased from early 2017. From December 2015 to January 2017, Stage-2, there was small expansion of the thermal anomaly area, but temperatures of the fumaroles remained about 96 °C (Figure 4).

4.3. Stage-3: January 2017 to February 2018

The geothermal anomaly area expansion rate increased from early 2017 and continued until early summer 2017 (Figure S1h–j). During the expansion of the thermal anomaly area, the jet fumaroles H and A were formed. The mud pot A was observed at the point A on 19 March (Figure 5a), and another pot appeared at F between 19 and 21 March (Figure 5c). Hydrothermal fluids continued to well at these pots and the jet fumarole H, 30 m east of the mud pot F, started roaring between 15 and 18 April. This activity around the fumarole H was reported in the "YamaReco" web site [51]. We surveyed the roaring jet fumarole H on 22 April 2017 (Figure 5d). Then the jet fumarole vent A with a 1.5 m diameter was first observed about 10 m away from the mud pot A on 5 May (Figure 5b). The area near the jet fumarole vent A was covered with a thin layer of light gray ash-sized altered material (ashy deposit) with block-sized altered host rock ejecta over a distance of a few meters. The surface on the southwestern area of the jet fumarole vent A became grayish-white.

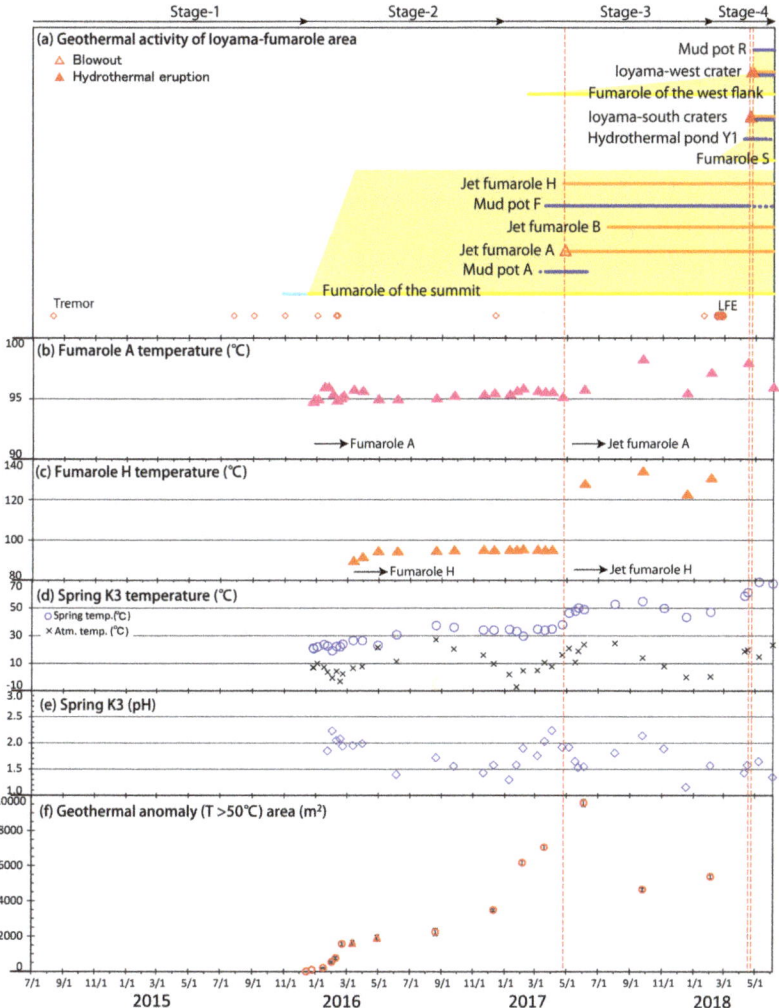

Figure 4. The sequence of geothermal activities at Ebinokogen Ioyama Volcano from July 2014 to June 2018. Geothermal locations are shown in Figure 3. (**a**) Yellow areas and bars denote the activity of Ioyama-fumarole area. Light blue line indicates a sulfur smell. Orange line denotes a jet fumarole. Blue line denotes a mud pot or hydrothermal pond. Tremor and low-frequency earthquake (LFE) based on the Japan Meteorological Agency (JMA) web site [50]; (**b**) Temperature observed at point A, before turning into a jet and the jet fumarole A; (**c**) Temperature observed at point H, before turning into a jet and the jet fumarole H; (**d**) Water temperature observed at the spring K3; (**e**) pH of water observed at the spring K3; (**f**) Geothermal anomaly (>50 °C) area in the summit of Ioyama. Triangles include larger error values.

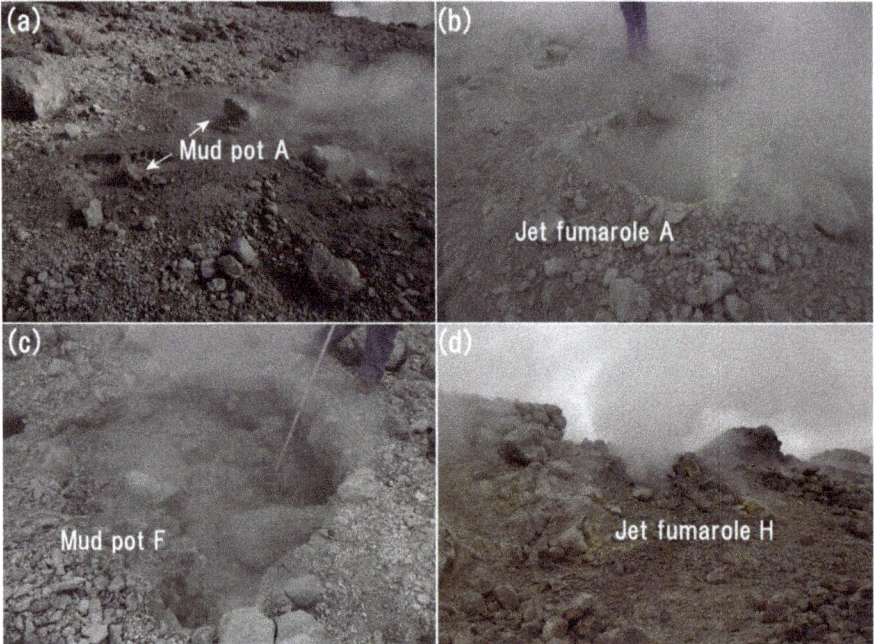

Figure 5. Details of geothermal activity at Ioyama fumarole area during Stage-3. (**a**) Mud pot A appeared just before/on 19 March 2017 (Photo on 19 March); (**b**) Jet fumarole A appeared on 26 April 2017 (Photo on 7 May); (**c**) Mud pot F appeared between 19 and 21 March 2017 (Photo on 22 April); (**d**) Jet fumarole H appeared between 15 and 18 April 2017 (Photo on 22 April).

That ashy deposit was found on green leaves and could be traced to about 200 m from the vent (Figure 6). The amounts of the deposit at two points around the jet fumarole vent A were 3.8 kg/m^2 and 3.4 kg/m^2. As the thickness of the deposit was nearly the same, the mass was estimated to be roughly 1 ton. The amount of deposit around the jet fumarole vent A coincided with the volume of the vent. We did not notice any change in the fumarole activity in this area until the morning of 26 April, according to the video footage of the Ioyama–south web camera maintained at the Japan Meteorological Agency (JMA) [50]. However, the footage showed a gray-white steam rising from the fumarole A at 11:29 am on 26 April 2017. The steam moved to the southwest from 10:20 am to 11:30 am, which coincided with the area covered with the ashy deposit. Therefore, it was concluded that the altered material ashy deposit was blown out from the jet fumarole vent A around 11:29 am on 26 April. The high rates of geothermal activity at the jet fumarole vents continued until July 2017. The temperature of the jet fumarole H was 134.2 °C on 24 September 2017. The geothermal anomaly area decreased from late summer or early autumn in 2017 (Figure S1k).

4.4. Stage-4: February 2018 to December 2018

The expansion of the geothermal anomaly area resumed around February 2018 (Figure 4). The fumarole S appeared about 30 m south of the Ioyama-fumarole area on 3 February 2018 (Figure 7a and Figure S1l). While the previous fumarole spots were named sequentially, beginning with A, this was named S.

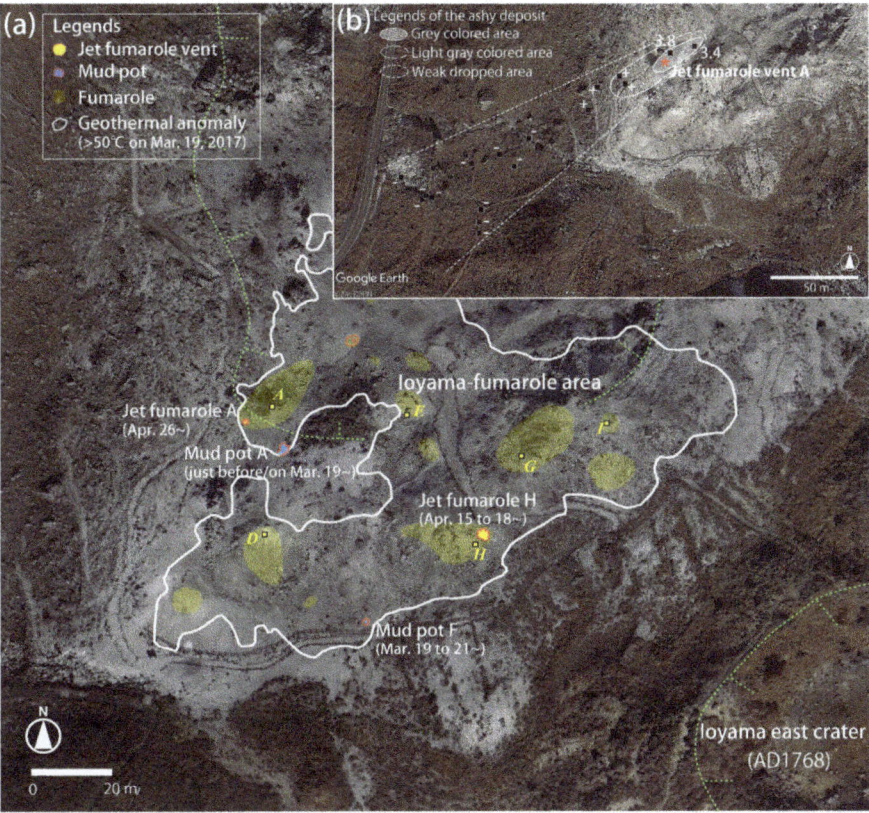

Figure 6. Drone and satellite map of the Ioyama-fumarole area. (**a**) Locations of mud pot A, jet fumarole vent A, mud pot F and jet fumarole vent H. Observation points are indicated by italic alphabet.; (**b**) Distribution of ashy deposit from the jet fumarole vent A on a Google Earth map [45]. +: coated, -: visible, two values of ashy deposit weight per a square (kg/m^2).

The outline of the eruption at the southern and western parts of the Ioyama-fumarole area in April 2018 (the 2018 Ebinokogen Ioyama eruption) was presented in our previous report [52]. The first sign, just 12 days before the eruption, was the appearance of the hydrothermal pond Y1 from around 2 am on 7 April 2018 [53]. The size of the pond Y1 was 9 m long and 5 m wide, and hydrothermal water had welled out and flowed downstream from the pond under observation on 16 April (Figure 7b). The temperature of the hydrothermal pond Y1 was 93.0 °C at that time. The activity of the fumarole S continued on 16 April, and the jet fumarole H was accompanied with vigorous roaring.

An eruption began on the south side of Ioyama at 3:39 pm on 19 April [53]. It took place first at the vent S5 near fumarole S, and almost immediately the vent S2 opened, expelling vigorous white steam moving about 30 m northeast from the vent S5 (Figure 3). This succession was captured in the NHK news video. Both vents appeared almost simultaneously within a span of one minute, and the vigorous steams developed into white to pale gray-white plumes with heights of 100 to 200 m. Subsequently, the plume from vent S5 heightened, and the basal jet part of the plume changed color intermittently from gray-white to brown. The size of the vent S2 also widened, and steams also rose from small vents around the vent S2. During an aerial survey with an airplane on the morning of 20 April, we found a new jetting fountain of dark-colored hydrothermal fluids at the vent S7 as a pond. The vents S1 to S7 (4–19 vents) were mapped via analysis of video camera images, and the details are reported separately [54]. The subsequent survey showed that the vents S1 to S3, the vents S4 to S6, and the vent S7 were merged later into craters. The vents S1 to S3 became the crater Y3, the vents S4 to S6 formed the crater Y2a, and the vent S7 formed the crater Y2b (Figure S2). These craters are collectively referred to as the Ioyama-south craters including the pond Y1. The deposit thickness of the 19 April 2018 eruption ejecta varied from a few centimeters to 45 cm of the Ioyama-south craters. It was composed of a bluish very fine ash-sized material layer at the bottom, a second layer of very fine to fine ash-sized material with coarse ash-sized material and lapilli-sized material, a third layer of fine to coarse ash-sized material with lapilli-sized material, and the uppermost layer of coarse ash-sized material and rounded lapilli to block sized material. The details of these layers will be reported subsequently. Host altered rocks of ballistics ejecta reaching 30 cm in diameter (φ) were distributed around the Ioyama-south craters as Y2a, Y2b and Y3 (Figure 8). An altered ejecta of the host rock of φ100 cm was found at the northern edge of the crater Y3, and an ejecta of φ70 cm was found at the southern edge of the crater Y2a. Altered host volcanic block and lapilli sized ejecta were distributed radially from the craters Y3 and Y2a, and we observed φ4 cm (diameter) altered ejecta at a distance of 140 m, and φ9 cm altered ejecta at a distance of the 125 m at the southeastern part of the crater Y2a.

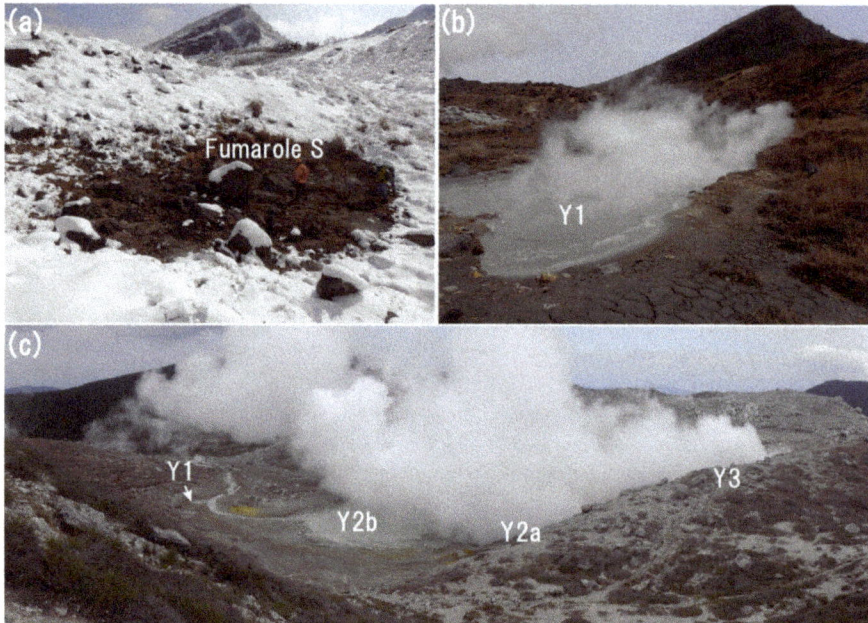

Figure 7. Photos showing the geothermal activity around Ioyama-fumarole area in 2018. (**a**) New fumarole S appeared at the south side of Ioyama-fumarole area (Photo on 3 February); (**b**) Hydrothermal pond Y1, 9 m long and 5 m wide that formed on 7 April (Photo on 16 April); (**c**) Distribution of craters of the steam driven eruption on 19 April (Photo on 4 June).

It was reported that the steaming could have started at a distance 500 m west of the Ioyama-south craters on the evening of 20 April [53]. A fissure vent composed of seven small hydrothermal explosion or steam vents (4–20 vents) was confirmed by later observations (Figure 8 and Figure S2). We did not notice any anomaly such as steaming around this area at about 4:30 pm on 20 April, and it is believed that the eruption occurred after 4:30 pm on 20 April [52]. Altered material ashy deposit and altered ejecta distribution around the two hydrothermal explosion vents W3 and W4 of the Ioyama-west crater were considered to be accompanied by a very small eruption (Figure 8). The time of deposition was considered the evening of 20 April. Photographs taken on the evening of 21 April showed that the large blocks near the Ioyama-west crater were covered with the ashy deposit. We observed the ashy deposit around the Ioyama-west crater, whose lower part was composed of fine to coarse ash-sized material, and the upper part was composed of fine ash-sized material. In addition, we observed altered ejecta comprising of different sizes of block and lapilli distributed at a distance of about 50 m from the Ioyama-west crater. The ejecta of $\varphi 8$ cm was deposited over a distance of about 25 m, while that of $\varphi 4$ cm was deposited about 40 m, and that of $\varphi 1$ cm went as far as about 50 m (Figure 8).

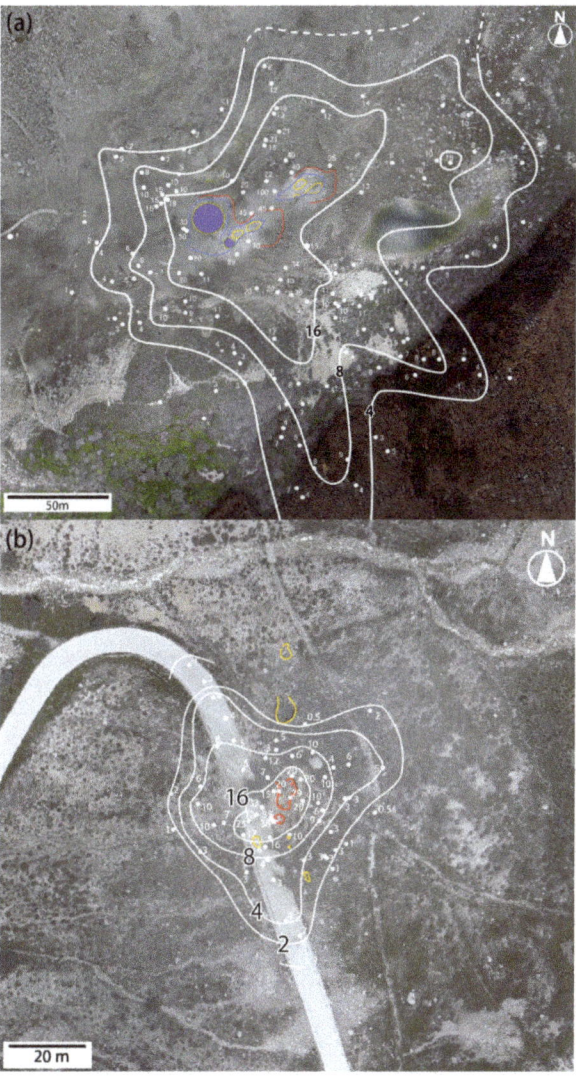

Figure 8. Distribution of ejected blocks and fragments from the Ioyama-south and west craters in April, 2018. (**a**) Block and fragment ejecta distribution from Ioyama-south craters using a drone map on a Google Earth map [45]; (**b**) Block and fragment ejecta distribution from the Ioyama-west crater. The value indicates the size of the diameter in centimeters.

Observations under a microscope showed that ashy deposits from both the Ioyama-south craters and the Ioyama-west crater were composed of altered materials and rounded crystals. Furthermore, we could not find any juvenile magmatic material under a microscope (Figure S3).

Other land features related to the April eruption were the crack and the mud pot R that formed south of the Fudoike Crater. Those were observed on 4 June 2018. The crack went across a road about 90 m southeast of the Fudoike Crater, and showed the rupture with a 3.5 cm south–up displacement and N44°E strike (Figure 3). Since we had not observed this crack before the eruption was presumed to have formed related to the eruptive activity. Furthermore, the mud pot R consisting of three small potholes, was formed to the south of the crack, and hydrothermal fluids were pouring out of one of them.

The vents S1 to S7 observed on 19 and 20 April. However, S1 and S7 subsequently became the three larger depression-shaped craters (Figures 3 and 7c). The crater Y2a was formed within two depressions extending from east–northeast to west–southwest, 37 m long and 15 m wide. Subsidence depth on the cliff of the Y2a varied in height ranging from 10 m to 3 m. The crater Y2b appeared almost circular but elongated slightly from east–northeast to west–southwest, 26 m long and 19 m wide. Subsidence depth on the cliff of the Y2b varied in height from 8 m to 2 m. The crater Y3 also elongated from east–northeast to west–southwest. Subsidence depth on the cliff of the Y3 varied in height from 12 m to 3 m (Table 1). We calculated elliptical shape areas of craters Y2a, Y3 and Y2b by using the long and short radii. Using the mean value of depths of the produced depressions, we calculated a total depression volume of approximately 8400 m^3 (min. 3500 m^3 to max. 13,400 m^3). It is assumed that three depression craters were formed between the time of the aerial survey on 20 April and 4 June, and that hydrothermal fluids continued to gush out from these craters. Furthermore, the ground uplifts and tilt–displacements continued until December 2018 [55].

Table 1. Topographical values of Ioyama-south craters.

Crater Name	Day of the Beginning	Diameter		Depression Depth		Depression Volume
		Long (m)	Short (m)	max (m)	min (m)	Mean (min–max) (m^3)
Y1	7 April, 2 am	9	5	–	–	–
Y2a	19 April, 3 pm	37	15	10	3	2800 (1300–4400)
Y3	19 April, 3 pm	33	19	12	3	3700 (1400–6000)
Y2b	19 to 20 April	26	19	8	2	1900 (800–3000)

5. Results of Geochemical Analyses

We analyzed ash-sized particles from the 2017 and 2018 events using XRD. Relative amounts (reference intensity ratios, RIR) in the XRD software of minerals are shown in Table 2.

Table 2. Relative intensities of hydrothermal minerals by the XRD analyses.

Sample Name	Alu	Kl	Cr	Qz	Py	Sm	S
20170426-ash	++++		+	+++			+
20180419-ash	+++	+	+	+++	+	+	+

Alu: alunite (soda), Kl: kaolinite, Cr: cristobalite, Qz: quartz, Py: pyrite, Sm: smectite, S: sulfur. Plus marks indicate relative intensities of the XRD peaks. ++++: ≥40% in weight, +++: ≥20%, ++: ≥10%, +: <10%.

An altered material ashy deposit sample from the April 2018 eruption (20180419-Ash) was collected on a shrub near the Ioyama-south craters, avoiding contamination by ground surface material. The sample was observed to contain significantly altered white colored fragments, light white to frosted clear-colored weakly altered fragments, and slightly altered black to gray to brown colored lithic fragments. It also contained rounded crystals of plagioclase and pyrite when observed under a microscope (Figure S3). Plagioclase, quartz, alunite (soda), kaolinite, cristobalite, pyrite, smectite and sulfur were detected by the XRD analysis (Table 2 and Figure S4). Plagioclase, pyroxenes and olivine are part of phenocrysts in volcanic rocks in this area, while quartz is not [32]. Therefore, quartz, alunite, kaolinite, cristobalite, pyrite, smectite, and sulfur in the ash-sized particles samples are considered to be the minerals produced by the hydrothermal activity or alteration. The RIR analysis showed quartz plus alunite as the major components of this sample, comprising of more than 70% in weight, and was followed by pyrite and plagioclase. The analysis also showed little amounts of kaolinite (about 10 wt%), cristobalite of about 5 wt% and smectite of about 1 wt%. These minerals are considered to have been produced primarily in acidic to neutral environments [56].

Ashy deposit from the April 2017 event (20170426-Ash) was collected from a large stone surface around the jet fumarole vent A, with minor contamination from ground surface material. The sample contained strongly and weakly altered fragments as well as slightly altered lithic fragments. Rounded plagioclase and pyroxenes were also observed using the microscope (Figure S3). Quartz, alunite (soda) cristobalite and sulfur were detected with the XRD analysis, as minerals associated with hydrothermal activity or alteration (Table 2 and Figure S4). The RIR analysis showed quartz plus alunite of nearly 90% in weight.

We collected fluid samples at the spring K3 on 16 and 21 February, and on 12 and 30 March 2016 and at the hydrothermal pond Y1 on 16 April 2018. We analyzed the abundances of ions of elements and the ratios of oxygen and hydrogen isotopes in those samples in Table S1. The water at the spring K3 in 2016 had δD values of −53‰ to −54‰ and $\delta^{18}O$ values of −8.0‰ to −8.3‰, and the compositions fall on the meteoric water lines with $\delta D = \delta^{18}O$ +10 to +20. However, the hydrothermal fluids sampled at the Y1 had a δD value of −31.3‰ and a $\delta^{18}O$ value of −0.0‰, which are values between those of Andesitic magmatic water (AMW) [57,58] and global meteoric water line (GMWL) [59] in Figure 9.

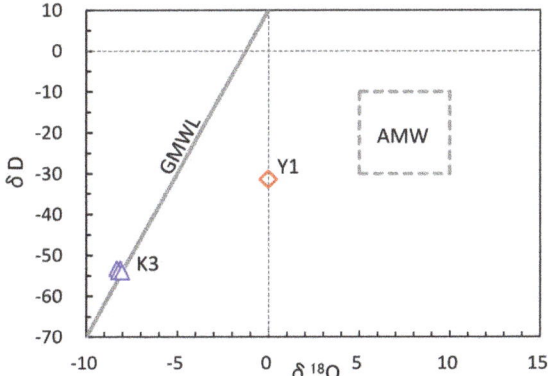

Figure 9. Oxygen and hydrogen isotope ratios of the spring K3 and the hydrothermal pond Y1 water. The "Andesitic magmatic water" (AMW) [57,58] and "global meteoric water line" (GMWL) [59] are also shown.

6. Discussions

6.1. Relationship between Observed Geothermal Activities and Underlying Geothermal System

Recently, MT inversion analysis showed that an extremely conductive layer of resistivity less than several Ω-m existed beneath 200 to 700 m depth (elevation 1100 m to 600 m) from the surface of the Ioyama [25]. This conductive layer was interpreted as composed of a hydrothermally altered clay-rich layer containing copious amounts of smectite [20]. Kagiyama et al. [60] also reported the presence of a conductive layer of 3- to 30-Ω-m resistivity with a thickness of 120 to 150 m beneath the area around the Ioyama. Inverse modeling, based on repeatedly operated and precise leveling survey data, revealed an inflation source at a depth of 600 to 700 m, just under the conductive layer [61]. Thus, it is suggested that the conductive layer behaved as a caprock preventing the up-flow of pressurized hydrothermal fluids [25]. This caprock, likely corresponding to a low permeable layer, can explain the activity sequence from Stage-1 to Stage-4. Furthermore, the geological setting identified Ebinokogen Ioyama Volcano can be considered to represent a Type 1 system described by Stix and Moor [11].

Another deeper pressure source, likely related to a magma reservoir, is present between 8 and 10 km below Ebinokogen. It appeared that this source was inflated before the eruption at Shinmoedake in 2011 and was deflated during the early large eruptions (subplinian and lava-accumulation stages) in 2011 [47]. Further, it was re-inflated after the early large eruption stages, and almost stagnated around January 2012 at Shinmoedake [62,63]. Seismicity then increased at the northern side of Karakunidake from December 2013 [48], implying the possibility that magmatic fluids started to be supplied from the deeper parts under Karakunidake to the shallow hydrothermal system under Ebinokogen Ioyama Volcano. The volcanic tremor event that took place on 20 August, 2014, may represent the beginning of the activation of the shallow hydrothermal system under the conductive caprock in Stage-1. As no manifestation was observed at the surface in this stage, it is hypothesized that the fluid-driven heating and pressurization of the shallow hydrothermal system beneath the Ioyama-fumarole area developed gradually (Figure 10). Incidentally, the deep system is the magmatic reservoirs in Kirishima Volcano Group [62,64,65].

The temperatures of fumaroles during Stage-2 were almost constant at around 96 °C (Figure 4), the boiling point temperature at an altitude of 1300 m. The relationship between the δD and the $\delta^{18}O$ ratio showed that the water from the spring K3 at this stage was derived from the meteoric water (Figure 9). Most likely the high-temperature hydrothermal fluids accumulated below a depth of 600 to 700 m in Stage-1 started to percolate through the conductive caprock layer during Stage-2. Thus, fumaroles began to appear at the ground surface (Figure 10).

The successive period was characterized by the appearance of mud pots and jet fumaroles between March and April 2017 in Stage-3. The stage was defined by a rapid expansion of the thermal anomaly area from early 2017, and a temperature rose in the spring K3 in early April (Figure 4). The mud pot A appeared just before/on 19 March, and the mud pot F appeared between 19 and 21 March (Figure 5). The jet fumarole H appeared about 30 m east of the mud pot F between 15 and 18 April, and the jet fumarole A appeared about 10 m northwest of the mud pot A on 26 April. Hydrothermal minerals of the blowout deposit from the jet fumarole A contained mostly quartz and alunite (soda) and almost no smectite (Table 1). The mineral assemblage of the blowout deposit well matches the surficial alteration near Ioyama, characterized by the alunite zone with quartz, cristobalite and tridymite [44]. In other cases, the lower limit of the alunite zone is at a 380 m depth in the Hatchobaru geothermal area in Kuju Volcano [66,67], as well alteration zones containing alunite develop in very shallow parts of Hakone and Ontake volcanoes [68,69]. The XRD analysis for a drilling core at Owakudani in Hakone showed that alunite was distributed up to a depth of 100 m from the surface [23]. A key information for estimating depth of hydrothermal fluids in the Ebino-fumarole area comes from the geothermal drilling carried in the 1950s, in which blowout occurred at two wells while drilling at depth between 18.3 and 18.4 m below the surface [41]. The hot groundwater (aquifer) at a depth of 20 m in the western part of Ioyama with the same topographical gradient is considered to have been at a 50 to 70 m depth below the summit of Ioyama in the 1950s. Currently, the height difference between the spring K3 and the summit is about 50 m. The temperature of the spring K3 at the time of the blowout event was relatively low, 40 °C, and it was probable that there was a hot confined aquifer located deeper than 50 m. Alunite is widely stable in a low pH environment, and its deposition or accumulation is known to form altered layers of low-permeability [70,71]. Recently, alunite filling among mineral grains was reported [72]. Therefore, the 2017 blowout event at the jet fumarole A with the products abundant in alunite can be considered as a phenomenon that originated from low pH hydrothermal liquids at a very shallow depth (mostly up to a depth of 100 m). Alunite, quartz, cristobalite, tridymite, kaolinite, smectite and halloysite were found on the surface around the Ioyama-fumarole area in 1950s [44]. It was probable that the alunite accumulation had formed low permeability layers (or spots) before the year 2008, under which developed pockets of hydrothermal fluids or confined groundwater (Figures 10 and 11).

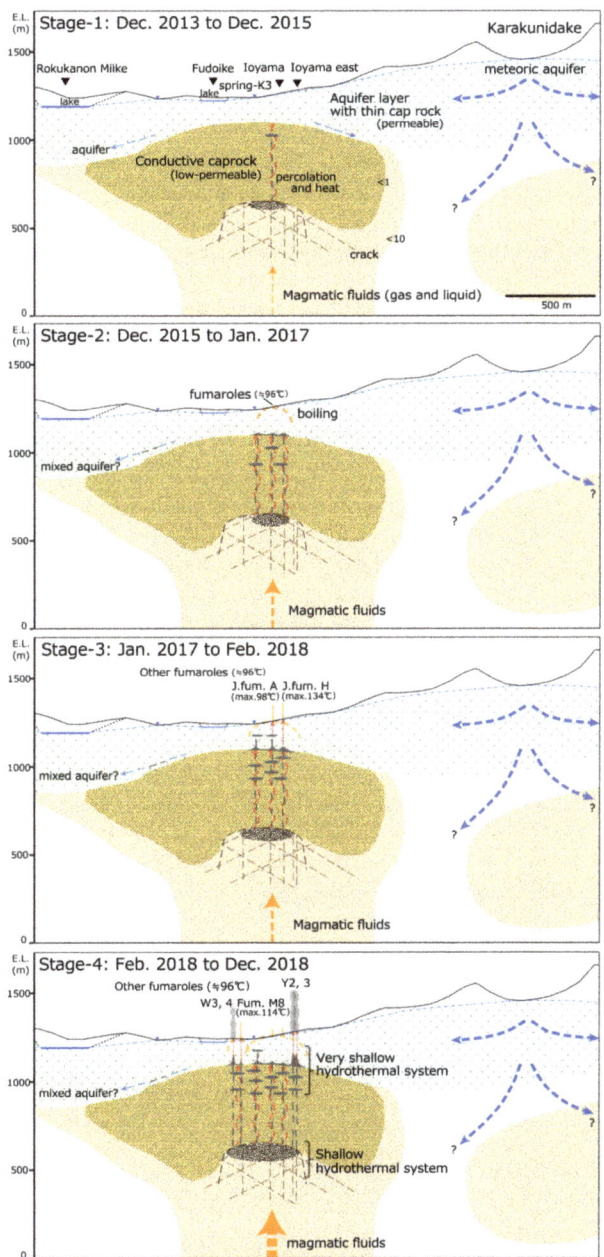

Figure 10. Conceptual sketch showing the different stages of geothermal activity in Ebinokogen Ioyama Volcano. Numerical values in Stage-1 denote Ω-m [25].

The high heat and gas emission released at the jet fumaroles H and A resulted in a decrease in the thermal anomaly area since August 2017. The temperature of the K3 spring rose again in February 2018, with the simultaneous appearance of the fumarole S. In this stage, a continuous uplift of the summit area was observed by the JMA tilt meter from 2018 mid-March. The Global Navigation Satellite System (GNSS) observations showed inflation prior to the 2018 eruption. The geophysical pressure source was estimated to be around 950 m altitude (about 350 m below the ground) [53]. The 2018 eruption deposits contained dominant quartz and alunite (soda) similar to those in the 2017 event deposits, and both probably came from similar hydrothermal environments. The smectite content of the 2018 eruption deposit was about 1 wt%, implying that the April 2018 eruption originated from a smectite depleted hydrothermal environment. The surface altered areas at the Ioyama contained minor amounts of the smectite mineral [44]. Less kaolinite (about 10 wt%) from the April 2018 eruption were derived from the hydrothermal region in an acidic environment. Part of the kaolinite or minor smectite may also have been taken from very shallow altered host rocks. The hydrothermal pond Y1 appeared on 7 April, and a large amount of hydrothermal fluids gushed out. The relationship between δD and $\delta^{18}O$ ratio of the water in the Y1 pond just before the eruption indicated that the pond liquid was a mixture of meteoric and magmatic liquids, creating an environment in which both end member liquids probably mixed. Thus, geological and geochemical observations show that the main excavation of the 2018 eruption may have been at a very shallow depth around the top of conductive layer. It is probable that the pressure source of about 350 m, indicated by low-resistivity implies the stagnation of magmatic–hydrothermal water. Seki et al. [73] suggested that a low-resistivity layer in the shallow part was formed due to the percolation of acidic hydrothermal fluid into the clayey lacustrine sediment at Midagahara Volcano in Tateyama. The resistivity >10 Ω-m was observed down to a depth of about 150 m (1150 m altitude) beneath Ioyama [60]. The similar interpretation is also possible here for the aquifer of hydrothermal fluids at depths of between 150 m and 350 m. We defined those systems above the 350 m depth as the very-shallow hydrothermal system (Figure 10). The main zone of acid alteration is the kaolinite zone without dickite which registered temperatures of up to 120 °C [13]. The temperature of 130 °C at the jet fumarole H in Stage-3 may suggest a leakage from this hydrothermal system (Figure 11).

The geology of Ioyama and the surrounding area consists of thin debris avalanche and thick welded pyroclastic deposits from Karakunidake [29,32]. The borehole survey at 1.1 km west of Ioyama [74] indicated that an upper boundary of the welded pyroclastic deposit was about 50 m depth. This deposit is exposed on the surface near the Ioyama-south craters. The elevation difference between the two sites is about 170 m. Assuming horizontally flat deposition, the welded pyroclastic deposit is distributed down to about 170 m depth under the Ioyama-south craters, corresponding to the distribution of a moderate resistivity layer down to 200 m depth. This welded pyroclastic deposit is interfingered pumice fall layers [75]. It is considered that pumice layers behave as permeable layers for ground water, where meteoric and magmatic liquids mix.

The conductive caprock layer with a low resistivity of <1 Ω-m at depths from 200 to 700 m was considered to have smectite as its main altered mineral [25]. In the laboratory experiments, higher contents of smectite achieved a lower resistivity [76]. Smectite exceeding 10% of maximum intensity by the XRD analyses was contained in the drilling core from the conductive layer in Hakone [23]. However, the lithological resistivity measured in the laboratory indicated a conductivity of about one order of magnitude higher than those in borehole observation. This difference was possibly affected by the pore fluid conductivity [24]. Additionally, the results of the effect of acidity on the conductivity were significant [77]. We will need further discussions about the morphology of the conductive layer.

For magma with a temperature of 870 °C such as the 2011 Shinmoedake silicic andesite magma [64] located at a depth of 600 to 700 m, the conductive, smectite-bearing caprock would not be observed around the inflation source because of the instability of smectite at temperatures above 200 °C [13]. It is known the high electrical conductivity is enhanced by the higher temperature e.g., [24,76]. Therefore, it was reasonable to assume that the main component of the inflation source was magmatic–hydrothermal fluids at a temperature slightly higher than 200 °C. Thus, we tentatively propose a hydrothermal reservoir system at the inflation source at around 600 to 700 m in depth (Figure 11). A high-temperature fumarole of 247 °C had previously developed at the summit of Ioyama in the 1970s [34], where gases probably originated from this depth.

6.2. A Kick Sign before the April 2018 Eruption

In order to forecast a steam-driven eruption, it is essential to understand how an aquifer of hydrothermal system is depressurized or heated and what triggers an eruption.

During the events of March to April 2017, the jet fumarole A blew out on 26 April, after the appearance of the mud pot A just before/on 19 March. The jet fumarole A blew out about a ten meter away from the mud pot A. Furthermore, welling-up of a large amount of fluids in the hydrothermal pond Y1 continued for 12 days immediately before the steam-driven eruption on 19 April, 2018 at Ioyama south craters (4–19 vents). The locations of the upwelling of hydrothermal fluids and of the eruption were about 60 m apart.

The characteristic feature in both the blowout and eruption is that the upwelling of hydrothermal fluids began just before the events. The technical report on drilling describes a phenomenon referred to as a "kick", in which fluid leaks into a borehole, which typically occurs before blowouts of pressurized fluid [3]. It was known that two boreholes blew out after reaching 18.3 to 18.4 m below the surface in the Ebino-fumarole area during a geothermal survey [41]. In this event, hot water welled up and steam blowout occurred after sudden increase of the borehole temperature. In the geothermal excavation, the pressurized hot water leaked, resulting in blowing through the wells. In addition, it is known that the cyclic behaviors of the lake water level and temperature in a hydrothermal system can be explained by the "two-phase boiling systems" phenomenon observed at the Inferno Crater Lake in New Zealand [78,79].

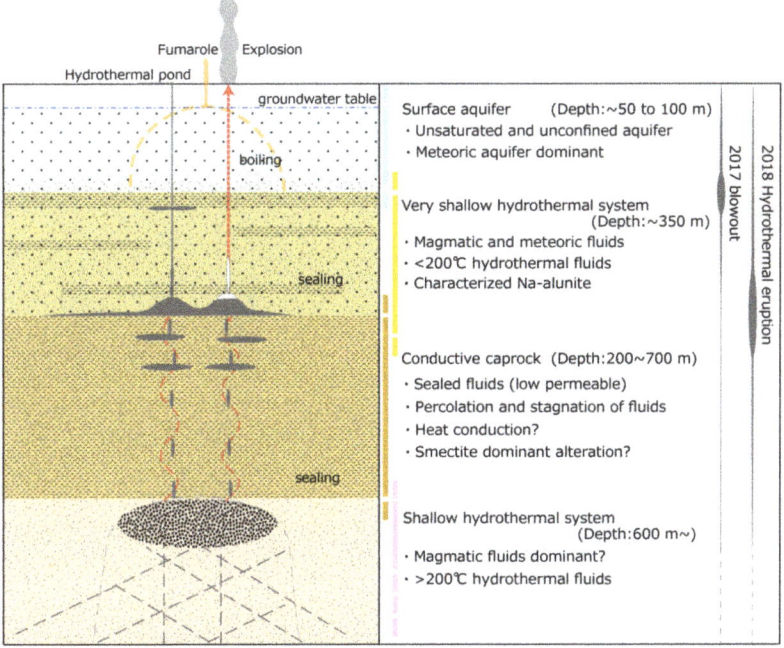

Figure 11. The schematic image of the shallow (ca. 600 m) and very shallow (ca. 350 m) systems beneath Ebinokogen Ioyama Volcano (See more details in Sections 6.1 and 6.2).

One question remains open about why hydrothermal fluids upwelling site and the explosion site were different. There appeared depressions of 3 to 10 m in the crater Y2a, 2 to 8 m in the crater Y2b, and 3 to 12 m in the crater Y3, a little time after the April 2018 eruption. The total depression volume was about 8400 m^3. And the depression volume of the Y2a and Y3 craters was about 6500 m^3 (Table 1). Those values are clearly larger than the ejecta volume of about 1500 m^3, assuming the density of the deposit as 1500 t [54] is 1000 kg/m^3. Therefore, the mechanism such as the crushing of the host rock around the conduit as in a maar formation [80] is not plausible. A hydrothermal dissolution [81] may be possible by continuous hydrothermal water outflow after the eruption. However, it is unclear if dissolution of a large volume in a short time. Therefore, we conclude that a certain amount of space had existed underground (Figure 11). The coupling between the geyser conduit and a laterally offset reservoir system [82] in the geyser eruption may be constructed under Ebinokogen Ioyama Volcano. However, the geyser system does not eject directly from a laterally offset reservoir, so it is difficult to apply this model for the April 2018 eruption. Likely, when the relatively low-temperature hydrothermal fluids in the upper layer leak and reduce the groundwater level, the resultant decreasing of hydrostatic pressure causes boiling to start. An instance of a similar trigger was seen when the dropping water levels at Gengissig lake in Kverkfjöll Volcano (Iceland) were observed to decompress the pressured aquifers, resulting in explosions due to boiling [15]. Additional explanation for the separation between the upwelling and eruption sites may originate from the fluid percolation pattern dictated by the deposit heterogeneity in the very-shallow hydrothermal system. Therefore, several mechanisms are possible for this eruption.

In summary, the pressurization of hydrothermal system is considered to be caused first as a bottom–up. The increase in fumarolic activity and the spring K3 temperatures from February 2018 indicated that the magmatic liquids were supplied to the very-shallow hydrothermal system. Second, hydrothermal fluids were observed to comprise mixed magmatic and meteoric waters up-welled at the Y1 just before the eruption in April 2018. We conclude that an initial bottom–up fluid pressurization (kick) destabilized the shallow part of the hydrothermal system, producing a boiling-front that penetrated downwards into the hydrothermal reservoir, followed by the explosion according to a top–down mechanism [12,83]. The expanding steam fragmented and dispersed the broken host rocks until the boiling decreased and stopped supplying sufficient energy to break and ejected rocks from the crater. Such mechanisms may explain the 2018 Ebinokogen Ioyama eruptions. Based on our observations, we also conclude that sudden injection and up-welling of hot magmatic fluids may cause an increase in temperature of very-shallow hydrothermal systems, and in turn produce changes in geothermal features at the surface. Thus, rapid increases in the geothermal manifestation over a short period of time, as in the case of Ebinokogen Ioyama Volcano, should be taken as a sign for caution in geothermal areas.

6.3. Issues of Steam-Driven Eruptions for the Volcanic Disaster Prevention

Information pertaining to the level of damage corresponding to steam driven eruptions originating in hydrothermal aquifer systems should be provided at tourist spots near and in geothermal areas. The magmatic–hydrothermal fluids storage system may have developed at two different levels in Ebinokogen Ioyama Volcano, contained by the conductive caprock layer (Figure 11).

Steam driven explosions from very-shallow levels are known to occur in some geothermal areas [84–87]. In a hydrothermal blowout event at the jet fumarole vent A, altered volcanic fragments of block size were scattered only a few meters from the vent. The magmatic hydrothermal eruption at Ioyama in April 2018 may have involved hydrothermal liquids shallower than 350 m depth. Based on past activity [32], larger steam-driven eruptions ejecting larger volumes of debris and producing tens-to-hundreds-meter-diameter craters at the Fudoike and Karakunidake north craters, occurred in Ebinokogen Ioyama Volcano. These explosions may involve greater depth, similarly to phreatic or phreato–magmatic eruptions, yet likely have a steam-driven dominant component e.g., [88–90]. The question remains, what may be the signs of bigger eruptions. The geophysically detected source of the 2014 phreatic eruption at Ontake Volcano, which scattered ballistics ejecta over 1 kilometer away [91], was thought to be from 600 to 1000 m deep [92,93]. Kato et al. [94] proposed that hot fluids infiltrated to cracks from the deeper magma chamber in middle September 2014 before the September 27 phreatic eruption in Ontake. The break of the seal of the conduit for hydrothermal fluids at a shallow level began just prior to the eruption [94], resulting in the decompression that led flashing of hydrothermal fluids from the top to downward. Recent research [95] supports this model by the finding of a small amount of juvenile material in the 2014 eruption products in Ontake. A steam-driven eruption at Sinmoedake Volcano in 2008 scattered ballistics as far as 800 m [31]. The geophysically detected source of the inflation and deflation in the eruption was estimated to be at a depth of 500 to 600 m below the volcano [96]. These are similar to the VLP events, 700 ± 200 m below the lake, in Kawah Ijen (Indonesia) and that of 850 ± 150 m below the crater at White Island (New Zealand) [97].

In case of Ebinokogen Ioyama in 2018, it took over two months for magmatic fluids to be supplied into shallow hydrothermal system before the eruption from the deep system, contrasting to about two weeks (11 to 16 days) in Ontake. In Ebinokogen Ioyama, the supplied amount and rate of magmatic fluids of the 2018 eruption may be much smaller than the 2014 eruption in Ontake. In summary, smaller events could be better understood as result of our study. And more energetic steam-driven eruptions may occur in this area if larger amount of magmatic fluid is injected rapidly and a wider area of the shallow hydrothermal system is destabilized. However, more studies are needed to improve our knowledge of triggers, mechanisms and potential precursors of steam-driven eruptions at Ebinokogen Ioyama. Additionally, we need to consider repetitive highly hazardous hydrothermal eruptions like that of Lake Okaro in New Zealand [12].

The technology to forecast very-shallow hydrothermal eruptions has developed by combining observations of hydrothermal phenomena such as increasing activity of geothermal areas, a welling-up kick, results of In-SAR and precise leveling observations. In the small eruption of Hakone 2015, In-SAR observed an uplift deformation area of approximately 200 m in diameter immediately before the eruption, along with a ground uplift of up to 30 cm. The vents appeared at the southern edge of deformation [98]. Locations of the fumarole S and the Y1 hydrothermal pond at Ebinokogen Ioyama may be possible sites for future eruption. A hydrothermal eruption occurred from the Ioyama-south craters on 19 April 2018. The next day, vents (Ioyama-west crater) formed 500 m away from the Ioyama-south craters. The impact area of the ejecta from the latter vents was as small as ~50 m. The outside of the alert level 2 area possibly impacted by ballistics, within a one-kilometer radius from the Ioyama crater, shown in the volcanic hazard map of Ebinokogen Ioyama [99,100], was not affected. Given that the tourist facility is relatively close to the potential venting area, in addition to addressing the 19 April vents, there is also the need to highlight the occurrence of the 20 April.

7. Conclusions

Geothermal activity resumed at Ebinokogen Ioyama Volcano in December 2015. We observed the very shallow hydrothermal event of a blowout and jet fumaroles from March to April 2017. This event was accompanied by a thermal anomaly expansion and rapid increase in the water temperature at the spring K3. The transformation of a hydrothermal mud pot A to jet fumarole A over one month following the blowout steaming event in 2017.

The magmatic hydrothermal eruptions took place in the Ioyama–south craters on 19 April and in the Ioyama-west crater on the evening of 20 April 2018. The increasing of geothermal anomaly phenomena before the eruption indicated that the magmatic liquid may be stored in a very-shallow hydrothermal system. Upwelling of hydrothermal water was observed to pour out at the hydrothermal pond Y1 just before the eruption in April 2018. The relationship of the δD and $\delta^{18}O$ ratio of this water in the Y1 indicated that the liquid was a mixture of meteoric and magmatic liquids. Lastly, the upwelling caused by a kick may have led to the eruption at about 60 m away from the upwelling center. Hydrothermal mineral analyses and geochemical observations suggest that the mixture was spouted from the very-shallow hydrothermal system (less than 350 m depth) at the top of the conductive caprock layer. In addition, the inflation source at 600 to 700 m depth was likely the shallow hydrothermal system (reservoir). Thus, we propose an underground hydrothermal structure consisting of two systems. The secondary eruptive event on the evening of 20 April, 2018 occurred 500 m away from the eruption site on 19 April, and it is necessary to investigate the formation mechanism and evaluation method for this event in more detail.

Supplementary Materials: The following are available online at http://www.mdpi.com/2076-3263/10/5/183/s1, Figure S1: The distribution of geothermal anomaly area over 50 °C in Ioyama fumarole area; Figure S2: Digital elevation models and photographs; Figure S3: Photomicrographs of ashy deposit grains; Figure S4: XRD analyses; Table S1: The oxygen and hydrogen isotope analyses of hydrothermal water from the Ioyama.

Author Contributions: Y.T.; geological and geothermal investigation, XRD analyses, project administration and writing—Original draft preparation, S.N.; supervision, geological investigation and writing—Review and editing. F.M.; drone investigation and project administration, T.H.; temperature investigation and geothermal monitoring, M.T. and A.I.; oxygen and hydrogen isotope analyses, T.M.; review and project administration, M.N.; geological investigation, J.F.; water investigation. All authors have read and agreed to the published version of the manuscript.

Funding: This research received no external funding.

Acknowledgments: The authors thank K. Ishihara, M. Ichihara, A. Takagi for offering helpful comments. The authors thank H. Shinohara, J. Itoh, N. Morikawa of GSJ and H. Suzuki, J. Teruya, A. Hasegawa of Nippon Koei for helping the field or laboratories. The Kirishima Nature Guide Club member as K. Haraguchi, T. Higashi, H. Yoshinaga, M. Baba, Y. Matsumura, T. Miyagawa, T. Nagatomo, S. Takada, S. Watanabe are thanked for helping with our field observations. We contacted the Kagoshima Meteorological Office of JMA when we entered the restricted area. In addition, permissions were obtained from the Ebino city and the Kyusyu Regional Forest Office at Miyakonojo during the investigation involving drones in the restricted access area. We thank J. Vandemeulebrouck, anonymous reviewers and J.C. Eichelberger editor for great constructive suggestions and helpful comments on this manuscript. We extend our heartfelt condolence to H. Itoh.

Conflicts of Interest: The authors declare no conflict of interest.

References

1. Barberi, F.; Bertagnini, A.; Landi, P.; Principe, C. A review on phreatic eruptions and their precursors. *J. Volcanol. Geotherm. Res.* **1992**, *52*, 231–246. [CrossRef]
2. Muffler, L.J.P.; White, D.E.; Truesdell, A.H. Hydrothermal explosion craters in Yellowstone National Park. *Geol. Soc. Am. Bull.* **1971**, *82*, 723–740. [CrossRef]
3. Browne, P.R.L.; Lawless, J.V. Characteristics of hydrothermal eruptions, with examples from New Zealand and elsewhere. *Earth Sci. Rev.* **2001**, *52*, 299–331. [CrossRef]
4. Mordret, A.; Jolly, A.D.; Duputel, Z.; Fournier, N. Monitoring of phreatic eruptions using interferometry on retrieved cross-correlation function from ambient seismic noise: Results from Mt. Ruapehu, New Zealand. *J. Volcanol. Geotherm. Res.* **2010**, *191*, 46–59. [CrossRef]
5. Yamamoto, T.; Nakamura, Y.; Glicken, H. Pyroclastic density current from the 1888 phreatic eruption of Bandai volcano, NE Japan. *J. Volcanol. Geotherm. Res.* **1999**, *90*, 191–207. [CrossRef]
6. Kitagawa, T. *Bakuhatsu Saigai No Kaiseki (Explosion Disaster Analysis)*; Nikkankogyoshinbunsha: Tokyo, Japan, 1980; pp. 277–324. (In Japanese)
7. Ogiso, C.; Uehara, Y. Experimental study of vapor explosions caused by the abrupt destruction of phase equilibrium. *J. Jpn. Soc. Saf. Eng.* **1985**, *24*, 192–198. (In Japanese)
8. Takashima, T.; Iida, Y. *Joukibakuhatsu No Kagaku (The Science of Vapor Explosion)*; Shoukabou: Tokyo, Japan, 1998; p. 172. (In Japanese)
9. Yamamoto, T. The pyroclastic density currents generated by the September 27, 2014 phreatic eruption of Ontake Volcano, Japan. *Bull. Geol. Surv. Japan.* **2014**, *65*, 117–127. (In Japanese) [CrossRef]
10. Oikawa, T.; Yoshimoto, M.; Nakada, S.; Maeno, F.; Komori, J.; Shimano, T.; Takeshita, Y.; Ishizuka, Y.; Ishimine, Y. Reconstruction of the 2014 eruption sequence of Ontake Volcano from recorded images and interviews. *Earth Planet. Space* **2016**, *68*, 489. [CrossRef]
11. Stix, J.; Moor, J.M. Understanding and forecasting phreatic eruptions driven by magmatic degassing. *Earth Planet. Space* **2018**, *70*, 83. [CrossRef]
12. Montanaro, C.; Cronin, S.; Scheu, B.; Kennedy, B.; Scott, B. Complex crater fields formed by steam-driven eruptions: Lake Okaro, New Zealand. *GSA Bull.* **2020**, 1–17. [CrossRef]

13. Stimac, J.; Goff, F.; Goff, C.J. Intrusion-Related Geothermal Systems. In *The Encyclopedia of Volcanoes*, 2nd ed.; Sigurdsson, H., Houghton, B., McNutt, S.R., Rymer, H., Stix, J., Eds.; Academic Press: Amsterdam, The Netherlands, 2015; pp. 799–822. [CrossRef]
14. Miyake, Y.; Osaka, J. Steam explosion of February 11th, 1995 at Nakanoyu hot spring, Nagano prefecture, central Japan. *Bull. Volcanol. Soc. Jpn.* **1998**, *43*, 113–121. (In Japanese) [CrossRef]
15. Montanaro, C.; Scheu, B.; Gudmundsson, M.T.; Vogfjörd, K.; Reynolds, H.I.; Dürig, T.; Strehlow, K.; Rott, S.; Reuschlé, T.; Dingwell, D.B. Multidisciplinary constraints of hydrothermal explosions based on the 2013 Gengissig lake events, Kverkfjöll volcano, Iceland. *Earth Planet. Sci. Lett.* **2016**, *434*, 308–319. [CrossRef]
16. Montanaro, C.; Scheu, B.; Cronin, S.J.; Breard, E.C.P.; Lube, G.; Dingwell, D.B. Experimental estimates of the energy budget of hydrothermal eruptions; application to 2012 Upper Te Maari eruption, New Zealand. *Earth Planet. Sci. Lett.* **2016**, *452*, 281–294. [CrossRef]
17. Hedenquist, J.W.; Henley, R.W. Hydrothermal eruptions in the Waiotapu geothermal system, New Zealand: Their origin, associated breccias, and relation to precious metal mineralization. *Econ. Geol.* **1985**, *80*, 1640–1668. [CrossRef]
18. Facca, A.; Tonani, F. The self-sealing geothermal field. *Bull. Volcanol.* **1967**, *30*, 271–273. [CrossRef]
19. Fournier, R.O. Hydrothermal processes related to movement of fluid from plastic into brittle rock in the magmatic-epithermal environment. *Econ. Geol.* **1999**, *94*, 1193–1211. [CrossRef]
20. Aizawa, K.; Ogawa, Y.; Ishido, T. Groundwater flow and hydrothermal systems within volcanic edifices: Delineation by electric self-potential and magnetotellurics. *J. Geophys. Res.* **2009**, *114*, B01208. [CrossRef]
21. Aizawa, K. Groundwater flow beneath volcanoes inferred from electric self-potential and magnetotellurics. *Bull. Volcanol. Soc. Jpn.* **2010**, *55*, 251–259. (In Japanese) [CrossRef]
22. Kanda, W.; Utsugi, M.; Tanaka, Y.; Hashimoto, T.; Fujii, I.; Hasenaka, T.; Shigeno, N. A heating process of Kuchi-erabu-jima volcano, Japan, as inferred from geomagnetic field variations and electrical structure. *J. Volcanol. Geotherm. Res.* **2010**, *189*, 158–171. [CrossRef]
23. Mannen, K.; Tanada, T.; Jomori, A.; Akatsuka, T.; Kikugawa, G.; Fukazawa, Y.; Yamashita, H.; Fujimoto, K. Source constraints for the 2015 phreatic eruption of Hakone volcano, Japan, based on geological analysis and resistivity structure. *Earth Planet. Space* **2019**, *71*, 1–20. [CrossRef]
24. Lévy, L.; Maurya, P.K.; Byrdina, S.; Vandemeulebrouk, J.; Sigmundsson, F.; Árnason, K.; Ricci, T.; Deldicque, D.; Roger, M.; Gibert, B.; et al. Electrical resistivity tomography and time-domain induced polarization field investigations of geothermal areas at Krafla, Iceland: Comparison to borehole and laboratory frequency-domain electrical observations. *Geophys. J. Int.* **2019**, *218*, 1469–1489. [CrossRef]
25. Tsukamoto, K.; Aizawa, K.; Chiba, K.; Kanda, W.; Ueshima, M.; Koyama, T.; Utsugi, M.; Seki, K.; Kishita, T. Three-dimensional resistivity structure of Iwo-yama volcano, Kirishima volcanic complex, Japan: Relationship to shallow seismicity, surface uplift, and a small phreatic eruption. *Geophys. Res. Lett.* **2018**, *45*. [CrossRef]
26. Fujinawa, A.; Kamoshida, T.; Tanase, A.; Tanimoto, K.; Nakamura, Y.; Kontani, K. Reconsideration of the 1900 explosive eruption at Adatara volcano, northeastern Japan. *Bull. Volcanol. Soc. Jpn.* **2006**, *51*, 311–325. (In Japanese) [CrossRef]
27. Sudo, Y. *Aso Ni Manabu (Learn. from Aso)*; Toukashobou: Fukuoka, Japan, 2007; p. 319. (In Japanese)
28. Kawanabe, Y.; Nogami, K. Kusatsu-Shiranesan. In *National Catalogue of the Active Volcanoes in Japan*, 4th ed.; Japan Meteorological Agency: Tokyo, Japan, 2013; pp. 1–25.
29. Imura, R.; Kobayashi, T. *Geological Map of Kirishima Volcano (Geological Map of Volcanoes 11)*; Geological survey of Japan: Tsukuba, Japan, 2001; pp. 1–8. (In Japanese)
30. Imura, R. Minor phreatic activity of Shinmoedake, Kirishima volcanoes, in 1991–1992. *Bull. Volcanol. Soc. Jpn.* **1992**, *37*, 281–283. (In Japanese) [CrossRef]
31. Geshi, N.; Takarada, S.; Tsustui, M.; Mori, T.; Kobayashi, T. Products of the August eruption of Shinmoedake Volcano, Kirishima Volcanic Group, Japan. *Bull. Volcanol. Soc. Jpn.* **2010**, *55*, 53–64. (In Japanese) [CrossRef]
32. Tajima, Y.; Matsuo, Y.; Shoji, T.; Kobayashi, T. Eruptive history of Ebinokogen volcanic area of Kirishima volcanoes for the past 15,000 years in Kyushu, Japan. *Bull. Volcanol. Soc. Jpn.* **2014**, *59*, 55–75. (In Japanese) [CrossRef]
33. Fukuoka District Meteorological Observatory; Kagoshima Local Meteorological Observatory; Miyazaki Local Meteorological Observatory. *Showa 34 Nen 2 Gatsu 17 Nichi No Kirishimayama Shinmoedake No Bakuhatsu (Eruption of Shinmoedake, Kirishima Volcano, on 17 February 1959)*; Fukuoka District Meteorological Observatory: Tokyo, Japan, 1959; p. 15. (In Japanese)

34. Kagiyama, T.; Uhira, K.; Watanabe, T.; Masutani, F.; Yamaguchi, M. Geothermal survey of the volcanoes Kirisima. *Bull. Earthq. Res. Inst.* **1979**, *54*, 187–210. (In Japanese)
35. Kagiyama, T. Eruption dominant volcanism vs. geothermal activity dominant volcanism—New aspect in volcanism. *J. Geotherm. Res. Soc. Jpn.* **2008**, *30*, 193–204. (In Japanese) [CrossRef]
36. McKee, K.; Fee, D.; Yokoo, A.; Matoza, R.S.; Kim, K. Analysis of gas jetting and fumarole acoustics at Aso volcano, Japan. *J. Volcanol. Geotherm. Res.* **2017**, *340*, 16–29. [CrossRef]
37. Heasler, H.P.; Jaworowski, C.; Foley, D. Geothermal systems and monitoring hydrothermal features. In *Geological Monitoring*; Young, R., Norby, L., Eds.; Geological Society of America: Boulder, CO, USA, 2009; pp. 105–140. [CrossRef]
38. Machida, H.; Arai, F. *Atlas of Tephra in and around Japan*; University of Tokyo Press: Tokyo, Japan, 2003; pp. 185–188. (In Japanese)
39. Kashmir3d, Ver.9.3.4; Sugimoto Tomohiko©. 2005. Available online: http://www.kashmir3d.com (accessed on 10 January 2020).
40. Okuno, M. Chronology of tephra layers in southern Kyushu, SW Japan, for the last 30,000 years. *Quat. Res.* **2002**, *41*, 225–236. (In Japanese) [CrossRef]
41. Geothermal Research Department, Geological survey of Japan. Studies on natural steam at Ebino hot spring in Miyazaki prefecture. *Bull. Geol. Surv. Jpn.* **1955**, *10*, 611–626. (In Japanese)
42. Oda, R. Kirishima kazan tiiki chishitu chousa houbun (Report of geological survey in Kirishima volcano area). *Bull. Imp. Earthq. Investig. Comm.* **1921**, *96*, 1–58. (In Japanese)
43. Funasaki, J.; Shimomura, M.; Kuroki, C. Document on geothermal activities at Ebino highland including Iou-yama, Kirishima volcano, since the Meiji era. *Quart. J. Seismol.* **2017**, *80*, 1–11. (In Japanese)
44. Kimbara, K.; Sakaguchi, K. Geology, distribution of hot springs and hydrothermal alteration zones of major geothermal areas in Japan. *Rept. Geol. Surv. Jpn.* **1989**, *270*, 1–428. (In Japanese)
45. *Google Earth Pro, 7.3.2.5776*; ZENRIN: Fukuoka, Japan, 2020.
46. Itoh, J.; Hamasaki, S.; Kawanabe, Y. Re-evaluation of explosive activities of Iwate Volcano in the last 10,000 years: Spatial and temporal relationship of phreatic and magmatic explosions. *J. Geol. Soc. Jpn.* **2018**, *124*, 271–296. (In Japanese) [CrossRef]
47. Nakada, S.; Nagai, M.; Kaneko, T.; Suzuki, Y.; Maeno, F. The outline of the 2011 eruption at Shinmoe-dake (Kirishima), Japan. *Earth Planet. Space* **2013**, *65*, 475–488. [CrossRef]
48. Kagoshima Meteorological Office, JMA. Fukuoka Regional Headquarters, JMA. Volcanic Activity of Kirishimayama Volcano—May–October 2014. *Rep. Coord. Comm. Predict. Volcan. Erupt.* **2016**, *119*, 213–259. (In Japanese)
49. Fukuoka Regional Headquaters, JMA. Volcanic Activity of Kirishimayama Volcano—June–October 2015. *Rep. Coord. Comm. Predict. Volcan. Erupt.* **2018**, *122*, 347–379. (In Japanese)
50. Japan Meteorological Agency Website. Available online: http://www.data.jma.go.jp/svd/vois/data/tokyo/volcano.html (accessed on 20 January 2019).
51. YamaReco. Available online: https://www.yamareco.com/ (accessed on 10 January 2019).
52. Tajima, Y.; Nakada, S.; Nagai, M.; Maeno, F.; Watanabe, A. Small eruption at the Ebinokogen Ioyama volcano of the Kirishima Volcano Group in April 2018. *Bull. Volcanol. Soc. Jpn.* **2019**, *64*, 147–151. (In Japanese) [CrossRef]
53. Kagoshima Local Meteorological Office, JMA. Regional Volcanic Observation and Warning Center, Fukuoka Regional Headquarters, JMA. Volcanic Activity of Kirishimayama Volcano—1 February 2018 –31 May 2018. *Rep. Coord. Comm. Predict. Volcan. Erupt.* **2018**, *130*, 213–284. (In Japanese)
54. Nagai, M.; Tajima, Y.; Maeno, F.; Nakada, S.; Furuzono, T.; Watanabe, A. Products of the 2018 eruption around Ioyama in Ebino kogen area, Kirishima Volcano Group, Southern Kyushu, Japan. In Proceedings of the Programme and Abstracts of the Volcanological Society of Japan 2018 Fall Meeting, Akita, Japan, 26 September 2018; p. 7. (In Japanese).
55. Kagoshima Local Meteorological Office, JMA. Regional Volcanic Observation and Warning Center, Fukuoka Regional Headquarters, JMA. Volcanic Activity of Kirishimayama Volcano—1 September 2017–31 January 2019. *Rep. Coord. Comm. Predict. Volcan. Erupt.* **2019**, *132*, 240–311. (In Japanese)
56. White, N.C.; Hedenquist, J.W. Epithermal gold deposits: Styles, characteristics and exploration. *SEG Newsl.* **1995**, *23*, 9–13.

57. Taran, Y.A.; Pokrovsky, B.G.; Dubik, Y.M. Isotopic composition and origin of water from andesitic magmas. *Dokl. Acad. Nauk. USSR* **1989**, *304*, 440–443.
58. Giggenbach, W.F. Isotopic shifts in waters from geothermal and volcanic systems along convergent plate boundaries and their origin. *Earth Planet. Sci. Lett.* **1992**, *113*, 495–510. [CrossRef]
59. Craig, H. Isotopic variations in meteoric waters. *Science* **1961**, *133*, 1702–1703. [CrossRef]
60. Kagiyama, T.; Yamaguchi, M.; Masutani, F.; Utada, H. VLF, ELF-MT survey around Iwo-yama, Kirishima Volcanoes. *Bull. Earthq. Res. Inst.* **1994**, *69*, 211–239. (In Japanese)
61. Morita, K.; Matsushima, T.; Yokoo, K.; Miyamachi, R.; Teguri, Y.; Fujita, S.; Nakamoto, M.; Shimizu, H.; Chiba, K.; Koga, Y.; et al. Vertical ground deformation of Iwoyama, Kirishima volcanoes measured by precise leveling survey (during June 2015–February 2017). In Proceedings of the Japan Geoscience Union Meeting 2018, Chiba, Japan, 20–24 May 2018.
62. Nakao, S.; Morita, Y.; Yakiwara, H.; Oikawa, J.; Ueda, H.; Takahashi, H.; Ohta, Y.; Matsushima, T.; Iguchi, M. Volume change of the magma reservoir relating to the 2011 Kirishima Shinmoe-dake eruption—Charging, discharging and recharging process inferred from GPS measurements. *Earth Planet. Space* **2013**, *65*, 505–515. [CrossRef]
63. Kagoshima Local Meteorological Office, JMA. Regional Volcanic Observation and Warning Center, Fukuoka Regional Headquarters, JMA. Volcanic Activity of Kirishimayama Volcano—June–October 2012. *Rep. Coord. Comm. Predict. Volcan. Erupt.* **2014**, *113*, 148–172. (In Japanese)
64. Suzuki, Y.; Yasuda, A.; Hokanishi, N.; Kaneko, T.; Nakada, S.; Toshitsugu, F. Syneruptive deep magma transfer and shallow magma remobilization during the 2011 eruption of Shinmoe-dake, Japan—Constraints from melt inclusions and phase equilibria experiments. *J. Volcanol. Geotherm. Res.* **2013**, *257*, 184–204. [CrossRef]
65. Aizawa, K.; Koyama, T.; Hase, H.; Uyeshima, M.; Kanda, W.; Utsugi, M.; Yoshimura, R.; Yamaya, Y.; Hashimoto, T.; Yamazaki, K.; et al. Three-dimensional resistivity structure and magma plumbing system of the Kirishima Volcanoes as inferred from broadband magnetotelluric data. *JGR Solid Earth.* **2014**, *119*, 198–215. [CrossRef]
66. Hayashi, M. Hydrothermal alteration in the Otake geothermal area, Kyushu. *J. Jpn. Geotherm. Energy Assoc.* **1973**, *38*, 9–46.
67. Kiyosaki, J.; Tanaka, K.; Taguchi, S.; Chiba, H.; Takeuchi, K.; Motomura, Y. Hypogene alunite from acid alteration zone in Hatchobaru geothermal field, Kyushu, Japan. *J. Geotherm. Res. Soc. Jpn.* **2006**, *28*, 287–297. (In Japanese) [CrossRef]
68. Hiraon, T. Hydrothermal alteration of volcanic rocks in the Hakone and northern Izu geothermal area. *Bull. Hot Springs Res. Inst. Kanagawa Pref.* **1986**, *17*, 73–166.
69. Minami, Y.; Imura, T.; Hayashi, S.; Ohba, T. Mineralogical study on volcanic ash of the eruption on September 27, 2014 at Ontake volcano, central Japan: Correlation with porphyry copper systems. *Earth Planet. Space* **2016**, *68*, 1841. [CrossRef]
70. Zimbelman, D.R.; Rye, R.O.; Breit, G.N. Origin of secondary sulfate minerals on active andesitic stratovolcanoes. *Chem. Geol.* **2005**, *215*, 37–60. [CrossRef]
71. Mayer, K.; Scheu, B.; Yilmaz, T.I.; Montanaro, C.; Gilg, H.A.; Rott, S.; Joseph, E.P.; Dingwell, D.B. Phreatic activity and hydrothermal alteration in the Valley of Desolation, Dominica, Lesser Antilles. *Bull. Volcanol.* **2017**, *79*, 82. [CrossRef]
72. Imura, T.; Ohba, T.; Nakagawa, M. Characteristics of hydrothermally altered minerals in volcanic products at Tokachidake volcano, central Hokkaido, Japan. *Jour. Geol. Soc. Jpn.* **2019**, *125*, 203–218. (In Japanese) [CrossRef]
73. Seki, K.; Kanda, W.; Ogawa, Y.; Tanbo, T.; Kobayashi, T.; Hino, Y.; Hase, H. Imaging the hydrothermal system beneath the Jigokudani valley, Tateyama volcano, Japan: Implications for structures controlling repeated phreatic eruptions from an audio-frequency magnetotelluric survey. *Earth Planet. Space* **2015**, *67*, 539. [CrossRef]
74. Akasaki, A.; Imura, R. Reconstruction of eruptive history of the central part of Kirishima Volcano from the drill core at Ebino Highland. In Proceedings of the Programme and abstracts of the Volcanological Society of Japan 2018 Fall Meeting, Akita, Japan, 26 September 2018; p. 217. (In Japanese).
75. Imura, R. Geology of Kirishima Volcano. *Bull. Earthq. Res. Inst. Univ. Tokyo.* **1994**, *69*, 189–209. (In Japanese)

76. Takakura, S. Influence of pore-water salinity and temperature on resistivity of clay-bearing rocks. *Butsuri Tansa.* **2009**, *62*, 385–396. (In Japanese) [CrossRef]
77. Byrdina, S.; Grandis, H.; Sumintadireja, P.; Caudron, C.; Syahbana, D.K.; Naffrechoux, E.; Gunawan, H.; Suantika, G.; Vandemeulebrouck, J. Structure of the acid hydrothermal system of Papandayan volcano, Indonesia, investigated by geophysical methods. *J. Volcanol. Geotherm. Res.* **2018**, *358*, 77–86. [CrossRef]
78. Vandemeulebrouck, J.; Stemmelen, D.; Hurst, T.; Grangeon, G. Analogue modeling of instabilities in crater lake hydrothermal systems. *JGR Solid Earth.* **2005**, *110*, B02212. [CrossRef]
79. Legaz, A.; Vandemeulebrouck, J.; Revil, A.; Kemna, A.; Hurst, A.W.; Reeves, R.; Papasin, R. A case study of resistivity and self-potential signatures of hydrothermal instabilities, Inferno Crater Lake, Waimangu, New Zealand. *Geophys. Res. Lett.* **2009**, *36*, L12306. [CrossRef]
80. White, J.D.L.; Ross, P.-S. Maar-diatreme volcanoes: A review. *J. Volcanol. Geotherm. Res.* **2011**, *201*, 1–29. [CrossRef]
81. Morgan, L.A.; Shanks, W.C.P.; Pierce, K.L. Hydrothermal processes above the Yellowstone magma chamber: Large hydrothermal systems and large hydrothermal explosions. *GSA Spec. Pap.* **2009**, *459*, 1–95. [CrossRef]
82. Vandemeulebrouck, J.; Sohn, R.A.; Rudolph, M.L.; Hurwitz, S.; Manga, M.; Johnston, M.J.S.; Soule, S.A.; McPhee, D.; Glen, J.M.G.; Karlstrom, L.; et al. Eruptions at Lone Star geyser, Yellowstone National Park, USA: 2. Constraints on subsurface dynamics. *JGR Solid Earth.* **2013**, *119*, 8688–8707. [CrossRef]
83. McKibbin, R.; Smith, T.A.; Fullard, L. Components and phases: Modelling progressive hydrothermal eruptions. *Anziam J.* **2009**, *50*, 365–380. [CrossRef]
84. Allis, R.G. Thermal history of the Karapiti area, Wairakei. *Geophys. Div. N. Z.* **1981**, *137*, 1–38.
85. Allis, R.G. The 9 April 1983 steam eruption at Crater of the Moon thermal area, Wairakei. *Geophys. Div. N. Z.* **1984**, *196*, 1–25.
86. Doke, R.; Harada, M.; Mannen, K.; Itadera, K.; Tatenaka, J. InSAR analysis for detecting the route of hydrothermal fluid to the surface during the 2015 phreatic eruption of Hakone Volcano, Japan. *Earth Planet. Space* **2018**, *70*, 63. [CrossRef]
87. Mannen, K.; Yukutake, Y.; Kikugawa, G.; Harada, M.; Itadera, K.; Tatenaka, J. Chronology of the 2015 eruption of Hakone volcano, Japan: Geological background, mechanism of volcanic unrest and disaster mitigation measures during the crisis. *Earth Planet. Space* **2018**, *70*, 68. [CrossRef]
88. Mastin, L.G. The roles of magma and groundwater in the phreatic eurptions at Inyo Craters, Long Valley Caldera, California. *Bull. Volcanol.* **1991**, *53*, 579–596. [CrossRef]
89. Valentine, G.A.; Graettinger, A.H.; Sonder, I. Explosion depths for phreatomagmatic eruptions. *Geophys. Res. Lett.* **2014**, *41*, 3045–3051. [CrossRef]
90. Valentine, G.A.; Sottili, G.; Palladino, D.M.; Taddeucci, J. Tephra ring interpretation in light of evolving maar–diatreme concepts: Stracciacappa maar (central Italy). *J. Volcanol. Geotherm. Res.* **2015**, *308*, 19–29. [CrossRef]
91. Kaneko, T.; Maeno, F.; Nakada, S. 2014 Mount Ontake eruption: Characteristics of the phreatic eruption as inferred from aerial observations. *Earth Planet. Space* **2016**, *68*, 1226. [CrossRef]
92. Maeda, Y.; Kato, A.; Terakawa, T.; Yamanaka, Y.; Horikawa, S.; Matsuhiro, K.; Okuda, T. Source mechanism of a VLP event immediately before the 2014 eruption of Mt. Ontake, Japan. *Earth Planet. Space* **2015**, *67*, 716. [CrossRef]
93. Takagi, A.; Onizawa, S. Shallow pressure sources associated with the 2007 and 2014 phreatic eruptions of Mt. Ontake, Japan. *Earth Planet. Space* **2016**, *68*, 119. [CrossRef]
94. Kato, A.; Terakawa, T.; Yamanaka, Y.; Maeda, Y.; Horikawa, S.; Matsuhiro, K.; Okubo, T. Preparatory and precursory processes leading up to the 2014 phreatic eruption of Mount Ontake, Japan. *Earth Planet. Space* **2015**, *67*, 111. [CrossRef]
95. Miyagi, I.; Geshi, N.; Hamasaki, S.; Oikawa, T.; Tomiya, A. Heat source of the 2014 phreatic eruption of Mount Ontake, Japan. *Bull. Volcanol.* **2020**, *82*, 1–17. [CrossRef]
96. Takagi, A. Ground deformation around the Shinmoedake crater before the 2011 Kirishimayama eruption. In Development of quantitative detection techniques of magma activity and improvement of evaluation of volcanic activity level. *Tech. Rep. Meteorol. Res. Inst.* **2013**, 146–151. (In Japanese)
97. Caudron, C.; Taisne, B.; Neuberg, J.; Jolly, A.D.; Christenson, B.; Lecocq, T.; Suparjan; Syahbana, D.; Suantika, G. Anatomy of phreatic eruptions. *Earth Planet. Space* **2018**, *70*, 168. [CrossRef]

98. Kobayashi, T.; Morishita, Y.; Munekane, H. First detection of precursory ground inflation of a small phreatic eruption by InSAR. *Earth Planet. Sci. Lett.* **2018**, *491*, 244–254. [CrossRef]
99. Volcanic Alert Levels of Kirishimayama (Around Ebinokogen (Ioyama)). Available online: https://www.data.jma.go.jp/svd/vois/data/tokyo/keikailevel.html (accessed on 10 January 2020).
100. Volcanic Disaster Prevention Map of Mt. Kirishima. Available online: https://www.pref.miyazaki.lg.jp/kiki-kikikanri/kurashi/bosai/bousai-kikikanri/kazan_bousai_map.html (accessed on 10 January 2020).

© 2020 by the authors. Licensee MDPI, Basel, Switzerland. This article is an open access article distributed under the terms and conditions of the Creative Commons Attribution (CC BY) license (http://creativecommons.org/licenses/by/4.0/).

Article

'Silent' Dome Emplacement into a Wet Volcano: Observations from an Effusive Eruption at White Island (Whakaari), New Zealand in Late 2012

Arthur Jolly [1,*], Corentin Caudron [2], Társilo Girona [3], Bruce Christenson [1] and Roberto Carniel [4]

1. GNS Science, Avalon, Lower Hutt 5010, New Zealand; B.Christenson@gns.cri.nz
2. ISTerre, Université Grenoble Alpes, Université Savoie Mont Blanc, 73000 Chambery, France; corentin.caudron@gmail.com
3. Jet Propulsion Laboratory, California Institute of Technology, Pasadena, CA 91109, USA; tarsilo.girona@jpl.nasa.gov
4. Laboratorio di Misure e Trattamento dei Segnali, DPIA, University of Udine, Udine, 33100 Friuli, Italy; carniel1965@gmail.com
* Correspondence: a.jolly@gns.cri.nz

Received: 2 April 2020; Accepted: 9 April 2020; Published: 14 April 2020

Abstract: The 2012–2016 White Island (Whakaari) eruption sequence encompassed six small explosive events that included one steam driven and five explosive phreato-magmatic eruptions. More enigmatic, a dome was observed at the back of the vent and crater lake in November 2012. Its emplacement date could not be easily determined due to persistent steam from the evaporating crater lake and because of the very low levels of discrete volcanic earthquakes associated with its growth. During this period, seismicity also included persistent tremor with dominant frequencies in the 2–5 Hz range. Detailed assessment of the tremor reveals a very slow evolution of the spectral peaks from low to higher frequencies. These gliding spectral lines evolved over a three-month time period beginning in late September 2012 and persisting until early January 2013, when the tremor stabilised. As part of the dome emplacement episode, the crater lake progressively dried, leaving isolated pools which then promoted persistent mud/sulphur eruption activity starting in mid-January 2013. We interpret the emplacement of the dome as a non-explosive process where the hot, mostly degassed, magma intruded slowly through the hydrothermal system in late September 2012 and cooled in a relatively quiet state. The tremor evolution might reflect the slow contraction of subsurface resonant cavities, which increased the pitch of the peak resonant frequency through time. Alternatively, spectral evolution might reflect a 'comb function' due to clockwork beating of the slowly cooling dome, although direct evidence of clockwork beats is not seen in the waveform data. Finally, it might represent frothing of the hydrothermal system ahead of the slowly propagating magma.

Keywords: dome emplacement; hydrothermal system; RSAM; tremor; gliding spectral lines; White Island; phreatic eruptions

1. Introduction

White Island (Whakaari in Te Reo Maori) is a frequently active and hazardous composite cone volcano with most of its relief lying below sea level within the south-eastern Bay of Plenty (Figure 1A). While White Island produces only small eruptions by global standards, the tragic 9 December 2019 event illustrates the possible societal impacts that relate to small eruptions at volcanoes that are frequented by tourists. Such eruptive activity was also present during the 2012–2016 period (Figure 1) which proceeded from the onset of unrest in mid-2011 [1,2], produced several well documented explosive eruptions [1,3,4] through 2012 and 2013, and culminated in April 2016 with a small phreatomagmatic event [2,5,6].

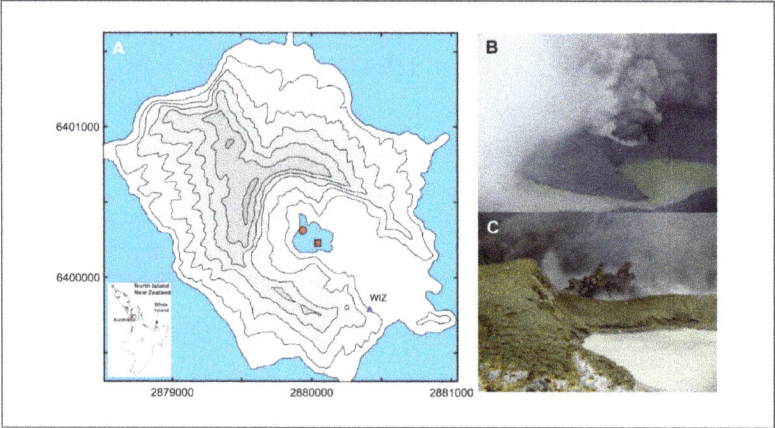

Figure 1. Map of White Island, located in the south-eastern Bay of Plenty (the topographic contour is 40 m). The single seismic station WIZ (blue triangle) and dome eruption vent (red circle) and south east eruption vent (red square) shown in (**A**). Ash venting during the 5 August post eruption ash venting period is shown for 9 August. The dome is shown in (**C**) and was taken 12 December. Note the remainder of the crater lake in the foreground. Photo (**B**) is unattributed courtesy of GeoNet, while (**C**) is by Brad Scott courtesy of GeoNet.

Eruptions at White Island are driven by persistently active magmatic degassing that is evident over the historical record [7]. This degassing requires long-term magmatic injection into shallow portions of the volcanic edifice, which may be accomplished via convective overturn within a conduit [8–10] or persistent injection of small batches of magma [11,12].

A small effusive dome was also emplaced on the back wall margin of the active lake filled eruption vent system, and hence this is an excellent example of magmatic propagation through a 'wet' volcano hydrothermal system. The dome effusion had significant consequences for the evolution of the hydrothermal system, promoting the drying of the lake, and a switch to persistent and well documented mud/sulphur eruption activity [13–16]. Dome forming eruptive activity is a frequent occurrence in volcanic systems with viscous magmas and have drawn significant scientific interest due to their persistent medium- to long-term hazards. These hazards are greatest when the domes are perched at elevation on unstable slopes of a volcanic edifice such as at volcanoes like Soufriere Hills, Montserrat [17]; Redoubt, Alaska [18]; or Merapi, Indonesia [19]. Domes may also form within positions of lower relief (e.g., Mt. St. Helens [20–22]), and in such cases, the eruptions can be emplaced through well-developed hydrothermal systems.

The hydrothermal system at White Island results from the interaction between magmatic heat and juvenile fluids with water from meteoric and oceanic sources that percolates into the crust [16]. Heat and fluids from the magma drive convective circulation that establish fluid phase transitions below the surface and a hydrostatically controlled phase equilibrium with its enclosing hydrothermal brine. In particular, at the magma-hydrothermal system interface, a single-phase gas enveloped by a two-phase fluid composed of liquid and gas bubbles dominates, whereas a single-phase liquid will eventuate at shallow levels and within the crater lake. These conditions are generally stable but can evolve in different ways depending on the position of magma within the system and externally modulating effects like rainfall (Figure 2).

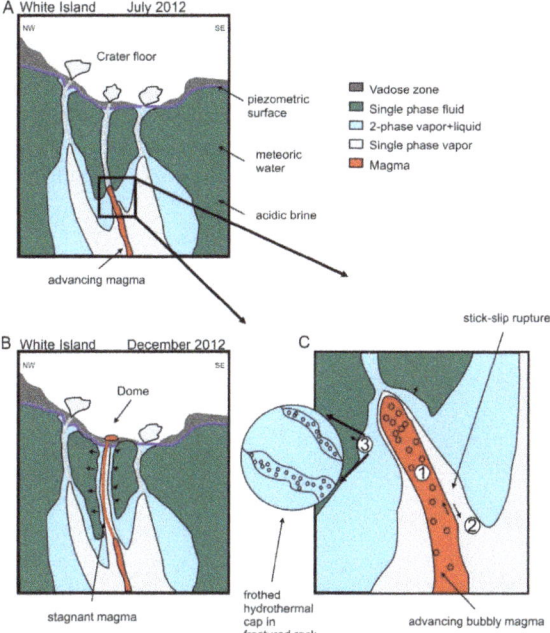

Figure 2. Schematic cross section depicting (**A**) the pre- and (**B**) post-extrusion scenarios for the dome. Acidic brine fluids underlie meteoric water/condensate lenses (both depicted green) and encapsulate the magmatic/fumarolic vent system (blue and grey). Magma ascends along the main conduit (designated by the red line in (**B**)), preceded by increased heat and gas flow which evaporates the two-phase vapour-liquid region to single-phase vapor along its way (grey). Three possible resonant systems are portrayed in the inset diagram (**C**). The bubble filled cavity (1). A clockwork stick-slip rupture mechanism (2), or the frothed expanding boiling front (3 and its inset) may produce harmonic signals discussed. The arrows in B show the expanded two-phase system that results from the propagating magma injection. See text.

From a seismological perspective, the shallow hydrothermal system may produce the full range of seismic observations seen at White Island and its global analogues like Kawah Ijen volcano, Indonesia [4,23]. Examples include discrete long-period (LP) [24], volcano-tectonic (VT) [2,25], very long period (VLP) seismicity [2], and persistent volcanic tremor [1]. Tremor is an important feature of volcanic seismicity [26]. It is characterised by a continuous, banded, or spasmodic signal, which is detectable only when exceeding the background seismic energy level [26]. Several data reduction methods [27] have been proposed to characterize its short- and long-term time evolution, that can lead to the identification of significant transitions between volcanic processes [28], sometimes triggered by external or internal events such as tectonic or volcano-tectonic (VT) seismic events [29]. Volcanic tremor can also occur precursory to an eruption [30–33]. However, the origins of tremor may vary at different volcanoes, and are often poorly understood; moreover, many different seismic sources can act at the same time and combine to produce the signal of interest [27].

Harmonic tremor, consisting of a fundamental frequency and evenly spaced overtones, is a common occurrence at volcanic systems worldwide [34,35], and is important both to determine possible source processes and as a monitoring tool. Changing spectral patterns in harmonic tremor (e.g., migration of the spectral lines, i.e., so-called gliding) is increasingly recognised at volcanoes [36–38] as an important short-term precursor that may occur several minutes prior to explosive activity.

Systematic evolution of spectra over longer time periods are less common; however, an assessment of such slowly evolving systems may also be important from a process and hazard standpoint [39].

It is surmised that movement of magma below or within a hydrothermal system should produce significant changes in seismic observations [40–42] (Figure 2). Hence, the extrusion of magma through a well-established hydrothermal system and onto the surface provides an excellent opportunity to re-examine remote monitoring data for hazards implications. The dome forming eruption at White Island is interesting in this context because its discovery was truly enigmatic. It was first documented by GeoNet on 11 December 2012 but was observed by White Island tour operators possibly a fortnight earlier (Volcanic Alert Bulletin WI-2012/16; https://www.geonet.org.nz). Surprisingly, its emplacement occurred without dramatic seismicity revealed on the permanent White Island seismic station (Figure 1A). Likewise, an evaluation of the White Island web camera system did not reveal the appearance of the small dome, mostly due to persistent obscuring steam plumes for the period from the initial onset of eruption activity on 5 August 2012 [1]. It also appeared to have been quite effusive, based on the smooth lumpy texture of the surface (Figure 1C). The dome, with an extent of 20–30 m in diameter and a height of 10–15 m, has a volume on the order of $1-7 \times 10^3$ m^3. Below, we retrospectively document the pre-eruption seismicity from the onset of eruptive activity from August to the observation of the dome in December 2012.

2. Seismic Data and Results

White Island seismic monitoring was composed of a single broadband seismic sensor at site WIZ (Figure 1A), which is comprised of a Guralp 3ESP seismometer and a Quanterra Q330 digitiser, sampling at 100 Hz and telemetering data in real time to the GeoNet data center. The archived seismic data were processed in two ways: (1) calculation of the real-time seismic amplitude measurement value (RSAM) by taking the time-series data from the vertical component sensor, correcting it to velocity, and computing the root mean square (RMS) amplitude within one minute long, non-overlapping windows (Figure 3A); and (2) computing the spectra via the fast-Fourier transform (FFT) for each one-minute window (Figure 3B). For the latter, we visualised the data by picking the peak amplitude from the spectra and monitoring that over the period of interest (before the first eruption to the end of the dome forming phase).

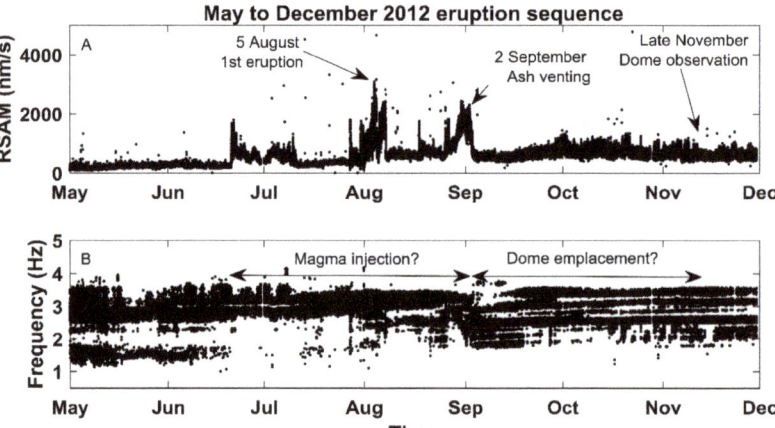

Figure 3. Real-time seismic amplitude measure (RSAM) computed from one-minute moving windows (**A**) and the maximum spectral peak amplitude (**B**) from Fast Fourier Transform of the same one-minute window. The 6 August 6 eruption and 2 September ash venting episode and the dome emplacement are marked (**A**) and the inferred injection of magma into the deeper hydrothermal system and emplacement of the dome are shown in (**B**).

The one-minute RSAM observations (Figure 3A) show the first eruption on 5 August 2012 (RSAM~2800 nm/s), as well as a persistent high amplitude signal in early September (~2100 nm/s), followed by sustained moderate level tremor (500–1000 nm/s) through the time of the dome observation in late November 2012. The spectral analysis (Figure 3B) shows that the peak frequency of tremor is generally focused within the ~1.0–4.0 Hz band, with specific and persistent peaks and notable evolutionary patterns. We note with particular interest the spectral changes associated with high amplitude tremor periods, including (1) a shift from broader spectrum tremor to 2.5–3.5 Hz in mid-June 2012; (2) the onset of a migration of the spectral peaks (termed gliding spectral lines in the literature [36]) in mid-July; (3) the onset of broader spectra tremor with slowly gliding spectral lines in early September; and (4) establishment of stable (non-gliding) persistent spectral peaks in late December (see red dots in Figure 4). These spectral changes of tremor are linked in time to specific aspects of volcanic activity, including: (A) Rapid volcanic lake fluctuations of about 4 m and minor geysering (observed in June/July); (B) the onset of the first strong eruption on 5 August; (C) persistent ash venting in early September; and (D) the first observation of the dome in late November.

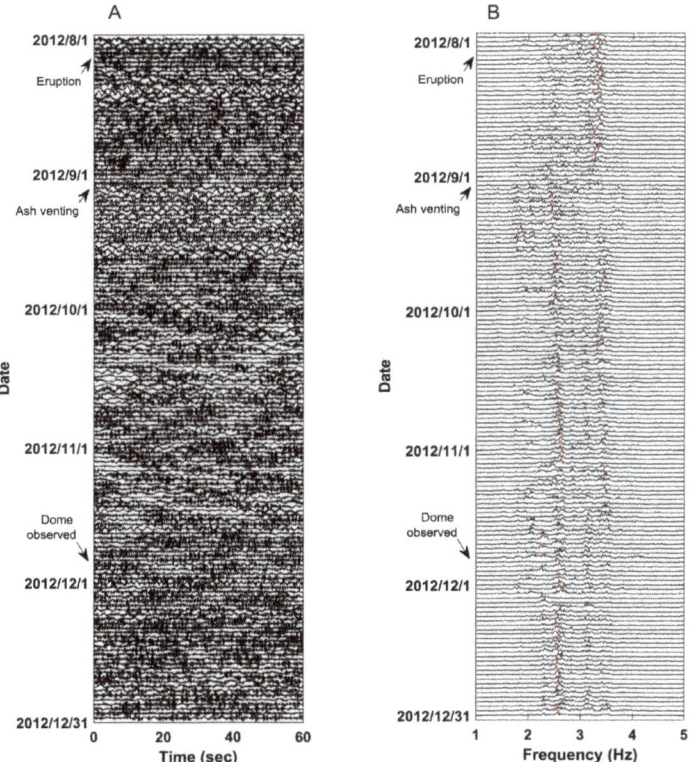

Figure 4. Example waveforms (**A**) and spectra (**B**) for a subset of the seven-month period of interest. The examples are extracted from the continuous seismic data and include one minute from the start of each Julian day (UT HH:MM 00:00 to 00:01). The initial eruption occurred on 5 August 2012 at NZST 04:55, and the waveforms are not shown. Note the change in peak frequency around the time of the secondary ashing on 2 September 2012 NZST. Also note the slow migration of peak frequencies until the observation of the dome (note red marks for peak spectral frequency). The crater was largely obscured by steam from the period of ashing to the observation of the dome.

To aid in interpreting the seismic observations, example waveforms (Figure 4A) and spectra (Figure 4B) are plotted from the first minute of each day for the period 1 July to 31 December 2012 (Figures 3 and 4 are presented in UTC). Although we realize that this procedure potentially produces an aliasing effect, the individual sample waveforms and spectra match closely a denser spectral analysis and confirm the observations in Figure 3. Hence, we regard the observations in Figures 3 and 4 as robust, illustrating both the longer-term features of the tremor and the slowly evolving migration from lower to higher spectral frequencies.

Specific spectral peaks become well established in late August 2012, which then perceptibly migrate towards higher frequencies over a three-month period. On ~ 29 August 2012, the specific peaks of interest are at 2.1 Hz (Figure 4B). These peaks shift to a stable frequency of 2.3 Hz by 30 November 2012 (Figure 4B), about the same period that the dome must have been emplaced based on the tremor and the first observation of the dome. A subsidiary peak is also observed, which weakly emerges at ~2.9 Hz after the ashing episode and persists after the first dome observation at a modestly higher frequency of ~ 3.2 Hz. There are also stationary spectral peaks observed at other specific frequencies (Figures 3 and 4B) as part of a possible harmonic pattern which began in early September and evolved until the dome was observed.

To assess possible harmonic patterns, we computed spectrograms using two-minute-long windows. Each time window is detrended with a mean and a linear function before tapering using a Hanning window (10% on each side). The frequency resolution is 0.0083 Hz. For each column of the spectrogram the resulting FFT amplitudes are color-coded and shown in Figure 5. After smoothing the results with a median of five days, we picked the central frequencies of continuous spectral lines: 1.4, 1.6, 1.8, 2.0, 2.2, 2.3, 2.7, 2.9, 3.2, 3.5, 4.9 Hz. We note that 2.2 Hz and 2.3 Hz seem to merge in September.

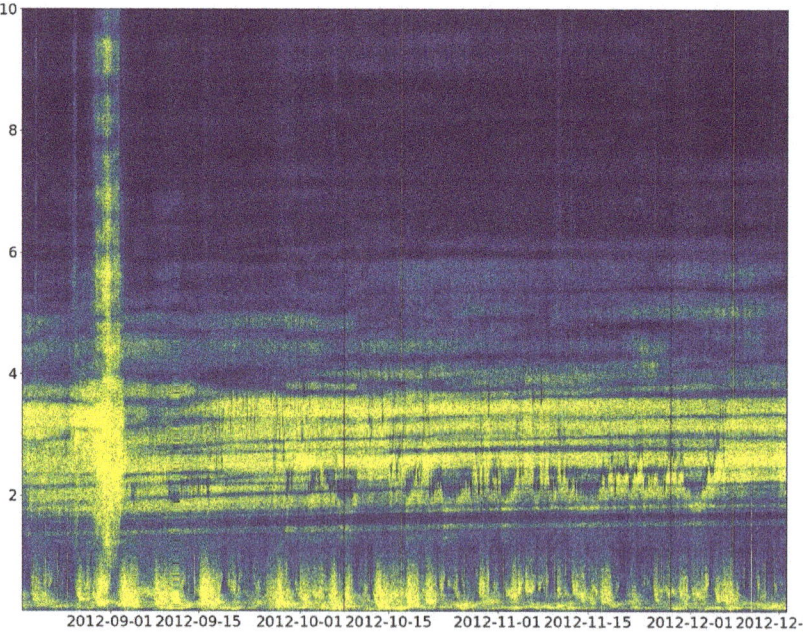

Figure 5. Spectrogram computed using two-minute-long windows after removing the mean, the trend and applying a cosine taper in each window. Yellow colours correspond to high amplitudes, whereas blue colours correspond to low amplitudes.

We then assess if the data were consistent with a harmonic tremor source. In fact, although peaks often appear to be equally spaced in frequency, the dominant frequency is rarely seen. We therefore developed an algorithm to automatically compute this possibly buried dominant frequency.

For each column of the spectrogram (i.e., 2 min), the algorithm determines all the peaks by looking for local maxima in the column. An histogram of these peaks is then built over a wider time window (e.g., 4 h or 12 h) by counting for each frequency how many times a peak is detected. Finally, we search for our dominant frequency in the 0.3–1.3 Hz range, and cumulate histogram values not only for that candidate dominant frequency but also for its potential harmonics. The cumulated value is then normalized (i.e. divided by the number of harmonics) because lower frequencies may yield higher numbers of harmonics and higher cumulated values.

For each column, we determine the potential dominant frequency, and we superimpose it, together with all its harmonics, on the spectrogram to assess its coherency with time. The results (Figure 6 for time windows of 4 h, and in Figure 7 for time windows of 12 h) are consistent with a harmonic oscillation over a narrow frequency band, hence confirming that the tremor could originate from a harmonic process. We must acknowledge that the observation is based on the sole single station observation and hence may include other influences such as path and structural effects in this complex volcanic edifice.

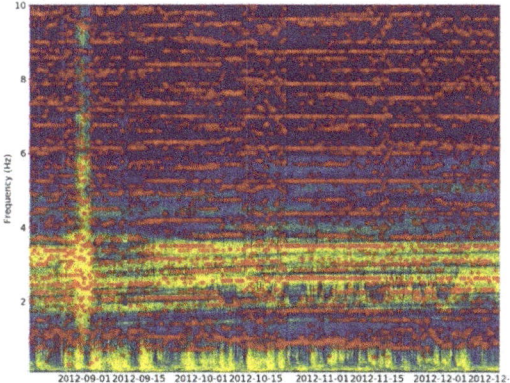

Figure 6. Same as Figure 5, but with potential dominant frequency and corresponding harmonics overlaid (red dots in 4-hr time windows).

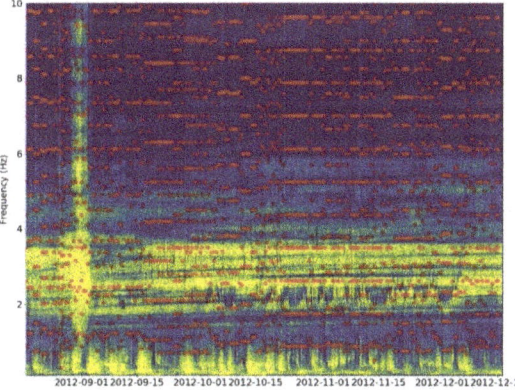

Figure 7. Same as Figure 6, but with potential dominant frequency and corresponding harmonics overlaid (12-hr time windows).

3. Discussion

Whilst the migration in spectral peaks is very small compared to other well-known examples like Soufriere Hills [36] and Redoubt [38], the spectral gliding observed for White Island occurs over a much longer time period of ~3–4 months. This timescale is more comparable to the one observed at Villarrica volcano [43], although in that case the frequency changes were more abrupt. In all of these cases, the spectral changes can be linked to eruptive activity at the volcano. The gliding frequencies at White Island are unusual in two respects. First, they persisted over several months, starting with the onset of low-level ash eruptions in early September 2012, and the observation of a dome over 2.5 months later in late November. Second, while the gliding spectral lines observed prior to large eruptive events (e.g., Soufriere Hills [36]; Redoubt [38]) seem to have persistent monotonically increasing frequencies (concave upward trending spectral peaks) for fundamental frequencies and overtones, the observations here include increasing spectral peak-lines with a reduced rate of change over time (i.e., a concave downward increase in frequency). This observation might suggest a different mechanism for the two processes (Figures 3B and 5).

Harmonic features in seismic data are often regarded as part of three potential processes, including fluid-elastic [39,40,44–54], frictional-induced [38,41], and permeable flow-controlled resonance [42,46]. In the first source process (see Figure 2C label (1)), fluid-elastic resonance involves the vibrations of cavities as a result of the motion of fluids in response to short-term perturbations (e.g., possibly the superposition of hybrid earthquakes thought to occur as magma proceeds through the glass transition at shallow depth [55]. In such a case, the different modes of vibration vary with the impedance contrast (i.e., with the properties of the fluid filling the cavity) and geometry [35,56–58]. In the second source process (Figure 2C label (2)), a frictional mechanism involves the generation of harmonic tremor by the superposition of highly regular repetition of mostly identical stick-slip earthquakes [41]. In such a case, the dominant frequencies vary with the elastic properties of the medium, stressing rates, and geometry, whereas overtones are the result of a Dirac comb effect. In the third source process (see Figure 2C label (3)), a permeable flow-controlled mechanism involves the spontaneous vibration of cavities while gas escapes toward the surface and is the result of the transient porous flow of magmatic/hydrothermal gases through the permeable medium that caps the cavity. In such a case, the pressure inside the cavity is governed by the equation of a non-linear oscillator, which reduces to a linear first-order harmonic oscillator for highly-fractured and thin ($\leq$100 m) caps [42]; hence, the specific frequencies depend on the permeability and porosity of the cap, gas properties, supply, and geometry of the cavity. In particular, a harmonic tremor may emerge through a Dirac comb effect when gas feeds the cavity regularly (e.g., through bubble clouds or foam collapse, [59–61]). Alternatively, a harmonic tremor may arise as a non-linear effect [46] or for thick ($\gtrsim$ 100 m) caps as result of the high-order terms controlling the pressure oscillations of the cavities [42].

At White Island, the observation of the dome in late November 2012 and the lack of strong seismicity associated with its appearance is enigmatic and can provide some constraints regarding the governing tremor mechanism. Persistent steam obscured the back of the crater lake and this might imply that the dome had been emplaced much earlier, possibly around the 2 September ash venting period shown in Figure 4. If this did mark the dome emplacement, then the slow cooling of the dome and conduit may have contributed to the spectral gliding within the tremor. For example, the slow progressive cooling and degassing might provide both the required seismic trigger mechanism and the higher impedance contrast cavity to produce harmonic tremor through fluid-elastic resonance. If a magmatic root remained beneath the dome, such a feature could hold exsolved gases within a bubble-rich root structure; Neuberg and O'Gorman [58] have shown how such a conduit could produce resonant tremor. If this conduit became progressively further degassed or if progressive top down solidification occurred, this could produce a smaller conduit with time. The slowly evolving changes in tremor might then reflect variations of the resonant root structure.

As an alternative, the slowly evolving spectral features could be regarded as a feature of the emplacement process if the dome were emplaced shortly before its observation in late November.

The 2–5 Hz tremor source process at White Island, thought to be dominantly within the hydrothermal system, is a long-term feature of the shallow volcanic system [1]. In this case, the modulation of the shallow seismicity might reflect a longer-term intrusion process at depth. If the sub-surface magma began its interaction with the shallow hydrothermal system about 2.5 months before the dome's observation, then the ascent rate of intrusion would have been very slow. We regard this alternative as less likely due to the apparent fluidity of the dome features, however it might be relevant to the spectral observations occurring prior to the 5 August explosive eruption (see Figure 3B).

Regarding the second source process (Figure 2), in order to produce clear overtones in the spectrogram, the repetition interval of seismic events must be very precise. However, we do not observe clear discrete earthquakes in the sequence (Figure 4A), so the application of a repeating clockwork pattern, and thus frictional-induced resonance, is not easy to invoke for White Island in this period. In any case, it might be possible for interactions of hydrothermal system fluids with the conduit walls to act as a source of persistent micro-seismicity. Whilst the occurrence of a regular clockwork pattern is not apparent in the waveform data (Figure 4A), it could still be part of the underlying tremor excitation process. This process might also be only weakly periodic but still interact with a nearby resonant cavity [62]. If, on the other hand, the dome was extruded immediately prior to its first observation in November, it is plausible that the tremor observed over the period September to November could be the result of interaction between the ascending magma column and the hydrothermal system through which it is moving. In Figure 2C labeled (3), we illustrate this process with the front of the ascending plug intruding into the overlying two-phase vapor–liquid region extant at the time. This process could explain the earlier June to September spectral features. Temperature gradients above the ascending magma are very high, ranging from an assumed sub-liquidus intrusion temperature of ~ 900 °C to the temperature of the vapour–liquid saturation curve. The overlying conduit, assumed to be a porous fractured medium, is two-phase over its entire length to the surface (see Figure 2C). This conduit could deliver non-condensable gas to an overlying compressible "gas pocket" created beneath a partial mineralogic seal. Such a seal system is argued to be present based on C/S ratios of vent emissions at the time [16].

Gas pockets trapped beneath permeable media or embedded in cracks within the dome may lead to spontaneous vibrations while gas escapes toward the surface [42] but also through thermal instabilities [54]. In these resonant gas pockets, gliding spectral behaviour results from changing lengths of the cavity, with shorter lengths generating progressively higher frequencies over the spectral range of 0.1 Hz to 15 Hz. It is possible that this could result from shortening of the conduit cavity during magma ascent, although this process cannot be definitively confirmed by the existing data. Whereas spectral gliding would be complete when the magma reaches the surface, a tremor would be ongoing during cooling and the ongoing interaction between the plug/dome with the hydrothermal system, as portrayed in Figure 2C.

We regard propagation and emplacement of the dome in early September as more likely due to the strong tremor and the relatively abrupt change of the spectral patterns (Figure 3). If so, then the perturbation of the hydrothermal system led to high levels of gas and steam discharge, which obscured the dome, and caused the observed spectral gliding. It also may have promoted the slow evaporation of the crater lake system, which was a central feature of the White Island crater vent for more than 10 years before the 2012–2016 eruption episode. Regardless of the timing and mechanism of emplacement and stabilisation, it is remarkable for a dome to be emplaced into an active wet hydrothermal system with such subtle changes of the observed tremor patterns.

4. Conclusions

We describe the visual and seismic observations related to the onset of eruptive activity at White Island, New Zealand, for the period June to December 2012. The period included an explosive eruption, ash venting, and a dome emplacement episode through a 'wet' volcano hydrothermal system. The period was marked by persistent elevated tremor amplitudes and slowly evolving gliding spectral lines. At the same time, the crater lake system progressively evaporated, which encompassed the full crater floor in June 2012 and progressed to isolated pools by the end of December 2012. Interestingly, this progressive lake loss promoted the next phase of mud/sulphur eruptions which began in mid-January 2013 [13–16] through the pre-existent Southeast vent system (Figure 2).

We regard the injection of magma into the shallow hydrothermal system as the key driver of the lake loss and shifts in tremor patterns in mid to late 2012. We interpret that the magmatic intrusion into the hydrothermal system began in early September and promoted the range of seismic observations documented here. Future work will be needed to examine the plausibility of the models outlined here with a rigorous analytical and numerical approach.

As a final note, the 9 December 2019 White Island eruption occurred within the timeframe of the development of this research and post-eruption unrest associated with that eruption is ongoing at the time of writing. Hence, this work does not include new insights from this most recent eruptive activity. At the conclusion of the present unrest period, a detailed retrospective assessment of the progression of unrest will be critical to determine similarities and differences for the two time periods. It is hoped that such an analysis will lead to ever-improving outcomes at difficult to monitor volcanoes like White Island. The present work illustrates how persistent shallow magma and well-developed hydrothermal systems may produce observations that may fit many possible conceptual models and produce challenging monitoring conditions. It is a central goal of the volcano science community to document such events and hence to improve monitoring outcomes for similar volcanic systems both in New Zealand and around the globe. This work will help to assess the viability of the various models within the context of the limited available data.

Author Contributions: Conceptualization, A.J., C.C., T.G.; B.C., and R.C.; Data curation, A.J.; Formal analysis, A.J., C.C., T.G., and R.C.; Investigation, A.J., C.C., T.G., B.C., and R.C.; Methodology, A.J., C.C., T.G. and R.C.; Visualization, A.J., C.C., T.G., B.C., and R.C.; Writing—original draft, A.J., C.C., T.G., B.C., and R.C.; Writing—review and editing, A.J., C.C., T.G., B.C., and R.C. All authors have read and agreed to the published version of the manuscript.

Funding: This research received no external funding.

Acknowledgments: This work used GeoNet seismic data which is freely available. All of the authors contributed toward manuscript and figure production as well as the data analysis and interpretation. Photos used in this study are from the GNS volcano database (VolcDB) and reflects the effort of the GeoNet team. A. Jolly and B. Christenson are supported by the Ministry of Business, Innovation and Employment (MBIE) Strategic Science Investment Fund (SSIF). T. Girona is supported by an appointment to the NASA Postdoctoral Program at the Jet Propulsion Laboratory, California Institute of Technology, administered by Universities Space Research Association under contract with the National Aeronautics and Space Administration. Chris Van Houtte and Tony Hurst reviewed an earlier version of the manuscript. We thank Alicia Hotovec-Ellis, William Chadwick, and a third reviewer for their useful comments which improved the manuscript.

Conflicts of Interest: The authors declare no conflict of interest.

References

1. Chardot, L.; Jolly, A.D.; Kennedy, B.; Fournier, N.; Sherburn, S. Using volcanic tremor for eruption forecasting at White Island volcano (Whakaari), New Zealand. *J. Volcanol. Geotherm. Res.* **2015**, *302*, 11–23. [CrossRef]
2. Jolly, A.D.; Lokmer, I.; Thun, J.; Salichon, J.; Fry, B.; Chardot, L. Insights into fluid transport mechanisms at White Island from analysis of coupled very long-period (VLP), long-period (LP) and high-frequency (HF) earthquakes. *J. Volcanol. Geotherm. Res.* **2017**, *343*, 75–94. [CrossRef]
3. Jolly, A.; Lokmer, I.; Christenson, B.; Thun, J. Relating gas ascent to eruption triggering for the April 27, 2016, White Island (Whakaari), New Zealand eruption sequence. *Earth Planets Space* **2018**, *70*, 177. [CrossRef]
4. Caudron, C.; Taisne, B.; Neuberg, J.; Jolly, A.D.; Christenson, B.W.; Lecocq, T.; Suparjanh; Syahbana, D.; Suantika, G. Anatomy of phreatic eruptions. *Earth Planets Space* **2018**, *70*, 168. [CrossRef]
5. Walsh, B.; Procter, J.; Lokmer, I.; Thun, J.; Hurst, A.W.; Christenson, B.W.; Jolly, A.D. Geophysical examination of the 27 April 2016 Whakaari/White Island, New Zealand, eruption and its implications for vent physiognomies and eruptive dynamics. *Earth Planets Space* **2019**, *71*. [CrossRef]

6. Kilgour, G.N.; Gates, S.; Kennedy, B.M.; Farquhar, A.; McSporran, A.; Asher, C. Phreatic eruption dynamics derived from deposit analysis: A case study from a small phreatic eruption from Whakaari/White Island, New Zealand. *Earth Planets Space* **2019**. [CrossRef]
7. Werner, C.A.; Hurst, A.W.; Scott, B.J.; Sherburn, S.; Christenson, B.W.; Britten, K.; Cole-Baker, J.; Mullan, B. Variability of passive gas emissions, seismicity, and deformation during crater lake growth at White Island Volcano, New Zealand, 2002–2006. *J. Geophys. Res.* **2008**, *113*, 15. [CrossRef]
8. Kazahaya, K.; Shinohara, H.; Saito, G. Excessive degassing of Izu-Oshima volcano: Magma convection in a conduit. *Bull. Volcanol.* **1994**, *56*, 207–216. [CrossRef]
9. Stevenson, D.S.; Blake, S. Modelling the dynamics and thermodynamics of volcanic degassing. *Bull. Volcanol.* **1998**, *60*, 307–317. [CrossRef]
10. Harris, A.; Carniel, R.; Jones, J. Identification of variable convective regimes at Erta Ale Lava Lake. *J. Volcanol. Geotherm. Res.* **2005**, *142*, 207–223. [CrossRef]
11. Cole, J.W.; Thordarson, T.; Burt, R.M. Magma Origin and Evolution of White Island (Whakaari) Volcano, Bay of Plenty, New Zealand. *J. Pet.* **2005**, *41*, 867–895. [CrossRef]
12. Girona, T.; Costa, F.; Schubert, G. Degassing during quiescence as a trigger of magma ascent and volcanic eruptions. *Sci. Rep.* **2015**, *5*, 18212. [CrossRef] [PubMed]
13. Jolly, A.D.; Kennedy, B.; Edwards, M.; Jousset, P.; Scheu, B. Infrasound tremor from bubble burst eruptions in the viscous shallow crater lake of White Island, New Zealand, and its implications for interpreting volcanic source processes. *J. Volcanol. Geotherm. Res.* **2016**, *327*, 585–603. [CrossRef]
14. Edwards, M.J.; Kennedy, B.M.; Jolly, A.D.; Scheu, B.; Jousset, P. Evolution of a small hydrothermal eruption episode through a mud pool of varying depth and rheology, White Island, NZ. *Bull. Volcanol.* **2017**, *79*. [CrossRef]
15. Schmid, D.; Scheu, B.; Wadsworth, F.B.; Kennedy, B.M.; Jolly, A.D.; Dingwell, D.B. A viscous-to-brittle transition in eruptions through clay suspensions. *Geophys. Res. Letts.* **2017**, *44*, 4806–4813. [CrossRef]
16. Christenson, B.W.; White, S.; Britten, K.; Scott, B.J. Hydrological evolution and chemical structure of a hyper-acidic spring-lake system on Whakaari/White Island. New Zealand. *J. Volcanol. Geotherm. Res.* **2017**, *346*, 180–211. [CrossRef]
17. Watts, R.B.; Herd, R.A.; Sparks, R.S.J.; Young, S.R. Growth Patterns and Emplacement of Andesitic Lava Dome at Soufriere, Hills, Volcano, Montserrat. In *The Eruption of Soufriere Hills Volcano, Montserrat, from 1995 to 1999*; Druitt, T.H., Kokelaar, B.P., Eds.; The Geological Society of London: London, UK, 2002; Volume 21.
18. Bull, K.F.; Anderson, S.W.; Diefenbach, A.K.; Wessels, R.L.; Henton, S.M. Emplacement of the final lava dome of the 2009 eruption of Redoubt Volcano, Alaska. *J. Volcanol. Geotherm. Res.* **2013**, *259*, 334–348. [CrossRef]
19. Hammer, J.E.; Cashman, K.V.; Voight, B. Magmatic processes revealed by textural and compositional trends in Merapi dome lavas. *J. Volcanol. Geotherm. Res.* **2000**, *100*, 165–192. [CrossRef]
20. Moore, J.G.; Lipman, P.W.; Swanson, D.; Alpha, T.R. Growth of lava domes in the crater, June 1980-January 1981. In *The 1980 Eruptions of Mt St. Helens*; Lipman, P.W., Mullineaux, D.R., Eds.; Geological Survey research Professor Paper; USGS: Washington, DC, USA, 1981; Volume 1250.
21. Swanson, D.A.; Dzurisin, D.; Holcomb, R.T.; Iwatsubo, E.Y.; Chadwick, W.W.; Casadevall, T.J.; Ewert, J.W.; Heliker, C.C. Growth of the lava dome at Mount St. Helens, Washington, 1981–1983. In *The Emplacement of Silicic Domes and Lava Flows*; Fink, J., Ed.; The Geological Society of America Special Paper; Geological Society of America: Boulder, CO, USA, 1987; Volume 212, pp. 1–16.
22. Vallance, J.W.; Schneider, D.J.; Schilling, S.P. Growth of the 2004–2006 Lava-Dome Complex at Mount St. Helens, Washington. *USGS Prof. Pap.* **2008**, *1750*, 169–208.
23. Caudron, C.; Syahbana, D.K.; Lecocq, T.; Van Hinsberg, V.; McCausland, W.; Triantafyllou, A.; Camelbeeck, T.; Bernard, A. Kawah Ijen volcanic activity: A review. *Bull. Volcanol.* **2015**, *77*, 16. [CrossRef]
24. Sherburn, S.; Scott, B.J.; Nishi, Y.; Sugihara, M. Seismicity at White Island volcano, New Zealand: A revised classification and inferences about source mechanism. *J. Volcanol. Geotherm. Res.* **1998**, *83*, 287–312. [CrossRef]
25. Carniel, R. Comments on the paper "Automatic detection and discrimination of volcanic tremors and tectonic earthquakes: An application to Ambrym volcano, Vanuatu" by Daniel Rouland; Denis Legrand; Mikhail Zhizhin; Sylvie Vergniolle. *J. Volcanol. Geotherm. Res.* **2010**, *194*, 61–62. [CrossRef]
26. Konstantinou, K.; Schlindwein, V. Nature, wavefield properties and source mechanism of volcanic tremor: A review. *J. Volcanol. Geotherm. Res.* **2003**, *119*, 161–187. [CrossRef]

27. Carniel, R. Characterization of volcanic regimes and identification of significant transitions using geophysical data: A review. *Bull. Volcanol.* **2014**, *76*, 848. [CrossRef]
28. Tárraga, M.; Martí, J.; Abella, R.; Carniel, R.; López, C. Volcanic tremors: Good indicators of change in plumbing systems during volcanic eruptions. *J. Volcanol. Geotherm. Res.* **2014**, *273*, 33–40. [CrossRef]
29. Carniel, R.; Tárraga, M. Can tectonic events change volcanic tremor at Stromboli? *Geophys. Res. Lett.* **2006**, *33*, 4–6. [CrossRef]
30. Chouet, B.A.; Matoza, R.S. A multi-decadal view of seismic methods for detecting precursors of magma movement and eruption. *J. Volcanol. Geotherm. Res.* **2013**, *252*, 108–175. [CrossRef]
31. Girona, T.; Huber, C.; Caudron, C. Sensitivity to lunar cycles prior to the 2018 eruption of Ruapehu volcano. *Sci. Rep.* **2018**, *8*, 1476. [CrossRef]
32. Ortiz, R.; García, A.; Marrero, J.M.; De la Cruz-Reyna, S.; Carniel, R.; Vila, J. Volcanic and volcano-tectonic activity forecasting: A review on seismic approaches. *Ann. Geophys.* **2019**, *62*. [CrossRef]
33. Caudron, C.; Girona, T.; Taisne, B.; Suparjan; Gunawan, H.; Kasbani; Kristianto. Change in seismic attenuation as a long-term precursor of gas-driven eruptions. *Geology* **2019**, *47*, 632–636. [CrossRef]
34. Sherburn, S.; Scott, B.J.; Hurst, A.W. Volcanic tremor and activity at White Island, New Zealand, July–September 1991. *N. Z. J. Geol. Geophys.* **1996**, *39*, 329–332. [CrossRef]
35. Neuberg, J. Characteristics and causes of shallow seismicity in andesite volcanoes. *Philos. Trans. R. Soc. Lond. A* **2000**, *358*, 1533–1546. [CrossRef]
36. Powell, T.W.; Neuberg, J. Time dependent features in tremor spectra. *J. Volcanol. Geotherm. Res.* **2003**, *128*, 177–185. [CrossRef]
37. Lesage, P.; Mora, M.M.; Alvarado, G.E.; Pacheco, J.; Métaxian, J. Complex behavior and source model of the tremor at Arenal volcano, Costa Rica. *J. Volcanol. Geotherm. Res.* **2006**, *157*, 49–59. [CrossRef]
38. Hotovec, A.J.; Prejean, S.G.; Vidale, J.E.; Gomberg, J. Strongly gliding harmonic tremor during the 2009 eruption of Redoubt Volcano. *J. Volcanol. Geotherm. Res.* **2013**, *259*, 89–99. [CrossRef]
39. Ripepe, M.; Gordeev, E. Gas bubble dynamics model for shallow volcanic tremor at Stromboli. *J. Geophys. Res.* **1999**, *104*, 10639–10654. [CrossRef]
40. Julian, B.R. Volcanic tremor: Nonlinear excitation by fluid flow. *J. Geophys. Res.* **1994**, *99*, 11859–11877. [CrossRef]
41. Dmitrieva, K.; Hotovec-Ellis, A.J.; Prejean, S.; Dunham, E.M. Frictional-faulting model for harmonic tremor before Redoubt Volcano eruptions. *Nat. Geosci.* **2013**, *6*, 652–656. [CrossRef]
42. Girona, T.; Caudron, C.; Huber, C. Origin of shallow volcanic tremor: The dynamics of gas pockets trapped beneath thin permeable media. *J. Geophys. Res.* **2019**. [CrossRef]
43. Ortiz, R.; Moreno, H.; Garcia, A.; Fuentealba, G.; Astiz, M.; Pena, P.; Sanchez, N.; Tarraga, M. Villarrica Volcano (Chile): Characteristics of the volcanic tremor and forecasting of small explosions by means of materials failure method. *J. Volcanol. Geotherm. Res.* **2003**, *128*, 247–259. [CrossRef]
44. Chouet, B. Excitation of a buried magmatic pipe: A seismic source model for volcanic tremor. *J. Geophys. Res.* **1985**, *90*, 1881–1893. [CrossRef]
45. Chouet, B. Long-period volcano seismicity: Its source and use in eruption forecasting. *Nature* **1996**, *380*, 309–316. [CrossRef]
46. Hellweg, M. Physical models for the source of Lascar's harmonic tremor. *J. Volcanol. Geotherm. Res.* **2000**, *101*, 183–198. [CrossRef]
47. Johnson, J.B.; Lees, J.M. Plugs and chugs—Seismic and acoustic observations of degassing explosions at Karymsky, Russia and Sangay, Ecuador. *J. Volcanol. Geotherm. Res.* **2000**, *101*, 67–82. [CrossRef]
48. Lane, S.J.; Chouet, B.A.; Phillips, J.C.; Dawson, P.; Ryan, G.A.; Hurst, E. Experimental observations of pressure oscillations and flow regimes in an analogue volcanic system. *J. Geophys. Res.* **2001**, *106*, 6461–6476. [CrossRef]
49. Balmforth, N.J.; Craster, R.V.; Rust, A.C. Instability in flow through elastic conduits and volcanic tremor. *J. Fluid Mech.* **2005**, *527*, 353–377. [CrossRef]
50. Fujita, E.; Araki, K.; Nagano, K. Volcanic tremor induced by gas-liquid two-phase flow: Implications of density wave oscillation. *J. Geophys. Res.* **2011**, *116*, B09201. [CrossRef]
51. Jellinek, A.M.; Bercovici, D. Seismic tremors and magma wagging during explosive volcanism. *Nature* **2011**, *470*, 522–526. [CrossRef]

52. Bercovici, D.; Jellinek, A.M.; Michaut, C.; Roman, D.C.; Morse, R. Volcanic tremors and magma wagging: Gas flux interactions and forcing mechanism. *Geophys. J. Int.* **2013**, *195*, 1001–1022. [CrossRef]
53. Lipovsky, B.P.; Dunham, E.M. Vibrational modes of hydraulic fractures: Inference of fracture geometry from resonant frequencies and attenuation. *J. Geophys. Res. Solid Earth* **2015**, *120*, 1080–1107. [CrossRef]
54. Montegrossi, G.; Farina, A.; Fusi, L.; de Biase, A. Mathematical model for volcanic harmonic tremors. *Sci. Rep.* **2019**, *9*, 14417. [CrossRef] [PubMed]
55. Neuberg, J.; Tuffen, H.; Collier, L.; Green, D.; Powell, T.W.; Dingwell, D. The trigger mechanism of low-frequency earthquakes on Montserrat. *J. Volcanol. Geotherm. Res.* **2006**, *153*, 37–50. [CrossRef]
56. Chouet, B. Resonance of a fluid-driven crack: Radiation properties and implications for the source of long-period events and harmonic tremor. *J. Geophys. Res.* **1988**, *93*, 4375–4400. [CrossRef]
57. Kumagai, H.; Chouet, B. Acoustic properties of a crack containing magmatic or hydrothermal fluids. *J. Geophys. Res.* **2000**, *105*, 25493–25512. [CrossRef]
58. Neuberg, J.; O'Gorman, C. A Model of the Seismic Wavefield in Gas-Charged Magma: Application to Soufrière Hills Volcano, Montserrat. In *The Eruption of Soufrière Hills Volcano, Montserrat, from 1995 to 1999*; Druitt, T.H., Kokelaar, B.P., Eds.; Geological Society of London: London, UK, 2002; Volume 21, pp. 603–609.
59. Manga, M. Waves of bubbles in basaltic magmas and lavas. *J. Geophys. Res.* **1996**, *101*, 17457–17466. [CrossRef]
60. Michaut, C.; Ricard, Y.; Bercovici, D.; Sparks, R.S.J. Eruption cyclicity at silicic volcanoes potentially caused by magmatic gas waves. *Nat. Geosci.* **2013**, *6*, 856–860. [CrossRef]
61. Spina, L.; Cimarelli, C.; Scheu, B.; Di Genova, D.; Dingwell, D.B. On the slow decompressive response of volatile- and crystal- bearing magmas: An analogue experimental investigation. *Earth Planet. Sci. Lett.* **2016**, *433*, 44–53. [CrossRef]
62. Jousset, P.; Neuberg, J.; Sturton, S. Modelling the time-dependent frequency content of low-frequency volcanic earthquakes. *J. Volcanol. Geotherm. Res.* **2003**, *128*, 201–223. [CrossRef]

© 2020 by the authors. Licensee MDPI, Basel, Switzerland. This article is an open access article distributed under the terms and conditions of the Creative Commons Attribution (CC BY) license (http://creativecommons.org/licenses/by/4.0/).

Article

Pressure Controlled Permeability in a Conduit Filled with Fractured Hydrothermal Breccia Reconstructed from Ballistics from Whakaari (White Island), New Zealand

Ben M. Kennedy [1,*], Aaron Farquhar [2], Robin Hilderman [2], Marlène C. Villeneuve [1,3], Michael J. Heap [4], Stan Mordensky [1], Geoffrey Kilgour [5], Art. Jolly [5], Bruce Christenson [5] and Thierry Reuschlé [4]

1. Department of Geological Sciences University of Canterbury, Private Bag 4800, Christchurch 8140, New Zealand; marlene.villeneuve@unileoben.ac.at (M.C.V.); stan.mordensky@gmail.com (S.M.)
2. Colorado College, Colorado Springs, Colorado, CO 80903, USA; farquharaaron@gmail.com (A.F.); robinhilderman@gmail.com (R.H.)
3. Chair of Subsurface Engineering, Montanuniversität Leoben, Erzherzog Johann-Straße 3, A-8700 Leoben, Austria
4. Université de Strasbourg, CNRS, Institut de Physique de Globe de Strasbourg UMR 7516, F-67000 Strasbourg, France; heap@unistra.fr (M.J.H.); thierry.reuschle@unistra.fr (T.R.)
5. GNS Science, 1 Fairway Drive, Lower Hutt 5011, New Zealand; G.Kilgour@gns.cri.nz (G.K.); a.jolly@gns.cri.nz (A.J.); B.Christenson@gns.cri.nz (B.C.)
* Correspondence: Ben.kennedy@canterbury.ac.nz

Received: 24 March 2020; Accepted: 8 April 2020; Published: 11 April 2020

Abstract: Breccia-filled eruption conduits are dynamic systems where pressures frequently exceed critical thresholds, generating earthquakes and transmitting fluids. To assess the dynamics of breccia-filled conduits, we examine lava, ash tuff, and hydrothermal breccia ballistics with varying alteration, veining, fractures, and brecciation ejected during the 27 April 2016 phreatic eruption of Whakaari/White Island. We measure connected porosity, strength, and permeability with and without tensile fractures at a range of confining pressures. Many samples are progressively altered with anhydrite, alunite, and silica polymorphs. The measurements show a large range of connected porosity, permeability, and strength. In contrast, the cracked samples show a consistently high permeability. The cracked altered samples have a permeability more sensitive to confining pressure than the unaltered samples. The permeability of our altered ballistics is lower than surface rocks of equivalent porosity, illustrating that mineral precipitation locally blocked pores and cracks. We surmise that alteration within the conduit breccia allows cracks to form, open and close, in response to pore pressure and confining pressure, providing a mechanism for frequent and variable fluid advection pulses to the surface. This produces temporally and spatially variable geophysical and geochemical observations and has implications for volcano monitoring for any volcano system with significant hydrothermal activity.

Keywords: alteration; porosity; eruption; fracture; permeability

---

1. Introduction

The tragic events on November 2019 at Whakaari (White Island; New Zealand), which killed over 20 people, highlight the need for a better understanding of the controls on phreatic eruptions. In particular, the behaviour of breccia-filled conduits remains a potentially controlling factor on the pressurization timescales and total pressure build-up. Many large volcanic eruptions begin with a phreatic/hydrothermal eruption that is responsible for clearing a pathway for magma to reach the surface [1]. Similarly, shallow intrusions and economically significant mineral resources are frequently associated with hydrothermal breccia sheets and pipes [2]. The physical properties of these breccia filled conduits is thus critical to volcano monitoring and economic mineral exploration.

The physical properties of the altered materials that fill hydrothermal conduits directly control (1) magma outgassing efficiency, (2) the build-up of pressure that can lead to explosive eruptions [3], and (3) subsequent fluid flow, hydrothermal alteration, and mineralization. Past studies of magmatic conduits and breccias reveal insights into magma flow, outgassing, and fragmentation processes [4–7]. Detailed studies of hydrothermal breccia filled conduits, however, are dispersed between the maar diatreme, hydrothermal mineralization, and petroleum basin literature [2,8–11] and are rarely applied to eruption models.

Until now, studies have described the physical properties of rocks collected from the crater floor of Whakaari to understand their mechanical behaviour and permeability [12,13] and to interpret fragmentation [14], however, these rocks may not be representative of the material filling the vent. Instead, the physical properties of volcanic ballistic samples can be used to provide insight into conduit processes [15]. Therefore, the ballistic field of the 2016 eruption [16] provides a unique opportunity to directly sample the pre-eruption conduit and the magmatic hydrothermal interface beneath the crater at Whakaari. Here, we combine the exceptional altered ballistic samples with established rock mechanical methods to make implications that can be directly applied to an active hazardous volcano.

Whakaari is an andesite stratocone, whose peak emerges 300 m from the Pacific Ocean 48 km off the east coast of New Zealand. The magmatism is the product of oblique western subduction of the Pacific plate under the Australian plate. Whakaari has been New Zealand's most active volcano in recent history. The most recent magmatic and hydrothermal eruption period has lasted from 1976 until present exhibiting a range of magmatic, phreatomagmatic, and hydrothermal eruptions [16,17]. Whakaari has undergone a complex sequence of unrest and minor eruptions since 2011. Geophysical [18–20], geochemical [21], and observational studies [22] reveal a complex interplay between a shallow magmatic system and surface hydrology with volcanic activity modulated by a variably permeable hydrothermal system. Magmatic gas is transferred [23] towards the surface via branching pathways to a series of vents and fumaroles [21]. This model is consistent with studies of hydrothermal vents elsewhere in New Zealand [24]. Quantifying the material properties and the gas transfer mechanisms is critical to developing realistic geophysical and fluid chemistry models and to understanding the hydrothermal magmatic interface with particular relevance to andesite volcanoes with well-developed hydrothermal systems.

Investigating the nature of the rocks that surround magma bodies allows for the interpretation of geophysical signals. For example, volcanic eruptions have been successfully forecast using volcanic earthquakes [25–29]. Earthquake types are inherently tied to the physical properties of the material they travel through [23,30] and the fluid within any cracks [31]. Similarly, models of deformation [20,32] and fluid flow [33] are reliant on the physical properties of conduits and volcanic edifice. During the current period of unrest at Whakaari, geophysical, geochemical, and geological phenomena seen at the surface are interpreted as a direct result of fluids travelling through variably altered volcanic rock.

Alteration affects the physical properties of volcanic rocks [34–37]. Alteration may drive the loss of porosity and permeability via mineral precipitation in pores and cracks [3,37,38] and/or cementation [39]. Reductions in porosity due to alteration can promote brittle behaviour [39]. Conversely, alteration may drive dissolution and pore creation with measurable porosity and permeability increase [40,41] and promote a ductile failure mode [38,42,43]. Additionally, microcracking is a critical textural consequence of alteration related processes. Fractures in volcanic environments typically form in response to tectonic, intrusive, hydration, or thermal forces [44–46], however, they can also form in response to specific thermally-driven mineral reactions and drive porosity/permeability changes [47,48]. Cracks are an important textural consideration in volcanic conduits, and their influence on, for example, the physical properties of rocks may be highly dependent on lithology and confining pressure, particularly in altered volcanic rocks [37,46,49–52].

Here, we present textural and petrological characteristics of volcanic ballistic samples from the 27 April 2016 hydrothermal eruption of Whakaari. We then present strength, porosity, and permeability variations as a function of confining pressure (i.e., depth). These data are compiled into a conceptual model in which transient cracks control fluid flow in an altered breccia-filled conduit.

2. Materials and Methods

2.1. Sampling and Sample Preparation

Ballistics were collected from various sites 15–200 metres from the rim of Whakaari's crater lake (Figure 1a) during several brief visits following the 27 April 2016 eruption by Kilgour, Farquhar, and Christenson on 2 May, 2 June, and 9 June, 2016 [53]. Three short sampling missions from three different scientists minimised individual exposure following protocol outlined in [54]. A quantitative analysis of a statistically relevant sample set of ballistic lithologies in the field via photo survey was not possible due to the partially buried and discoloured nature of the ballistics and the limited time for systematic sampling within the Health and Safety Risk considerations of GNS Science and University of Canterbury (Figures 1c and 2) [16]. Further, our sampling was biased towards ballistics that were suitable for cutting and drilling (>10 cm max. diameter) (Figure 2). We also include some additional (nonballistic) measurements made on samples of lava, ash tuff, and sulfur cemented tuff collected from surfaces exposed in 2015 (location details provided in [12,13].

Figure 1. (**a**) Whakaari's crater lake with ballistic collection area highlighted; (**b**) location of Whakaari north-east of the central North Island of New Zealand; (**c**) typical ballistic crater showing discoloration and partial burial of the ballistic.

Figure 2. Photographs showing examples of collected ballistic samples from the 2016 eruption at Whakaari cut in two to reveal interior, (**a**) Unaltered lava, (**b**) altered lava, (**c**) Ash Tuff, (**d**) Hydrothermal breccia, (**e**) hydrothermal breccia of altered and unaltered lava.

The collected blocks were subsampled by drilling 20 mm diameter cylinders nominally 20–40 mm long. After porosity and permeability were measured on these intact samples, electrical tape was wrapped around the cores and they were loaded diametrically in compression until the appearance of a through-going tensile macrofracture (i.e., a "Brazilian" indirect tensile test). The tape ensured that the two sample halves remained in contact upon failure, so that the permeability of the fractured samples could be measured in the laboratory (following the method described in [49]).

An ideal methodology would have allowed all samples to have been analysed using all methods. However, this was not possible, due to (1) the timing of availability of samples and experimental facilities across four institutions, three countries, and three student projects, (2) the destructive nature of some of the methodologies (e.g., XRD and accidental sample destruction during drilling), (3) available time for analysis. Our focus was on characterising the permeability and porosity of the rocks that comprised the conduit and encompassing the major breccia lithologies. We made 60 permeability measurements at 3 MPa confining pressure following the standard procedure at University of Canterbury, and a further 25 measurements were done at 1 MPa (following the standard procedure at the University of Strasbourg); all of these had a corresponding connected porosity measurement (Table S1 supplemental data). Nine samples were selected for the two series of time-consuming variable confining pressure experiments: one series without fractures and then the same samples used in the second series with tensile fractures (Table S2 Supplemental data). A subset of 17 of these were chosen for uniaxial compressive strength (UCS) tests (Table S3 Supplemental data). Individual sample numbers and lithological groupings are reported in the supplemental data.

Fifteen thin sections were selected for SEM analysis, and 18 samples were chosen for XRD analysis. These samples were selected to represent the range of lithologies and styles of alteration and allow some overlap of techniques on the same sample. We focus our presentation here on representative samples and textures.

2.2. SEM and XRD Preparation and Methods

Thin sections were prepared to dimensions of 25 × 47 mm, ground to 30 μm thick, and mounted without cover slips. Most samples contained variability in alteration degree, and whenever possible, thin sections were prepared so as to include both sides of these mineralogical boundaries. Thin sections of eight samples were analyzed using a JEOL-IT300 Scanning Electron Microscope (SEM). This was to determine the mineral composition of the rock and the infill in open pore spaces. Backscatter Electron Detector Contrast (BED-C) images and Energy Dispersive X-ray Spectroscopy (EDS) analyses were obtained with images collected at various scales ranging from 60× to 600× magnification, although our results here display all 100× images. The weight-percent ratios of elements within points of each thin section were measured using the EDS detector in order to calculate mineralogy. Compositional element maps (EDS) and lithological structures (BED-C) were produced using the computer program AZtec.

The mineralogy of rock powders and also clay separates were analysed using a Philips PANalytical X'Pert Pro X-ray diffraction (XRD) machine, using X-rays at 2θ angles of 7–70°. Dislodged electrons were received through a 1/4° slit. Data were received into the PANalytical computer software Data Collector and then processed to determine mineralogy in the PANalytical computer software High Score.

Two factors were combined in order to determine the most likely mineralogy of each sample. (1) The d-spacing peaks of the minerals within the sample were cross-referenced within the mineral database in High Score to generate a list of possible minerals. (2) Their likelihood of presence within the volcanic-hydrothermal system in combination with existing SEM data. The most abundant peaks for each sample (>50%) were considered a dominant mineral.

In order to determine which clays are present in the ballistic samples, the clays were physically separated from powders onto glass slides. Ten grams of each sample were measured on a precision scale and mixed with 200 mL of distilled water. This was mixed in a Waring blender for two minutes at high power. After mixing, the liquid and fine fraction of the mixture was rapidly decanted into a beaker. This sample was then placed into a 50 mL centrifuge tube and centrifuged at 750 rpm for two minutes. The centrifuged sample was quickly decanted into a second 50 mL tube, which was also centrifuged. This liquid was decanted, and the clay fraction at the bottom of the tube was suctioned out by pipette and placed onto a glass slide. This was placed in the oven at 90 °C for at least one hour, until the sample was dry and ready for XRD analysis. When running clay separation slides, 2θ angles of 5–65° and a 1/8° slit were used.

2.3. Physical Laboratory Measurements

Skeletal volume measurements were undertaken using an AccuPyc II 1340 pycnometer. We used helium and ultra-high purity nitrogen for the measurements performed at the Institut de Physique du Globe at the University of Strasbourg (IPGS; France) and the University of Canterbury (New Zealand), respectively. These values were used to calculate connected porosity using the bulk sample volume.

Samples (20 × 20 mm) of unaltered intact lava (WI20 series) and fractured lava (WI20 series) and high-permeability ballistic samples (20 mm in diameter and nominally 20–40 mm in length) were measured at the IPGS. Permeability was measured using a benchtop gas (nitrogen) permeameter (see Farquharson et al., 2016; Heap and Kennedy, 2016). The oven-dry samples were first allowed to equilibrate for one hour at the target confining pressure of 1 MPa (the confining fluid used was also nitrogen gas). We used the pulse-decay technique to measure the intact lava samples and the steady-state method to measure the fractured lava and porous ballistic samples. For the pulse-decay measurements, a pore pressure differential of 0.2 MPa (measured using a Keller pressure transducer) was first imposed on the sample for a duration of one hour. The valve to the gas bottle was then closed and the decay of the upstream pore pressure (within a known volume) across the sample was recorded as a function of time. These data were then used to assess permeability using Darcy's law and to check for any ancillary corrections, such as the Forchheimer and Klinkenberg corrections, which were applied on a case-by-case basis [55]). These pulse-decay measurements required the Klinkenberg correction. For the steady-state measurements, steady-state volumetric flow rates were measured using a Bronkhorst gas flowmeter for six different pressure differentials. Steady-state was ensured by waiting for the volumetric flow rate and differential pressure to stabilize before measurements were recorded. These data were then used to assess permeability using Darcy's law and to check for any ancillary corrections, such as the Forchheimer and Klinkenberg corrections, which were applied on a case-by-case basis. These steady-state measurements required the Forchheimer correction.

Samples (20 × 20 mm) of unaltered intact lava (WI20 series) and fractured lava (WI20 series) were also measured at IPGS up to a confining pressure of 30 MPa. These experiments used either the pulse-decay (for low-permeability measurements) or the steady-state method (for high-permeability measurements), using the techniques described above. For these measurements, water was used as the confining fluid and argon gas was used as the pore fluid. The confining pressure was only increased to the next increment after the permeability of the sample was measured to be the same on two consecutive days.

Low-permeability 20 mm diameter by nominally 20–40 mm length cores of lava, altered lava, altered ash tuff, hydrothermal breccia, and fumarolic sulfur flow (Table 1) were measured in a Core Laboratories Pulse Decay Permeameter-200 and analyzed using the software PDP V2.75 at the University of Canterbury Rock Mechanics Laboratory. Each oven-dried core was slid into a permeameter sleeve, and a hand pump was used to move Penrite ISO46 hydraulic oil into a surrounding chamber which set the confining pressure to 3 MPa. This pressure corresponds to the confining pressure that the rock experienced within the shallow subsurface [13]. The system was then left to equilibrate for at least thirty minutes. The calculations of connected porosity, mass, dimensions of the core, and atmospheric conditions such as temperature and barometric pressure values acted as a necessary input into the PDP software before running each set of tests. After entering these data, the computer automatically pressurized the chamber from above with high purity nitrogen gas, creating a pressure differential of 10 PSI between the top of the sample and the bottom. The rate at which the sample equilibrated to matching pressures above and below the sample was used by the permeameter calculator to determine the permeability of the sample, as described by [40].

Table 1. X-ray diffraction data for the samples studied.

Sample Type and Numbers	XRD Peaks Identified
Unaltered Lava (n = 3), unaltered portions from samples 1-2, 2-2, 2-4	Labradorite, Anorthoclase, Anhydrite, Alunite, Pyrite, Cristobalite, Zaherite *
Altered Lava (n = 4) altered portions from samples 2-1, 2-4, 3-2, 161-8,	Albite, Opal Anhydrite Alunite Cristobalite, Zaherite *, Gyrolite *
Altered Ash Tuff (n = 5) 3-3, 1-1, 3-1, 3-4, 3-7	Cristobalite Alunite, Tridymite Rectorite *, Montmorillonite *,
Hydrothermal Breccia (n = 5) 2615-4, 2-3, 163-2b, 161-22, 3-5 Surge matrix	Alunite, Anhydrite, Cristobalite, Pyrite, Zaherite *, Gyrolite * Cristobalite, Labrodorite

* Clay found using clay-separation method, not performed on all XRD samples.

The permeability of the fractured low porosity rock cores was measured in the steady-state gas permeameter at the University of Canterbury. The cores dwelled at 1 MPa confining pressure for 30 min in a Hoek cell and a compression frame as high purity nitrogen gas flowed through the sample at 4 bar to equilibrate the pressure. The volumetric gas flow rate was increased five times without exceeding 500 mL/min to reduce the risk of potential damage to the flow meter from broken fragments of the core entering the downstream tube and to reduce the deformation of the fractures themselves. The Forchheimer correction was applied to correct for flow inertia (as described in [55]).

Nine samples were tested under varied confining pressure, which was increased to mimic the environment of increasing depth of the conduit of Whakaari. These variable pressures simulate different hydrostatic pressures at Whakaari from about 300 m to 3 km depth and follow the same methodology as above and described in [40]. The Forchheimer correction was applied at each of these confining pressure steps.

2.4. Mechanical Laboratory Measurements

Select oven dried core samples from a wide range of lithologies and degrees of alteration (Table 1) were placed in a Tecnotest KE300/ECE compression loading frame to measure their uniaxial compressive strength (UCS), where $\sigma_1 > 0$ and $\sigma_2 = \sigma_3 = 0.1$ MPa (atmospheric). To measure UCS, a 20 mm diameter x 40 mm long core sample was placed between the machine's two platens, which then deformed the sample at a constant displacement rate of 0.03 mm/s. Macroscope failure of the sample was signalled by the formation of a through-going fracture, corresponding to a rapid decrease in stress. The peak axial stress achieved during the experiment was taken as the UCS.

3. Results

3.1. Lithologies

We focus our analysis on relatively unaltered lava from the surface, ballistics (unaltered lava, altered lava, altered tuff, and hydrothermal breccias), and sulfur cemented tuff from surface fumaroles [13]. The relatively unaltered lava blocks had some combination of very thin alteration rinds (Figures 2a and 3a) and central portions that were unaltered allowing unaltered cores to be subsampled (Figure 2b). Textural and compositional analyses reveal low porosity with unaltered phenocrysts of pyroxene, plagioclase, and iron oxides, and microlites of plagioclase and pyroxene, typical of Whakaari lava [56]. Porosity is observed to be mainly created by open macro- and microcracks (Figure 3a).

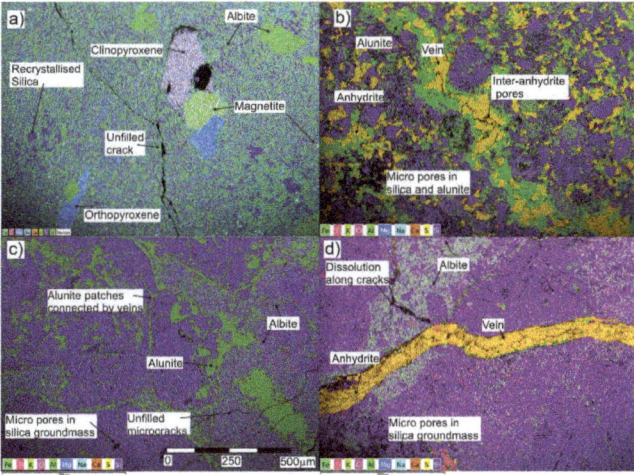

Figure 3. Scanning Electron Microscopy EDS (Energy Dispersive X-ray Spectroscopy) element maps of (**a**) relatively unaltered lava; (**b**) altered lava; (**c**) altered lava; (**d**) altered lava. Note that the scale of each image is the same as in (**c**).

Altered lava cores were drilled from yellow or white ballistics, from partial or complete alteration rinds thicker than 2 cm (Figure 2b). Alteration of the albite microlite rich lava groundmass has resulted in a crystallised microporous groundmass of silica polymorphs. Macrofractures and macropores are enlarged by dissolution and completely replaced, forming anhydrite and alunite veins, pockets (Figure 3b,d), or more diffuse areas partially replaced with alunite (Figure 3c). Larger pores contain coarser recrystallized grains of anhydrite (Figure 3b).

Ballistics of pyroclastic rocks were also found: altered ash tuff, lapilli tuff and tuff breccia (Figure 2c,d). These ballistics were completely altered to a yellow or white colour rather than just on their surfaces. Such pervasive alteration makes clasts and matrix hard to distinguish in hand specimen, due to their similar texture and mineralogy. The matrix/cement consists of <1 mm recrystallized silica polymorphs (Figure 4a), large areas of alunite (Figure 4a), and local areas of anhydrite (Figure 4b) and some clay minerals identified in XRD. Clasts are tuff, or occasional recrystallized coarser grained lava with albitic patches (Figure 4b). Clasts and matrix are relatively porous with pores existing between recrystallized grains (Figure 4c,d).

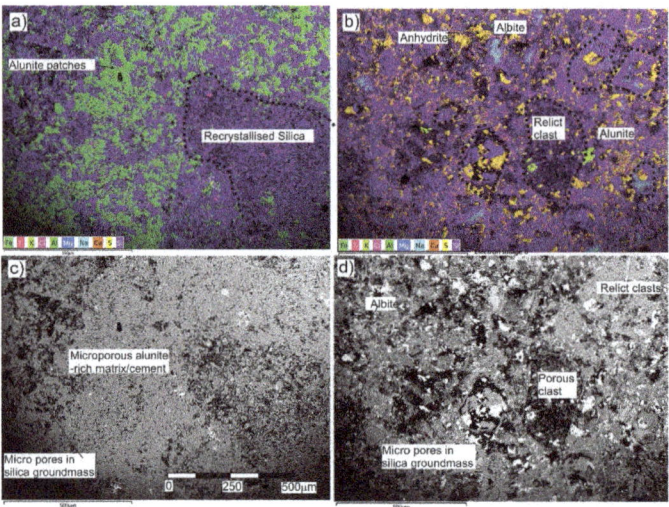

Figure 4. Scanning Electron Microscopy EDS (Energy Dispersive X-ray Spectroscopy) element maps and backscatter images, respectively, of (**a**) and (**c**) ash tuff matrix of a hydrothermal breccia from block 2-3; (**b**) and (**d**) ash tuff block 3-3.

Many ballistics consist of large clasts with only patches of matrix and were significantly mineralised and as such are described here as hydrothermal breccias. A large proportion of these samples are recrystallized and are frequently made up of more than 50% hydrothermal minerals such as alunite or anhydrite (Figure 2d,e and Figure 5b,c). Elemental sulfur is rare but does occur in small patches (Figure 5b). Texturally characteristic cristobalite with fishscale style fractures also occurs, while porous cristobalite is also seen in some vein margins (Figure 5c,d). Multiple generations of alunite and anhydrite veins are seen producing in-situ brecciation of clasts. Fractures are also observed within veins, or parallel to pre-existing fractures, and at the boundary between veins and stronger silicic clasts (Figure 5f). It is also worth noting the contrasting porosity in the vuggy anhydrite, the microporous lava, and the characteristic fishscale cracked cristobalite.

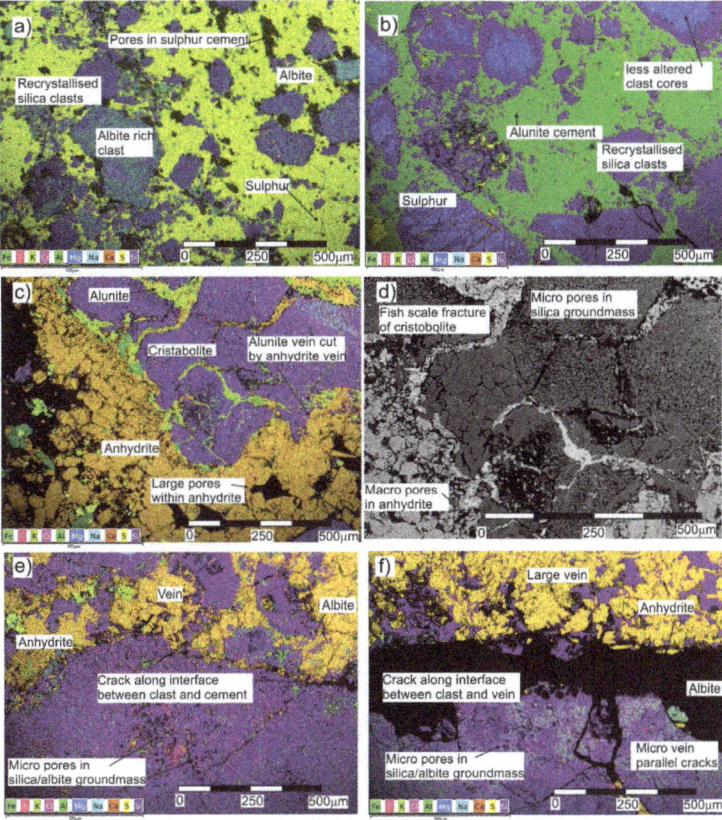

Figure 5. Scanning Electron Microscopy EDS (Energy Dispersive X-ray Spectroscopy) element maps of (**a**) EDS stacked image of sulfur cemented ash tuff; (**b**); EDS stacked image of alunite cemented portion of hydrothermal breccia (**c**) EDS element map of hydrothermal breccia; (**d**) same sample as (**c**) in backscatter to highlight porosity distribution (**e**) EDS stacked image of hydrothermal breccia (**f**) EDS stacked image of open vein in hydrothermal breccia.

These samples are additionally compared against data and textural observations of sulfur encrusted and cemented ash tuff from the surface. These yellow rocks are dominantly matrix supported and cemented with sulfur and contain angular ash and lapilli sized clasts of silica polymorphs and albite and irregular rounded pores resembling vesicles (Figure 5a).

3.2. Porosity and Permeability

Porosity for the entire sample set varies from a couple of percent up to ~60%, and permeability varies from ~4×10^{-19} to ~4×10^{-15} m^2 (Figure 6). Although there is a general trend of increasing permeability with increasing porosity, as observed in previous studies on the permeability of volcanic rocks [57,58], there is also substantial scatter within and between lithologies (Figure 6). For example, the permeability of samples with a porosity of ~40% can vary from ~2×10^{-16} to ~4×10^{-15} m^2 (Figure 6a). The relatively unaltered lava ballistics generally have lower porosity and permeability than the altered lava, altered ash tuff, and sulfur flow (Figure 6a). Compared to rocks collected from the surface (data from [13,59]; shown in grey on Figure 6a), the ballistic samples (yellow symbols on Figure 6a) generally have lower porosity and a narrower range of permeability. The samples with experimentally created tensile fractures are 4–5 orders of magnitude more permeable than the unfractured rocks at both confining pressures of 1 and 3 MPa (Figure 6a,b). Irrespective of the initial permeability, the permeabilities of the fractured samples are very similar at low confining pressures (~10^{-12} m^2 at 1 and 3 MPa; Figure 6a,b).

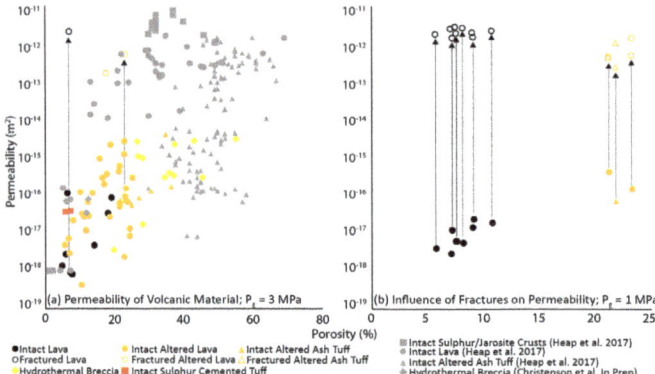

Figure 6. Permeability as a function of porosity at (**a**) confining pressure 3 MPa (this study) and 1 MPa, and (**b**) confining pressure 1 MPa showing the wide variability of matrix permeability (solid black points are unaltered intact lava from the edifice; yellow are altered intact ballistic samples collected at 3 MPa confining pressure) and low variability of fracture permeability (open points).

As confining pressure increases, the matrix permeability of the rocks decreases (Figure 7). A decrease in permeability as a function of increasing confining pressure has been previously reported in laboratory studies on volcanic rocks (e.g., [60,61]). The intact (i.e., without tensile fracture) altered samples have a higher permeability than the unaltered lava (Figure 7). An increase in confining pressure from 1 to 30 MPa leads to a decrease in permeability of ~1 order of magnitude in the unaltered lava and fractured unaltered lava, a decrease of 1-2 orders of magnitude in the unfractured altered lava, and a decrease in permeability of 2–4 orders of magnitude in the fractured altered lava (Figure 7). Importantly, our data show that the permeability reduction as confining pressure is greater in the fractured altered samples than in the fractured unaltered samples (Figure 7).

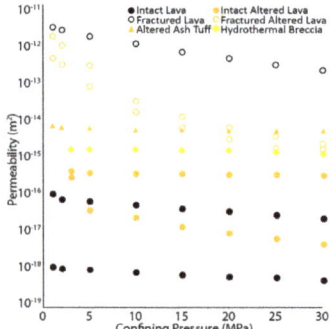

Figure 7. Permeability as a function of confining pressure showing the different impacts of confining pressure on altered and unaltered, fractured and unfractured samples.

Generally, our data show that uniaxial compressive strength (UCS) varies greatly over the range of porosity we tested. When combined with the other data from Whakaari [12] a clearer trend becomes evident in which UCS decreases as a function of increasing porosity (Figure 8), in agreement with previous studies on the strength of volcanic rocks (e.g., [62–64]). The observed decrease in UCS as a function of porosity is not linear (Figure 8). The compressive strength data show a consistently low strength for samples containing >10% porosity (Figure 8). Low porosity unaltered lava is the strongest sublithology, while altered rocks are both more porous and weaker, and the sulfur cemented tuff is significantly weaker than the lava samples of similar porosity.

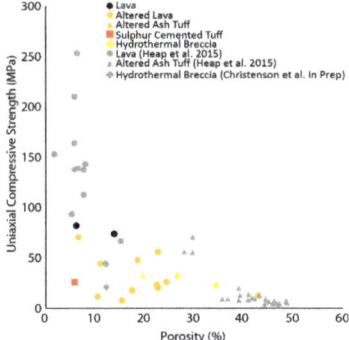

Figure 8. Uniaxial compressive strength (UCS) as a function of porosity for samples from this study and from [13,59].

3.3. Results Summary

The ballistics erupted from Whakaari are andesitic lavas, breccias, and tuffs. These rocks are all variably altered, containing silica polymorphs, alunite, and anhydrite. Many clasts show concentric alteration rinds. Anhydrite and alunite are common in pockets and veins, and they contain intercrystalline porosity, evidence for multiple cracking, veining, brecciation events, and cristobalite precipitation. In the tuff ballistics, it is difficult to distinguish matrix and clasts as both are recrystallized with silica polymorphs and alunite producing an overprinted microporosity. XRD results (Table 1) support the SEM interpretations, and additional analysis from the surge deposit shows dominantly cristobalite and plagioclase with little evidence of anhydrite and alunite. Consistent with previous data from surface rocks, tuff and altered lava ballistics are more porous and permeable than the less altered lavas. The permeability of the altered outer rind is generally higher when compared to the unaltered core of the same sample. When tensile fractures were created in cores, permeability increased by 4-6 orders of magnitude. However, the permeability of the fractured altered rocks decreases more than that of the fractured unaltered lava as confining pressure increases. Our results hint at the critical role fractures may play in the permeability of the altered rock that likely dominate the conduit at Whakaari.

4. Discussion

Our new laboratory data show that the permeability of our ballistic samples increases as a function of increasing porosity (Figure 6a), as observed in previous studies on the permeability of volcanic rocks (e.g., [57,58,65]) and porous sedimentary rocks (e.g., [66]). We also observe that a single tensile fracture in a laboratory sample can increase permeability by many orders of magnitude (Figure 6b). Similar to previous experiments [49] showed similar increases to permeability following the formation of a macroscopic fracture. Experiments performed at elevated pressures show that permeability decreases as a function of increasing confining pressure (Figure 7). We interpret this here as a result of the closure of pre-existing microcracks and high-aspect ratio pores within the samples, as concluded by previous studies that measured the permeability of microcracked volcanic rocks (e.g., [60,61]) and microcracked granite (e.g., [67]) as a function of increasing confining pressure. Finally, our uniaxial data show that the uniaxial compressive of our ballistic samples decreases as a function of increasing porosity (Figure 8), in agreement with previous studies on the strength of volcanic rocks (e.g., [62–64]).

Our new data allow us to assess the structure and composition of the breccia-filled conduit at Whakaari. The ballistic samples illustrate four distinct lithologies: (1) generally relatively unaltered lava blocks with thin alteration rinds, (2) completely or dominantly altered lava blocks, (3) completely altered tuffs, and (4) hydrothermal breccias. These lithologies represent the polylithic and variable breccia that comprise the conduit (Figure 9a).

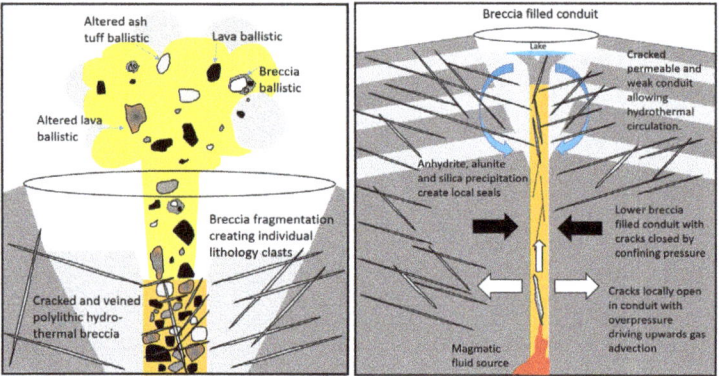

Figure 9. Proposed model of Whakaari conduit (**a**) near the surface and (**b**) eruption.

During an eruption, porous, weaker rocks are fragmented into smaller grain sizes [68] and therefore may be underrepresented in the block-sized ballistic samples. Soon after the 2016 eruption, the ash-sized component of the surge deposit had a generally yellow colour; however, our XRD analysis of the surge deposits indicates a dominantly altered cristobalite-rich composition. This indicates that the most fragmented component of the surge had a dominantly cristobalite mineralogy, and this likely comprised the finer grained dominantly tuff matrix of conduit-filling breccia which fragmented to form the surge and ballistics. Cristobalite was also observed in a dacitic bomb erupted at Whakaari in 1999 and is an inevitable product of shallow vapour phase alteration [69].

Several ballistic samples show evidence of brecciation and hydrothermal veining, with many samples also displaying anhydrite and alunite-bearing altered rims (Figures 3 and 5). The ballistic sample lithologies parallel those discussed in previous studies of the edifice of Whakaari: (1) altered tuffs composed of dominantly silica polymorphs and alunite and (2) relatively unaltered coherent lavas and breccias dominated by plagioclase and pyroxene (e.g., [12,14]. However, our data adds a third and fourth additional lithology, less common at the surface but dominant in the ballistics: (3) altered lavas dominated by anhydrite, and (4) hydrothermal breccias. We only found trace amounts of sulfur precipitation in the ballistics, indicating that sulfur-precipitated tuff may not be common in the subsurface.

We thus propose a model of a breccia-filled conduit containing clasts of dominantly altered lava and tuff in a matrix of relict ash-sized pyroclasts now replaced by silica polymorphs, alunite and anhydrite (Figure 9a). Similar to [13], we envisage that the conduit is surrounded by less altered coherent beds of lava, lava breccia, and tuff. Textural evidence in the ballistics and observations at the surface suggest that macrofractures are ubiquitous within the conduit and throughout the edifice, although these may be locally filled with alteration minerals such as anhydrite and alunite to form veins (Figure 9b). We acknowledge that some of the fractures present in the ballistics may be created during eruption, however, the overprinting of multiple fracture events recording in veining and mineralisation (Figure 5) show that fractures also develop pre-eruption in the conduit. Therefore, conduit itself may locally resemble a mineralised cockade breccia [35] but also contain facies similar to the heterolithic unbedded country rock breccia of maar diatremes [9] and the conduit beneath Unzen volcano [70]. Several vents were identified in the 2016 eruption [16] which support a model of a breccia filled vent system that diverges at shallow levels similar to that envisaged by [24].

5. Implications for Fluid Flow Monitoring and Eruption

At Whakaari, degassing occurs at distinct fumaroles, beneath the lake [21,22] and diffuse across the crater floor [71]. Macro- and microcracks are common in rocks collected at the surface [13]. Based on our data and previous studies [49], we estimate a fractured rock mass permeability of ~10^{-13} to ~10^{12} m^2 at the surface, away from open fumaroles. Our data also show that, for fractured samples, permeability in altered material is highly confining pressure-dependent, whereas the permeability in unaltered material is much less confining pressure-dependent. This relationship implies that permeability in the (highly fractured, altered) conduit is more confining pressure-dependent than the surrounding (less fractured, unaltered) edifice.

The effective pressure on any given part of the subsurface varies and is an interplay between depth (confining pressure) and pore pressure. Volcanoes in a state of unrest have variable pore pressures that frequently exceed confining pressures driving fluid flow, if pore pressure is greater than the tensile strength of the rock this can cause tensile failure (e.g., [12]) and if combined with a decompression event can drive fragmentation and eruption [14]. This is seen dramatically at Whakaari, which has been in an extended period of unrest and minor eruption since 2011. Outgassing and lake properties (temperature, chemistry, and level) vary significantly [21] and frequently shift on timeframes of hours to days, implying highly variable pore pressure, including periods where pore pressure exceeds confining pressure.

During recent unrest there were a range of surface phenomena, with several episodes of lake draining, mud fountaining and star bursting, ash venting [22] (Edwards et al., 2018), directed ballistic bursts and surges [16,18] (Chardot et al., 2014; Kilgour et al., 2019), and the appearance of a lava dome. Geophysics and satellite imagery show minor localised deformation and slope instability [32], significant tremor, and very long period (VLP) earthquakes [18,19]. Taken as a whole, these data reveal a model of shallow intrusions contributing time-variable heat and mass to the surface, along fumaroles and through eruptive vents. This model is consistent with the interpretation of gas advection as a mechanism to explain measured VLP activity [23]. Our interpretation of the eruptive vents as hydrothermally-cemented altered breccias with confining pressure-dependent permeability provides a mechanism to facilitate some of this time-dependent cycling behaviour.

Our observations indicate that cracks and veins are common in the ballistic rocks—something that we would expect from a conduit environment subjected to changing compressive and tensile stresses associated with intrusion and fluid movements. Our data suggest that pre-existing cracks in weaker hydrothermally mineralised zones easily close at confining pressures relevant for conduit processes. In unaltered rocks fractures may not fully close as asperities on crack surfaces may prop them open, maintaining permeability even at higher confining pressure. On long timescales of weeks to months these fractures allow eventual alteration and mineral replacement along these fluid flow pathways, and minerals may eventually crystallise within them to form veins. In contrast in the altered breccia-filled conduit, the weaker strength of hydrothermally mineralised fracture margins allow the asperities to be crushed, thus allowing closure under increased confining pressures, even on short timescales. This allows cracks to open and close as pressurized pore fluids open cracks and travel towards the surface (fluid advection; [23]). Once a fluid pulse has passed, pore pressures are reduced and the confining pressure recloses the fractures. In this manner, fractures can close instantaneously and reopen on timescales associated with gas advection. Hence, the opening and closing of the cracks is controlled by the accumulation of gas, sufficient to overcome confining pressure and generate a pore pressure. This allows time-variable advection of fluids, explaining the rapidly changing surface phenomena outlined in [22] and also the time variable tremor and very long period earthquakes described in ([18,23].

Effective pressure-controlled crack closure and permeability reduction are unlikely to build sufficient pore pressure to drive Whakaari's explosive events. Our data provide ample additional evidence for repeated fracture creation and mineral precipitation. Hydrothermally altered material in the conduit is generally weaker than the edifice forming rocks (Figure 8) and with tensile strengths of

3–5 MPa [59]. The generally weak and porous nature of the altered tuff and hydrothermally altered breccia implies it has a low fragmentation threshold [72] and would be susceptible to fragmentation due to phase-change related pressure fluctuations [14]. The dominance of the cristobalite in the surge matrix implies that it is composed of altered ash recycled from fine fragmentation of a porous tuff matrix [14]. There have been several different locations of vents on the lake floor [16] and the conduits are likely branching close to the surface with variable depths and types of mineralisation and resultant fragmentation histories similar to those envisaged at Okaro [24]. The frequency of explosive processes is on the order of weeks to years [18], which is sufficiently long to allow minerals to precipitate in veins and between breccia clasts in the conduits, resealing them. The exact conditions for this process are explained in [59]. Similar timescales for pressurisation as a result of alteration-induced reductions to permeability were proposed by [3]. Similar processes have been envisaged at other volcanoes [73]. However, here we offer an explanation for geophysical, geochemical, and visible changes that can occur in seconds to hours.

In conclusion, we provide evidence for a breccia-filled conduit of altered lava and tuff clasts dominantly cemented by alunite and anhydrite. The permeability of this altered material is susceptible to rapid variation in effective pressure allowing highly time-variable fluid advection, outgassing, and geophysical changes at the surface.

Supplementary Materials: An excel spreadsheet with all our data used for Figures 6–8 are available online at http://www.mdpi.com/2076-3263/10/4/138/s1.

Author Contributions: Conceptualization, B.M.K., M.C.V. and M.J.H.; methodology, B.M.K., M.C.V. and M.J.H.; formal analysis, A.F., R.H., S.M., M.C.V. and T.R.; investigation and experimentation, A.F., R.H., S.M., M.J.H., and T.R.; writing—original draft preparation, B.M.K., A.F., R.H., M.C.V., and M.J.H.; writing—review and editing, B.M.K., A.F., M.C.V., M.J.H., G.K., A.J., and B.C.; supervision, B.M.K., M.C.V. and M.J.H.; project administration, B.M.K.; funding acquisition, Sampling on the island was carried out by A.F., G.C., B.C. All authors have read and agreed to the published version of the manuscript.

Funding: This research was funded by New Zealand Ministry of Business, Innovation and Employment (MBIE) Strategic Science Investment Fund. M.J.H acknowledges an Erskine Teaching Fellowship awarded by the University of Canterbury.

Acknowledgments: We would like to acknowledge and thank White Island Tours, Frontiers Abroad, and the Green Room in aiding with the good times of data collection. Stephanie Gates and Ame McSporran also contributed useful discussion and data not directly used in the paper. Geoff Kilgour Bruce Christenson and Art Jolly were supported by the New Zealand Ministry of Business, Innovation and Employment (MBIE) Strategic Science Investment Fund. Shaun Mucalo assisted with SEM analysis.

Conflicts of Interest: The authors declare no conflict of interest. The funders had no role in the design of the study; in the collection, analyses, or interpretation of data; in the writing of the manuscript, or in the decision to publish the results.

References

1. Gardner, C.A.; White, R.A. Seismicity, gas emission and deformation from 18 July to 25 September 1995 during the initial phreatic phase of the eruption of Soufriere Hills Volcano, Montserrat. *Geol. Soc. Lond. Mem.* **2002**, *21*, 567–581. [CrossRef]
2. Norton, D.L.; Cathles, L.M. Breccia pipes, products of exsolved vapor from magmas. *Econ. Geol.* **1973**, *68*, 540–546. [CrossRef]
3. Heap, M.J.; Troll, V.R.; Kushnir, A.R.; Gilg, H.A.; Collinson, A.S.; Deegan, F.M.; Darmawan, H.; Seraphine, N.; Neuberg, J.; Walter, T.R. Hydrothermal alteration of andesitic lava domes can lead to explosive volcanic behaviour. *Nat. Commun.* **2019**, *10*, 1–10. [CrossRef] [PubMed]
4. Schauroth, J.; Wadsworth, F.B.; Kennedy, B.; von Aulock, F.W.; Lavallée, Y.; Damby, D.E.; Dingwell, D.B. Conduit margin heating and deformation during the AD 1886 basaltic Plinian eruption at Tarawera volcano, New Zealand. *Bull. Volcanol.* **2016**, *78*, 12. [CrossRef] [PubMed]
5. Stasiuk, M.V.; Barclay, J.; Carroll, M.R.; Jaupart, C.; Ratté, J.C.; Sparks, R.S.J.; Tait, S.R. Degassing during magma ascent in the Mule Creek vent (USA). *Bull. Volcanol.* **1996**, *58*, 117–130. [CrossRef]
6. Tuffen, H.; Dingwell, D. Fault textures in volcanic conduits: Evidence for seismic trigger mechanisms during silicic eruptions. *Bull. Volcanol.* **2005**, *67*, 370–387. [CrossRef]

7. Wadsworth, F.B.; Kennedy, B.M.; Branney, M.J.; von Aulock, F.W.; Lavallée, Y.; Menendez, A. Exhumed conduit records magma ascent and drain-back during a Strombolian eruption at Tongariro volcano, New Zealand. *Bull. Volcanol.* **2015**, *77*, 71. [CrossRef]
8. White, J.D.; Ross, P.S. Maar-diatreme volcanoes: A review. *J. Volcanol. Geotherm. Res.* **2011**, *201*, 1–29. [CrossRef]
9. Lefebvre, N.S.; White, J.D.L.; Kjarsgaard, B.A. Unbedded diatreme deposits reveal maar-diatreme-forming eruptive processes: Standing Rocks West, Hopi Buttes, Navajo Nation, USA. *Bull. Volcanol.* **2013**, *75*, 739. [CrossRef]
10. Zhang, J.Q.; Li, S.R.; Santosh, M.; Luo, J.Y.; Li, C.L.; Song, J.Y.; Lu, J.; Liang, X. The genesis and gold mineralization of the crypto-explosive breccia pipe in the Yixingzhai gold region, central North China Craton. *Geol. J.* **2019**. [CrossRef]
11. Omosanya, K.O.; Eruteya, O.E.; Siregar, E.S.; Zieba, K.J.; Johansen, S.E.; Alves, T.M.; Waldmann, N.D. Three-dimensional (3-D) seismic imaging of conduits and radial faults associated with hydrothermal vent complexes (Vøring Basin, Offshore Norway). *Mar. Geol.* **2018**, *399*, 115–134. [CrossRef]
12. Heap, M.J.; Kennedy, B.M.; Pernin, N.; Jacquemard, L.; Baud, P.; Farquharson, J.I.; Scheu, B.; Lavallée, Y.; Gilg, H.A.; Letham-Brake, M.; et al. Mechanical behaviour and failure modes in the Whakaari (White Island volcano) hydrothermal system, New Zealand. *J. Volcanol. Geotherm. Res.* **2015**, *295*, 26–42. [CrossRef]
13. Heap, M.J.; Kennedy, B.M.; Farquharson, J.I.; Ashworth, J.; Mayer, K.; Letham-Brake, M.; Reuschlé, T.; Gilg, H.A.; Scheu, B.; Lavallée, Y.; et al. A multidisciplinary approach to quantify the permeability of the Whakaari/White Island volcanic hydrothermal system (Taupo Volcanic Zone, New Zealand). *J. Volcanol. Geotherm. Res.* **2017**, *332*, 88–108. [CrossRef]
14. Mayer, K.; Scheu, B.; Gilg, H.A.; Heap, M.J.; Kennedy, B.M.; Lavallée, Y.; Letham-Brake, M.; Dingwell, D.B. Experimental constraints on phreatic eruption processes at Whakaari (White Island volcano). *J. Volcanol. Geotherm. Res.* **2015**, *302*, 150–162. [CrossRef]
15. Kennedy, B.; Spieler, O.; Scheu, B.; Kueppers, U.; Taddeucci, J.; Dingwell, D.B. Conduit implosion during Vulcanian eruptions. *Geology* **2005**, *33*, 581–584. [CrossRef]
16. Kilgour, G.N.; Gates, S.; Kennedy, B.; Farquhar, A.; McSporran, A.; Asher, C. Phreatic eruption dynamics derived from deposit analysis: A case study from a small, phreatic eruption from Whakāri/White Island, New Zealand. *Earth Planets Space* **2019**, *71*, 36. [CrossRef]
17. Houghton, B.F.; Nairn, I.A. The 1976–1982 Strombolian and phreatomagmatic eruptions of White Island, New Zealand: Eruptive and depositional mechanisms at a 'wet' volcano. *Bull. Volcanol.* **1991**, *54*, 25–49. [CrossRef]
18. Chardot, L.; Jolly, A.D.; Kennedy, B.M.; Fournier, N.; Sherburn, S. Using volcanic tremor for eruption forecasting at White Island volcano (Whakaari), New Zealand. *J. Volcanol. Geotherm. Res.* **2015**, *302*, 11–23. [CrossRef]
19. Jolly, A.; Lokmer, I.; Christenson, B.; Thun, J. Relating gas ascent to eruption triggering for the April 27, 2016, White Island (Whakaari), New Zealand eruption sequence. *Earth Planets Space* **2018**, *70*, 1–15. [CrossRef]
20. Fournier, N.; Chardot, L. Understanding volcano hydrothermal unrest from geodetic observations: Insights from numerical modeling and application to White Island volcano, New Zealand. *J. Geophys. Res. Solid Earth* **2012**, *117*. [CrossRef]
21. Christenson, B.W.; White, S.; Britten, K.; Scott, B.J. Hydrological evolution and chemical structure of a hyper-acidic spring-lake system on Whakaari/White Island, NZ. *J. Volcanol. Geotherm. Res.* **2017**, *346*, 180–211. [CrossRef]
22. Edwards, M.J.; Kennedy, B.M.; Jolly, A.D.; Scheu, B.; Jousset, P. Evolution of a small hydrothermal eruption episode through a mud pool of varying depth and rheology, White Island, NZ. *Bull. Volcanol.* **2017**, *79*, 16. [CrossRef]
23. Jolly, A.D.; Chardot, L.; Neuberg, J.; Fournier, N.; Scott, B.J.; Sherburn, S. High impact mass drops from helicopter: A new active seismic source method applied in an active volcanic setting. *Geophys. Res. Lett.* **2012**, *39*. [CrossRef]
24. Montanaro, C.; Cronin, S.; Scheu, B.; Kennedy, B.; Scott, B. Complex crater fields formed by steam-driven eruptions: Lake Okaro, New Zealand. *GSA Bull.* **2020**. [CrossRef]
25. Chouet, B.A. Long-period volcano seismicity: Its source and use in eruption forecasting. *Nature* **1996**, *380*, 309–316. [CrossRef]

26. McNutt, S.R. Seismic monitoring and eruption forecasting of volcanoes: A review of the state-of-the-art and case histories. In *Monitoring and Mitigation of Volcano Hazards*; Springer: Berlin/Heidelberg, Germany, 1996; pp. 99–146.
27. Sparks, R.S.J. Forecasting volcanic eruptions. *Earth Planet. Sci. Lett.* **2003**, *210*, 1–15. [CrossRef]
28. Brenguier, F.; Shapiro, N.M.; Campillo, M.; Ferrazzini, V.; Duputel, Z.; Coutant, O.; Nercessian, A. Towards forecasting volcanic eruptions using seismic noise. *Nat. Geosci.* **2008**, *1*, 126–130. [CrossRef]
29. Bell, A.F.; Greenhough, J.; Heap, M.J.; Main, I.G. Challenges for forecasting based on accelerating rates of earthquakes at volcanoes and laboratory analogues. *Geophys. J. Int.* **2011**, *185*, 718–723. [CrossRef]
30. Bean, C.J.; De Barros, L.; Lokmer, I.; Métaxian, J.P.; O'Brien, G.; Murphy, S. Long-period seismicity in the shallow volcanic edifice formed from slow-rupture earthquakes. *Nat. Geosci.* **2014**, *7*, 71–75. [CrossRef]
31. Clarke, J.; Adam, L.; Sarout, J.; van Wijk, K.; Kennedy, B.; Dautriat, J. The relation between viscosity and acoustic emissions as a laboratory analogue for volcano seismicity. *Geology* **2019**, *47*, 499–503. [CrossRef]
32. Hamling, I.J. Crater Lake controls on volcano stability: Insights from White Island, New Zealand. *Geophys. Res. Lett.* **2017**, *44*, 11–311. [CrossRef]
33. Todesco, M.; Rinaldi, A.P.; Bonafede, M. Modeling of unrest signals in heterogeneous hydrothermal systems. *J. Geophys. Res. Solid Earth* **2010**, *115*. [CrossRef]
34. Pola, A.; Crosta, G.; Fusi, N.; Barberini, V.; Norini, G. Influence of alteration on physical properties of volcanic rocks. *Tectonophysics* **2012**, *566*, 67–86. [CrossRef]
35. Sruoga, P.; Rubinstein, N. Processes controlling porosity and permeability in volcanic reservoirs from the Austral and Neuquén basins, Argentina. *AAPG Bull.* **2007**, *91*, 115–129. [CrossRef]
36. Wyering, L.D.; Villeneuve, M.C.; Wallis, I.C.; Siratovich, P.A.; Kennedy, B.M.; Gravley, D.M.; Cant, J.L. Mechanical and physical properties of hydrothermally altered rocks, Taupo Volcanic Zone, New Zealand. *J. Volcanol. Geotherm. Res.* **2014**, *288*, 76–93. [CrossRef]
37. Mordensky, S.P.; Villeneuve, M.C.; Kennedy, B.M.; Heap, M.J.; Gravley, D.M.; Farquharson, J.I.; Reuschlé, T. Physical and mechanical property relationships of a shallow intrusion and volcanic host rock, Pinnacle Ridge, Mt. Ruapehu, New Zealand. *J. Volcanol. Geotherm. Res.* **2018**, *359*, 18–20. [CrossRef]
38. Mordensky, S.P.; Heap, M.J.; Kennedy, B.M.; Gilg, H.A.; Villeneuve, M.C.; Farquharson, J.I.; Gravley, D.M. Influence of alteration on the mechanical behaviour and failure mode of andesite: Implications for shallow seismicity and volcano monitoring. *Bull. Volcanol.* **2019**, *81*, 44. [CrossRef]
39. Heap, M.J.; Gravley, D.M.; Kennedy, B.M.; Gilg, H.A.; Bertolett, E.; Barker, S.L. Quantifying the role of hydrothermal alteration in creating geothermal and epithermal mineral resources: The Ohakuri ignimbrite (Taupō Volcanic Zone, New Zealand). *J. Volcanol. Geotherm. Res.* **2020**, *390*, 106703. [CrossRef]
40. Cant, J.L.; Siratovich, P.A.; Cole, J.W.; Villeneuve, M.C.; Kennedy, B.M. Matrix permeability of reservoir rocks, Ngatamariki geothermal field, Taupo Volcanic Zone, New Zealand. *Geotherm. Energy* **2018**, *6*, 2. [CrossRef]
41. Farquharson, J.I.; Wild, B.; Kushnir, A.R.; Heap, M.J.; Baud, P.; Kennedy, B. Acid-induced dissolution of andesite: Evolution of permeability and strength. *J. Geophys. Res. Solid Earth* **2019**, *124*, 257–273. [CrossRef]
42. Heap, M.J.; Farquharson, J.I.; Baud, P.; Lavallée, Y.; Reuschlé, T. Fracture and compaction of andesite in a volcanic edifice. *Bull. Volcanol.* **2015**, *77*, 55. [CrossRef] [PubMed]
43. Siratovich, P.A.; Heap, M.J.; Villeneuve, M.C.; Cole, J.W.; Kennedy, B.M.; Davidson, J.; Reuschlé, T. Mechanical behaviour of the Rotokawa Andesites (New Zealand): Insight into permeability evolution and stress-induced behaviour in an actively utilised geothermal reservoir. *Geothermics* **2016**, *64*, 163–179. [CrossRef]
44. Ogata, K.; Senger, K.; Braathen, A.; Tveranger, J. Fracture corridors as seal-bypass systems in siliciclastic reservoir-cap rock successions: Field-based insights from the Jurassic Entrada Formation (SE Utah, USA). *J. Struct. Geol.* **2014**, *66*, 162–187. [CrossRef]
45. von Aulock, F.W.; Nichols, A.R.L.; Kennedy, B.M.; Oze, C. Timescales of texture development in a cooling lava dome. *Geochim. Cosmochim. Acta* **2013**, *114*, 72–80. [CrossRef]
46. Saubin, E.; Kennedy, B.; Tuffen, H.; Villeneuve, M.; Davidson, J.; Burchardt, S. Comparative field study of shallow rhyolite intrusions in Iceland: Emplacement mechanisms and impact on country rocks. *J. Volcanol. Geotherm. Res.* **2019**, *388*, 106691. [CrossRef]
47. Siratovich, P.; Villeneuve, M.; Cole, J.; Kennedy, B.; Bégué, F. Saturated heating and quenching of three crustal rocks and implications for thermal stimulation of permeability in geothermal reservoirs. *Int. J. Rock. Mech. Min.* **2015**, *80*, 265–280. [CrossRef]

48. Mordensky, S.; Kennedy, B.; Villeneuve, M.; Lavallée, Y.; Reichow, M.; Wallace, P.; Siratovich, P.; Gravley, D. Increasing the Permeability of Hydrothermally Altered Andesite by Transitory Heating. *Geochem. Geophys. Geosyst.* **2019**, *20*, 5251–5269. [CrossRef]
49. Heap, M.J.; Kennedy, B.M. Exploring the scale-dependent permeability of fractured andesite. *Earth Planet. Sci. Lett.* **2016**, *447*, 139–150. [CrossRef]
50. Lamur, A.; Kendrick, J.E.; Eggertsson, G.H.; Wall, R.J.; Ashworth, J.D.; Lavallée, Y. The permeability of fractured rocks in pressurised volcanic and geothermal systems. *Sci. Rep.* **2017**, *7*, 6173. [CrossRef]
51. Eggertsson, G.H.; Lavallée, Y.; Kendrick, J.E.; Markússon, S.H. Improving fluid flow in geothermal reservoirs by thermal and mechanical stimulation: The case of Krafla volcano, Iceland. *J. Volcanol. Geotherm. Res.* **2018**, *391*. [CrossRef]
52. Mordensky, S.P.; Villeneuve, M.C.; Farquharson, J.I.; Kennedy, B.M.; Heap, M.J.; Gravley, D.M. Rock mass properties and edifice strength data from Pinnacle Ridge, Mt. Ruapehu, New Zealand. *J. Volcanol. Geotherm. Res.* **2018**, *367*, 46–62. [CrossRef]
53. Farquhar, A. Ballistic Analysis Inferring Subsurface Hydrothermal Alteration and Mineralogical Seal Control on Eruptions at Whakaari Volcano, New Zealand. Colorado College: Colorado Springs, CO, USA, 2018; unpublished.
54. Deligne, N.I.; Jolly, G.E.; Taig, T.; Webb, T.H. Evaluating life-safety risk for fieldwork on active volcanoes: The volcano life risk estimator (VoLREst), a volcano observatory's decision-support tool. *J. Appl. Volcanol.* **2018**, *7*, 7. [CrossRef]
55. Heap, M.J.; Kushnir, A.R.; Gilg, H.A.; Wadsworth, F.B.; Reuschlé, T.; Baud, P. Microstructural and petrophysical properties of the Permo-Triassic sandstones (Buntsandstein) from the Soultz-sous-Forêts geothermal site (France). *Geotherm. Energy* **2017**, *5*, 26. [CrossRef]
56. Cole, J.W.; Thordarson, T.; Burt, R.M. Magma origin and evolution of White Island (Whakaari) volcano, Bay of plenty, New Zealand. *J. Pet.* **2000**, *41*, 867–895. [CrossRef]
57. Farquharson, J.; Heap, M.J.; Varley, N.R.; Baud, P.; Reuschlé, T. Permeability and porosity relationships of edifice-forming andesites: A combined field and laboratory study. *J. Volcanol. Geotherm. Res.* **2015**, *297*, 52–68. [CrossRef]
58. Wadsworth, F.B.; Vasseur, J.; Scheu, B.; Kendrick, J.E.; Lavallée, Y.; Dingwell, D.B. Universal scaling of fluid permeability during volcanic welding and sediment diagenesis. *Geology* **2016**, *44*, 219–222. [CrossRef]
59. Christenson, B.W.; Kennnedy, B.M.; Reyes, A.G.; Farquahar, A.; Heap, M.J.; Henley, R.W. Permeability reduction and other processes leading to phreatic eruptions from wet volcanic systems: Insights from the 27 April 2016 eruption from White Island, New Zealand. *Geophys. Res. Abstr.* **2019**, *21*, 1.
60. Vinciguerra, S.; Trovato, C.; Meredith, P.G.; Benson, P.M. Relating seismic velocities, thermal cracking and permeability in Mt. Etna and Iceland basalts. *Int. J. Rock Mech. Min. Sci.* **2005**, *42*, 900–910. [CrossRef]
61. Nara, Y.; Meredith, P.G.; Yoneda, T.; Kaneko, K. Influence of macro-fractures and micro-fractures on permeability and elastic wave velocities in basalt at elevated pressure. *Tectonophysics* **2011**, *503*, 52–59. [CrossRef]
62. Al-Harthi, A.A.; Al-Amri, R.M.; Shehata, W.M. The porosity and engineering properties of vesicular basalt in Saudi Arabia. *Eng. Geol.* **1999**, *54*, 313–320. [CrossRef]
63. Heap, M.J.; Xu, T.; Chen, C.F. The influence of porosity and vesicle size on the brittle strength of volcanic rocks and magma. *Bull. Volcanol.* **2014**, *76*, 856. [CrossRef]
64. Schaefer, L.N.; Kendrick, J.E.; Oommen, T.; Lavallée, Y.; Chigna, G. Geomechanical rock properties of a basaltic volcano. *Front. Earth Sci.* **2015**, *3*, 29. [CrossRef]
65. Kushnir, A.R.; Martel, C.; Bourdier, J.L.; Heap, M.J.; Reuschlé, T.; Erdmann, S.; Komorowski, J.C.; Cholik, N. Probing permeability and microstructure: Unravelling the role of a low-permeability dome on the explosivity of Merapi (Indonesia). *J. Volcanol. Geotherm. Res.* **2016**, *316*, 56–71. [CrossRef]
66. Bourbie, T.; Zinszner, B. Hydraulic and acoustic properties as a function of porosity in Fontainebleau sandstone. *J. Geophys. Res. Solid Earth* **1985**, *90*, 11524–11532. [CrossRef]
67. Darot, M.; Guéguen, Y.; Baratin, M.L. Permeability of thermally cracked granite. *Geophys. Res. Lett.* **1992**, *19*, 869–872. [CrossRef]
68. Kueppers, U.; Scheu, B.; Spieler, O.; Dingwell, D.B. Fragmentation efficiency of explosive volcanic eruptions: A study of experimentally generated pyroclasts. *J. Volcanol. Geotherm. Res.* **2006**, *153*, 125–135. [CrossRef]

69. Schipper, C.I.; Mandon, C.; Maksimenko, A.; Castro, J.M.; Conway, C.E.; Hauer, P.; Kirilova, M.; Kilgour, G. Vapor-phase cristobalite as a durable indicator of magmatic pore structure and halogen degassing: An example from White Island volcano (New Zealand). *Bull. Volcanol.* **2017**, *79*, 74. [CrossRef]
70. Goto, Y.; Nakada, S.; Kurokawa, M.; Shimano, T.; Sugimoto, T.; Sakuma, S.; Hoshizumi, H.; Yoshimoto, M.; Uto, K. Character and origin of lithofacies in the conduit of Unzen volcano, Japan. *J. Volcanol. Geotherm. Res.* **2008**, *175*, 45–59. [CrossRef]
71. Bloomberg, S.; Rissmann, C.; Mazot, A.; Oze, C.; Horton, T.; Gravley, D.; Kennedy, B.; Werner, C.; Christenson, B.; Pawson, J. Soil Gas Flux Exploration at the Rotokawa Geothermal Field and White Island, New Zealand. In Proceedings of the Thirty Sixth Workshop on Geothermal Reservoir Engineering, Stanford, CA, USA, 30 January–1 February 2012; Volume 30.
72. Spieler, O.; Kennedy, B.; Kueppers, U.; Dingwell, D.B.; Scheu, B.; Taddeucci, J. The fragmentation threshold of pyroclastic rocks. *Earth Planet. Sci. Lett.* **2004**, *226*, 139–148. [CrossRef]
73. Caudron, C.; Taisne, B.; Neuberg, J.; Jolly, A.D.; Christenson, B.; Lecocq, T.; Syahbana, D.; Suantika, G. Anatomy of phreatic eruptions. *Earth Planets Space* **2018**, *70*, 168. [CrossRef]

© 2020 by the authors. Licensee MDPI, Basel, Switzerland. This article is an open access article distributed under the terms and conditions of the Creative Commons Attribution (CC BY) license (http://creativecommons.org/licenses/by/4.0/).

Article

A CO$_2$-Driven Gas Lift Mechanism in Geyser Cycling (Uzon Caldera, Kamchatka)

Alexey V. Kiryukhin [1,2,*] and Gennady Karpov [1]

[1] Institute of Volcanology & Seismology FEB RAS, Piip, 9, 683006 Petropavlovsk-Kamchatsky, Russia; karpovga@kscnet.ru
[2] Kronotsky Federal Nature Biosphere Reserve, Ryabikova 48, 684000 Elizovo, Russia
* Correspondence: avkiryukhin2@mail.ru

Received: 11 March 2020; Accepted: 12 May 2020; Published: 14 May 2020

Abstract: Here, we report on a new geyser (named Shaman) formed in the Uzon caldera (Kronotsky Federal Nature Biosphere Reserve, Russia) in autumn 2008 from a cycling hot Na-Cl spring. The geyser is a pool-type CO$_2$-gas lift driven. From 2012 to 2018, the geyser has shown a rather stable interval between eruptions (IBE) from 129 to 144 min with a fountain height up to 4 m, and the geyser conduit has gradually enlarged. In 2019, the Shaman geyser eruption mode significantly changed: cold water inflow from the adjacent stream was re-directed into the geyser conduit and the average IBE decreased to 80 min. We observed two eruptive modes: a cycling hot spring (June 2019) and a cycling geyser (after June 2019). Bottom-hole temperature recording was performed in the geyser conduit to understand its activity. The TOUGH2-EOS2 model was used to reproduce the obtained temperature records and estimate geyser recharge/discharge parameters in both modes. Modeling shows that a larger cold inflow into the conduit causes a switch from cycling geyser to hot cycling spring mode. It was also found that the switch to cycling geyser mode corresponds to a larger mass of CO$_2$ release during the time of the eruption.

Keywords: geyser; Uzon; CO$_2$; TOUGH2; modeling

1. Introduction

A thorough analysis of world-wide geyser distributions and functionality performed by Hurwitz and Manga [1] showed that geysers mostly occur within high-temperature hydrothermal systems hosted in active silicic volcanic areas. Dissolved gas sample data from Yellowstone National Park (USA) suggest that the presence of magma-derived CO$_2$ as well as N$_2$ derived from air-saturated meteoric water reduced the near-surface saturation temperature in geyser conduits [2]. Thermodynamic calculations suggest that the dissolved CO$_2$ and N$_2$ modulate the dynamics of geysers and may trigger hydrothermal geyser eruptions when recharged into shallow reservoirs at high concentrations [2].

Cold CO$_2$ gas-lift geyser behavior has been observed at abandoned oil and gas boreholes [3]. A recent studies of geysers in Iceland using submersible cameras showed detailed images from two of the largest geysers at depths exceeding 20 m: near the surface, the conduit of the geysers were near-circular, but at a depth of 9–12 m, the shape changed into a crack-like elongated fissure [4]. Experimental investigation of the importance of the conduit shape and cold water inflow for the type of discharge (geyser, fumaroles or boiling pool) have been reported [5].

Analysis of some particular cases from the Valley of Geysers (Kamchatka) highlights the two factors: (i) the possibility of high porosity-permeability conduit formation due to fast rhyolite/thermal water chemical interaction in thermal spring discharge areas [6,7]; and (ii) the importance of CO$_2$-gas lift for driving geyser eruptions [8,9].

Nevertheless, partial destruction of geysers in the Valley of Geysers (03 June 2007) and the almost simultaneous creation of a new geyser in the same caldera (2008) further raises the questions of how, where, and why geysers exist. Why, despite similar hydrogeological conditions in Uzon and in the Valley (Figure 1), have most geysers formed in the Valley? What are the properties of the conduits where geyser behavior is preferred? How is geyser fountain power generated in underground reservoirs?

This paper examined a newly created geyser in Uzon (Appendix A, Figures A1–A8) as an example to address the above-mentioned issues. Another critical issue is whether a new Uzon geyser is a precursor of an on-going diatreme (Figure 2), which may eventually cause a catastrophic event in future. Thus, the modeling only focuses on the very last duration of the observations.

2. Geological Setting

The age of the Uzon–Geysernaya caldera (Figure 1) is estimated to be 39,600 ± 1000 years according to the radiocarbon dating of soil samples below the caldera-forming ignimbrites [10]. Uzon–Geysernaya pre-caldera deposits comprise dacite–andesite tuffs and lavas that are 40,000–140,000 years old (corresponding geological indexes are αQ_3^{1-2}, αQ_3^3, and Q_3^3 ust). Initially, this caldera was an isolated hydrological basin, where volcanogenic and sedimentary lake deposits were formed (Q_3^4). These deposits, which have thicknesses up to 400 m near the caldera rim, are represented by layered pumice tuffs and minor breccias and conglomerates. Caldera lake deposits are overlain by 15,000–20,000-year-old rhyolite to dacite lavas, which formed large domes and adjacent lava flows up to 100–150 m thick (corresponding geological indexes are ξQ_3^4 and $\alpha \xi Q_3^4$). Approximately 9000 to 12,000 years ago, the southeastern wall of the caldera was eroded by the Shumnaya and Geysernaya Rivers, initiating the drainage and formation of intensive hydrothermal discharge in the Geysernaya River basin by 5000–6000 years ago [6].

The current magmatic activity in the Uzon–Geysernaya caldera was detected by uplift with an amplitude of up to 15 cm from 1999 to 2003 (around 4 cm per year), which was identified by radar interferometry data analysis [11]. The source of these deformations is supposed to be a partially molten magma body under the northeastern part of the Uzon–Geysernaya caldera.

On June 3, 2007, a catastrophic landslide occurred in the Valley of Geysers, Kamchatka [6]. Within a few minutes, 20×10^6 m^3 of rocks shifted 2 km downstream in the Geysernaya River, creating a dam with Podprudnoe Lake and burying more than 23 geysers. After this partial loss of geysers in the Valley of Geysers from the catastrophic landslide in 2007, a new geyser appeared at 652 m.a.s.l. in the Uzon caldera in the autumn of 2008. The new geyser site is located in an active hydrothermal area (East of Uzon) composed of deeply altered caldera lake deposits. This area included several magma-hydrothermal eruption vents (Figures 1 and 2) such as a 1.5 km maar of Dalnee Lake (8–10 kY old, Figure A10 in the Appendix A), and craters enclosing Khloridnoe Lake, Fumarolnoye Lake, and Bannoe Lake (all are less than 1.5 kY old). Bannoe Lake (Figure A9 in Appendix A) hosted the most recent phreatic eruption in 1989, which created a crater ($4 \times 3 \times 8$ m) at the bottom [12]. Shallow sills and the formation of geomechanical conditions may be associated with isotope signed magmatic CO_2 discharge observed in this area (see Section 3 below).

At the time of creation, this geyser was named Mutny (Muddy in Russian) because it erupted pieces of clay rocks (Appendix A, Figures A1 and A2). The geyser conduit was logged down to 2.3 m, with a diameter of 30 cm at a 60 cm depth. A bi-modal (12 and 20 min) interval between eruptions (IBE) was observed [13]. Geyser pool temperature varied from 64.4–94.9 °C, with a maximum temperature achieved 1–1.5 min before the eruption. Duration of eruptions were around of 25 s and the height of erupted fountains was up to ~4 m. It was also noted at that time, that the water level in the geyser conduit lowered by 60 cm after the eruption, then erupted water flowed back from the pool into the conduit.

A different name, Shaman, has been proposed for this geyser in the literature [14–16], so this name was adopted. It was noted that Shaman arose within a cycling chloride sodium hot spring with a temperature of 80 °C, geyser pool area of 4.5×6 m with two terraces with 40 cm elevation differences (terraces reflected water level in the pool before and after eruptions). After the switch from a hot spring to a geyser, its chemical composition remained relatively stable (Cl: 1223–1454 ppm; Na: 812–941 ppm) at least until 2009, while the temperature of eruption increased to 99 °C.

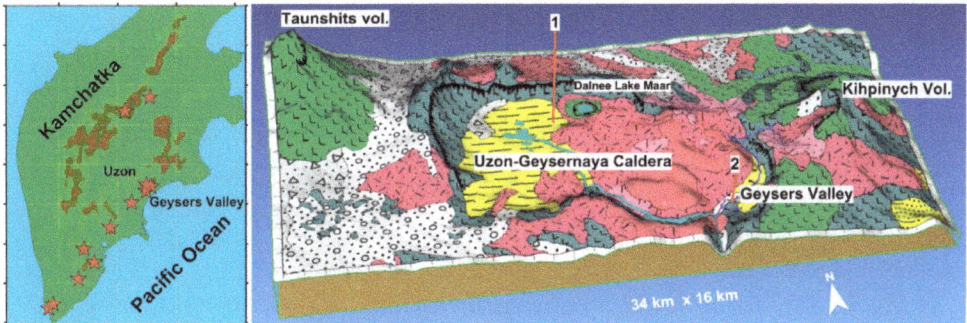

Figure 1. Kamchatka high temperature geothermal systems (**left**) and schematic 3D view of the Uzon–Geysernaya caldera (**right**). The geological units are shown by the colors: alluvial and glacial deposits (light gray), caldera lake deposits (Q_3^4) (yellow), rhyolite–dacite extrusions (ξQ_3^4 and $\alpha \xi Q_3^4$) (pink), pre-caldera tuffs and sedimentary deposits (gray), and basalt-andesite lavas (green) (αQ_3^{1-2}, αQ_3^3, and Q_3^3 ust). Geysers under monitoring: 1—Shaman geyser area, 2—Bolshoy, Velikan geysers area.

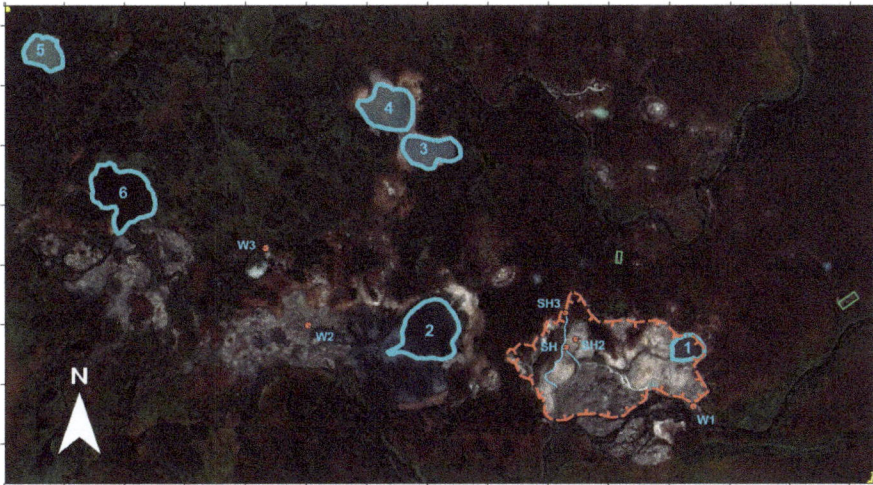

Figure 2. Shaman geyser area. Air-photo (Google Earth, 19 September 2017) scheme of the East Uzon thermal field, where geyser Shaman is located. SH—Shaman geyser (Figures A1–A8 in Appendix A), SH2—hot CO_2 spring adjacent to geyser (Figure A11 in Appendix A), SH3–hot spring where warm–cold creek started bypassing Shaman. Since 2019, a significant fraction of creek flow has diverted into the Shaman geyser conduit (Figure A6). Diatreme shapes: 1—Bannoe Lake (Figure A9 in Appendix A); 2—Chloridnoe Lake; 3,4—Lake Vosmerka; 5—Seroye Lake; 6—Utinoye Lake, Red hatched area—Shaman geyser "on-going" diatrem. Axes grid scale—100 m.

3. Gas-Chemistry Composition and Interval Between Eruption (IBE) History of the Shaman Geyser

In Figure 3, we show the time variation of the average IBE of this new geyser in the period 2012–2019. The average IBE for August 2012–May 2017 was rather stable in the range from 129 to

144 min, but in 2018, the average IBE dropped to 88 minutes, then to 80 minutes in August 2019. The chemical history of the Shaman geyser showed a significant dilution over time of the major components (Cl, Na, K, Ca, and H_3BO_3 decreased by a factor of ~3), pH decreased from 6.4 to 4.6; nevertheless, SO_4 remains rather stable (Table 1). There is a trend of stable water isotopes ($\delta D‰$, $\delta^{18}O‰$) decreasing during the time period of 2015–2018. Gas composition was relatively stable with the dominance of CO_2 (80% vol on average) (Table 2), which resulted in the lowering of the boiling temperature in the geyser pool to 82.7 °C (Figure A4 in Appendix A). It is worth noting the clear $\delta^{13}C‰$ signatures of magmatic CO_2 (varies from −6.8‰ to −3.4‰) and abiotic-thermogenic CH_4 (varies from −30.4‰ to −22.9‰) [17] in the gas components of the Shaman geyser (Table 2), and the N_2/Ar ratios (varies from 50 to 492) also exceeds the atmospheric saturated water ratio, pointing to non-atmospheric origin of some N_2. The NH_4 increase on 24.08.2019 may have been caused by the dissolution of NH_3 generated by the reaction $N_2 + 3H_2 \rightarrow 2NH_3$, if H_2 generation in the magma–hydrothermal system increased at that time. Gas samples taken from both Shaman and from Shaman-2 on 02.05.2018 and the sample from Shaman-2 on 27.08.2019 were strongly contaminated with air. Concurrently, the surficial hydrological conditions changed: the temperature variations of the creek water, recharged from spring SH3, and rain/snow-melting (Figure 2, Figure A3 in Appendix A) partially re-directed (Figure A6 in Appendix A) their water flow into the Shaman pool, triggered by pool area subsidence (Figure A6 in Appendix A). The Shaman geyser conduit also significantly enlarged at that time (Figure A5 in Appendix A). Eruptions styles of the Shaman geyser in 2014 and 2019 are also shown in Figures A7 and A8 in Appendix A.

In April 2019, the Shaman geyser eruption style changed when compared to the regular cycling styles observed previously (match vertical images row of 06.09.2014 and 21.04.2019). During the period from 2008 to 2018, a single major eruption was observed, but in April 2019, multiple (from three to four) eruptions took place in a few minutes of geyser discharge (Figure A8). Then, in June 2019, the Shaman geyser eruptions ceased altogether. Only bubbling was observed at regular times when eruptions were expected. Nevertheless, in July 2019, the geyser cycling style (multiple eruption styles of three to five) was re-established and the geyser activity continued (Figures A7 and A8 (12.09.2019)).

Table 1. Chemical parameters of the Shaman geyser water (in ppm). Chemical composition of pre-Shaman hot spring (2006), sampling in 2006–2009 was performed by G.A. Karpov, and sampling in 2015–2019 was performed by A.V. Kiryukhin. Chemical analysis was performed in Central Chemical Lab Institute of Volcanology & Seismology Far East Branch Rassia Academy of Scieces (IVS FEB RAS). Isotope analysis was performed by P.O. Voronin using LGR IWA 35EP. Notes: SH—Shaman geyser, SH3—hot spring where warm–cold creek started bypassing Shaman (see Figure 2).

##	Data	pH	HCO$_3$	Cl	SO$_4$	NH$_4$	Na	K	Ca	Mg	H$_3$BO$_3$	SiO$_2$	T$_{Na-K}$	T$_{SiO2}$	δD ‰	δ^{18}O ‰
SH	2006	5.2	6	1365	365	4	859	72	45	0.20	366	411	169.4	235.2		
SH	2009	5.9	68	1454	231	3	1007	61	50	1.20	412	472	139.3	247.2		
SH	2009	6.9	56	1418	231	1	941	62	48	2.40	408	255	146.7	197.3		
SH	27.07.15	6.3	38	837	158	35	491	47	37	4.0	228	286	181.9	205.8	−102.7	−11.5
SH	28.07.17	6.1	55	682	192	15	478	43	10	3.3	161	219	176.2	186.5	−105.7	−12.1
SH	02.05.18	4.7	10	521	221	13	379	23	20	6.2	131	204	139.8	181.4	−111.9	−13.5
SH	02.05.18	5.6	21	695	231	15	493	35	13	5.5	176	241	153.0	193.3	−108.7	−13.4
SH	22.04.19	3.6		425	259	15	308	32	27	4.4	176	241	191.8	193.3		
SH	24.08.19	5.5	1	532	240	98	298	45	17	4.4	219	358	236.8	223.6		
SH3	26.08.19	4.4		266	207	4	201	25	32	15	47	394	212.4	231.5		

Table 2. Gas composition (vol. %) of the Shaman geyser and adjacent gas-boiling hot spring Shaman-2 (Figure A11 in Appendix A). Gas sampling in 2009, 2012, 2014, and 05.09.2019 was performed by G.A. Karpov, and gas sampling in 2018–2019 was performed by A.V. Kiryukhin and N.B. Zhuravlev using the sampling method (RU patent # 195670)). Chemical analysis was performed at the Central Chemical Lab. IVS FEB RAS, isotope analysis was performed by V.Y. Lavrushin and B.G. Pokrovsky in GIN RAS (δ^{13}C‰).

Geyser	Date	He	H_2	Ar	O_2	N_2	CO_2	CH_4	H_2S	δ^{13}C CO_2	δ^{13}C CH_4
Shaman	18.09.2009	0.0004	4.12	0.01	0.08	3.15	89.6	1.83	0.8		
Shaman	25.07.2012	<0.001	0.68	0.206	1.96	15.4	77.9	2.47			
Shaman	26.09.2014	<0.007	0.017	0.007	0.01	3.4	93.6	2.24			
Shaman	02.05.2018	0.0004	0.005	0.81	17.0	70.0	12.1	0.04		−6.7	
Shaman-2	02.05.2018	0.0003	0.23	0.66	12.6	56.2	29.2	1.05			
Shaman	22.04.2019	0.0006	0.25	0.14	1.46	11.5	83.5	3.1			
Shaman	27.08.2019	0.0005	0.001	0.13	0.93	10.3	86.2	2.4		−3.4	−22.9
Shaman-2	27.08.2019	0.0004	0.002	0.56	10.9	52.7	34.9	1.0		−15.8	
Shaman	05.09.2019	0.0004	2.03	0.014	0.83	9.83	84.1	2.4		−6.76	−30.4

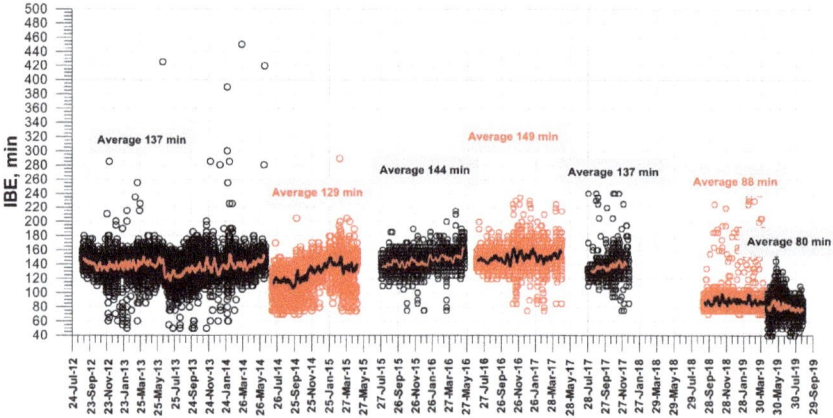

Figure 3. IBE history of the Shaman geyser during the time period from 2012 to 2019. Note: black and red symbols show different times when the temperature loggers were replaced on site.

4. Bottom-Hole Temperature Tests and Sampling of Shaman Geyser in 2019

We performed two tests to measure the bottom hole transient temperature at the Shaman geyser. In both cases, a HOBO U12-015 temperature logger was used. Test #1 was carried out during the period from 25.06.2019 18:00 to 29.06.2019 18:00 (Figure 4). During the first part of test #1 (from 25.06.2019 18:00 to 27.06.2019 18:00), a 62 mm diameter iron pipe was installed in the geyser conduit to ensure that the logger reached the bottom. At that time, the pipe reached the depth of 286 cm, and then the temperature logger was installed inside the pipe at the bottom. Two days after the iron pipe was removed from the geyser conduit and the temperature logger was installed alone on a cable at the same depth of 286 cm (the second part of test #1 lasted from 27.06.2019 18:00 to 29.06.2019 18:00). These operations gave us some idea of the shape and geometry of the Shaman geyser conduit: (1) Penetrated depth to the bottom was 286 cm; (2) the geyser bottom was joined to some sub-horizontal cavern, from where the geyser was recharged by hot water and gas; (3) the geyser conduit was conical with a top diameter from 55 to 65 cm, medium diameter 17 cm at the depth of 180 cm, and a bottom diameter at least of 12.5 cm at the depth of 272 cm; and (4) the maximum bottom-hole temperature of 98.6 °C is significantly lower than the boiling temperature of pure water at the given hydrostatic pressure (104.8 °C). During this time, 73 cycles of temperature with an average interval of 78.4 min were observed, but no fountain eruptions took place, only intense gas bubbling.

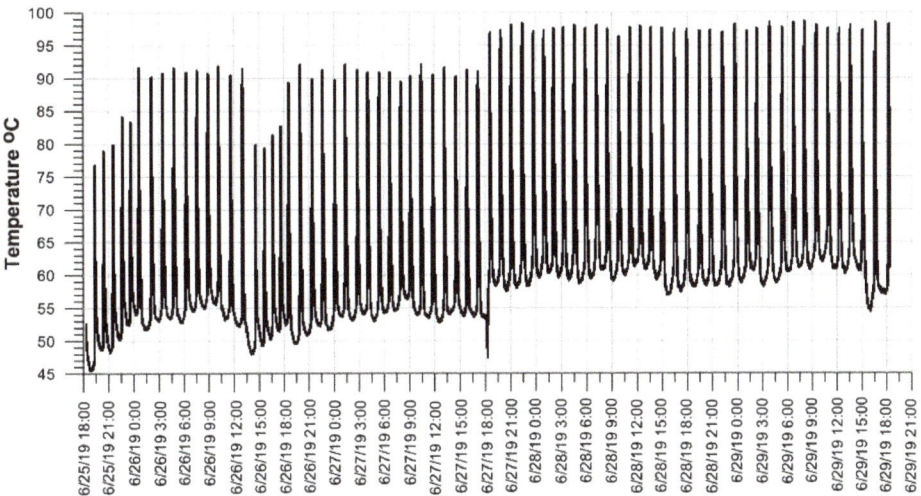

Figure 4. Shaman geyser temperature cycling records obtained at the bottom of the conduit (depth of 2.9 m) during the time period of observations from 25.06.2019 18:00 to 29.06. 2019 18:00 (73 eruptions, 78.4 min). Note that a 5 °C temperature positive shift after the casing tube was removed on 27 June 2019 18:00.

Test #2 was a response to a Shaman geyser eruption that started in mid-July 2019. We used the same logger configuration and depth as in the second part of test #1. Test #2 was performed from 24.07. 2019 09:00 to 28.07. 2019 15:00 (Figure 5). During that time, we observed 77 eruptions with an IBE of 82.6 min. Each eruption was characterized by three to five sub-eruptions with a duration of 5–7 min, followed by a 1 m water level drawdown in a conduit. The eruption period temperature records were clearly in contrast to non-eruption ones with a minimum bottom hole temperature rise from 55 to 65 °C (Figures 4 and 5). A 10 °C temperature increase in a geyser bottom during day-time (Figure 5) was also observed when the water temperature in the surface recharge creek rose. It is worth noting that the maximum bottom-hole temperature of ~99.4 °C was significantly lower than the boiling temperature of pure water at a given hydrostatic pressure (104.8 °C).

During the period from 24.08.2019 to 29.08.2019, the following parameters of the Shaman geyser were observed: (1) Gas rate 2.5 g/s of CO_2 at the top of the conduit area (we used a sampling method with a down-hole probe connected by a stainless steel tube to a pump on a ground surface (RU patent # 195670)); (2) In the time between eruptions, a cold water inflow about 0.25 l/s (volume method of measurement, simply measuring with a 3 L bucket the volume of water per time) into the geyser conduit took place from the adjacent creek (T = 35 °C, pH 6.45); (3) Surface geyser pool temperature 80 °C (IR survey, see Figure A4 in Appendix A), geyser pool pH 5.53–5.6; (4) Tracer test using a mass of m = 1 kg NaCl (diluted in a 5 L bottle of geysers water before injection) shows an active volume of the Shaman geyser system of 3.8 m^3. Cl concentration after eruption and before injection was C_1 = 581.4 10^{-3} kg/m^3, Cl concentration after salt brine injection and just before the next eruption was C_2 = 737.9 10^{-3} kg/m^3, thus the volume V of the conduit was determined by the formula V = 0.60684 m/(C_2–C_1) = 3.8 m^3, where 0.60684 is the weight fraction of Cl in NaCl (tracer test methodology to estimate the geyser's conduit volume was described in Section 3.3.2 [8]). This estimate is based on the assumption that injected salt is homogeneously distributed in all of the active geyser volume (conduit and adjacent caverns) during time between subsequent eruptions, thus this is a lower volume limit estimate; and (5) Opal deposition from geyser brine was found (on a logger surface) as a main secondary phase (Sergeeva A.V., pers. com. 2019).

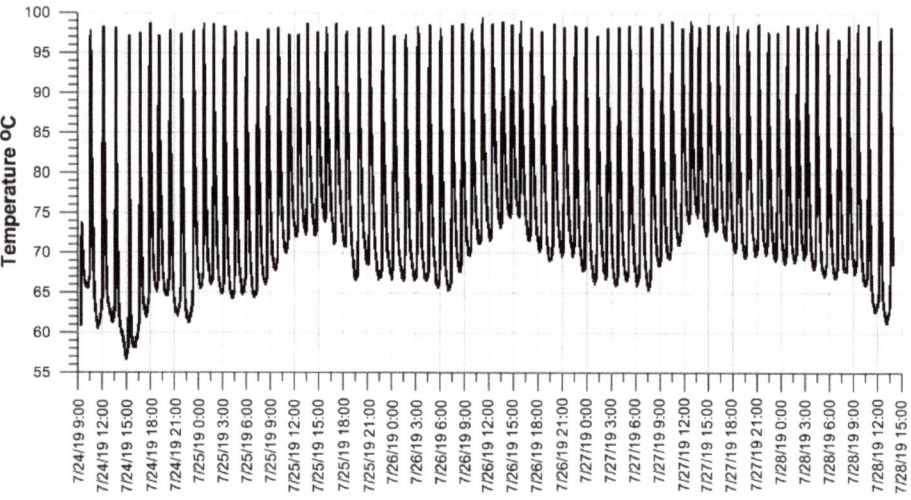

Figure 5. Shaman geyser temperature cycling records obtained at the bottom of the conduit (depth of 2.9 m) during the time period of observations from 24.07.2019 09:00 to 28.07.2019 15:00 (77 eruptions, IBE = 82.6 min).

5. Conceptual Model of Geyser Shaman Cycling

Based on the information above, a conceptual model of the Shaman geyser cycling in 2019 is shown in Figure 6. The geyser conduit has a 3 m depth and 0.3 m radius connected to the reservoir-cavern (yellow in a Figure 6) fed by two-phases (hot water + CO_2) from a deeper reservoir (red arrows in a Figure 6). An active volume of the geyser system was estimated as no less than 3.8 m^3, which includes the conduit volume 0.8 m^3 ($\pi r^2 H$) and reservoir-cavern active volume of 3 m^3. If we assume a cavern-reservoir porosity of 0.5 (as a maximum of the Uzon–Geysernaya caldera tuff unit porosity, which ranges from 0.16 to 0.52, see Section 2.8.1 [6]), then its actual volume should be 6 m^3. The isometric shape of subsidence around the geyser conduit pointed to radial-cylindrical geometry, thus the cavern-reservoir radius should be 2 m if its thickness of 0.5 m is assumed. Surface boundary conditions must include constant atmospheric pressure on the top (average value for the Uzon caldera conditions is 0.93 bars). There is a possibility of self-discharge conditions and cold water injection from the adjacent creek (0.25 kg/s and 35 °C as estimated in August 2019, although colder temperature values like 30 °C may be assumed in July 2019). Mass output from 5 min of fountain discharge must be applied to within an 80 min interval of cycling. Bottom CO_2 recharge was estimated at the top as 2.5 g/s and hot water recharge was 0.25 kg/s, if a water/gas mass rate ratio ~100 (observed during gas sampling) was used. Hot water recharge enthalpy may be slightly above 420 kJ/kg to match the bottom-hole maximum temperature records of 99.4 °C. Host rock for the active geyser system might be low permeability hydrothermally altered rocks.

Integrated Figure A7 in the Appendix A may be used as an illustration of the 5 min fountain eruption stage of the geyser, which was approximated as 5 min constant flow discharge, followed by the start of cold water injection in the geyser conduit.

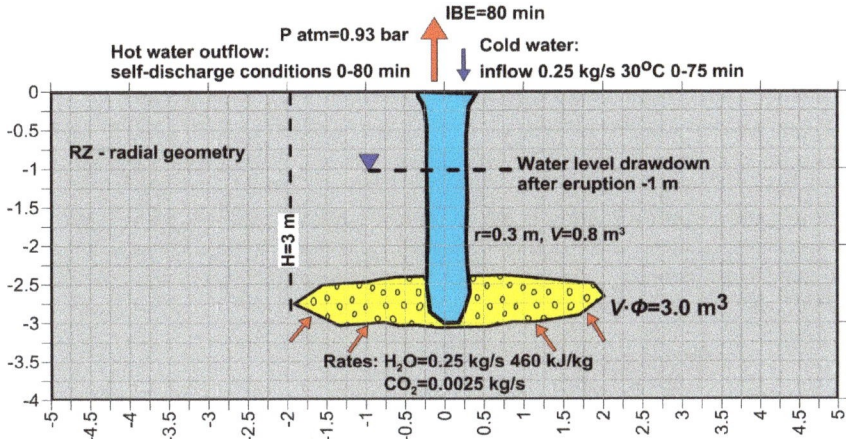

Figure 6. Shaman geyser conceptual model of heat and mass recharge/discharge conditions, conduit (blue) and recharge reservoir (yellow) geometry.

6. TOUGH2-EOS2 Modeling of Shaman Cycling Conditions

6.1. Model Setup

The TOUGH2 program with EOS2 fluid module of state (water + CO_2) [18] was used to reproduce the thermal–hydrodynamic processes in a geyser conduit and adjacent geothermal reservoir. The EOS2 fluid module describes single-phase and 2-phase conditions with two fluid components (H_2O and CO_2). In single-phase conditions, primary variables are P (pressure), T (temperature), and P_{CO2} (partial CO_2 pressure). In 2-phase conditions, the primary variables are P_g (gas phase pressure), S_g (gas saturation), and P_{CO2} (partial CO_2 pressure). In both cases, P_{CO2} is calculated using Henry's law using the mole fraction of CO_2 in a liquid phase $XCO2_{liq}$ and K_H (Henry constant, non-linearly dependent of temperature) as input parameters. The CO_2 component travels between phases according to the CO_2 solubility defined by Henry's law. Thus, if hot compressed single phase CO_2-saturated water is going upward to a lower pressure discharge point, then it eventually comes into 2-phase conditions (when P minus P_{CO2} becomes less than saturation pressure at a given temperature (this is a check point in the TOUGH2 program to switch fluid phase conditions). Then, large volume changes may take place due to CO_2 release from a dissolved state in a liquid to the gas phase.

A simple radial–cylindrical geometry with hot water feed and surface discharge/recharge conditions was used to describe the functionality of the Shaman geyser. Figure 7 shows the cross-section of this model including the numerical grid and boundary conditions.

Model geometry was assigned as a cylinder 3 m high with a 2 m radii, and a radial–cylindrical RZ-grid was used accordingly. Regular spacing in the Z direction (six elements) and non-regular spacing in the radial direction (0.3 m, 10 × 0.17 m) were applied.

Model zonation includes three domains: (i) the geyser conduit domain (CHAN) assigned to axial elements; (ii) adjacent reservoir domain (RESER) properties were assigned to the bottom layer of the model; and (iii) host–rock domain properties were assigned to the rest model elements (ROCK). All of the above model domain properties are shown in Table 3.

Initial conditions in the top model layer elements (ROCK) were assigned as fixed temperature T = 10 °C and pressure P = 0.93 bars. Other model elements' initial conditions were assigned as a result of natural state modeling with a constant flow/recharge boundary conditions.

Boundary conditions were defined in the following order: Hot water rate of 0.25 kg/s with enthalpy of 460 kJ/kg and CO_2 rate of 0.0025 kg/s, with the same enthalpy assigned in the bottom layer. Surface discharge was assigned as a "well on deliverability" [17] in a top element of the geyser conduit (production index PI = 1 × 10^{-10} m^3, bottom-hole pressure 0.95 bar, atmospheric pressure plus 0.2 m water column).

Another assumption of the model was the time-dependent cycling of sources and sinks. Cold water downflows were assigned at the bottom element of the geyser conduit with a rate q (unknown estimated parameter) and enthalpy h (unknown estimated parameter) as a time-dependent source active during the time interval [(k − 1) × IBE, 75 min + (k − 1) × IBE], where k is a cycle number (k = 1,2,3 ... 10).

Geyser eruptions were assigned in a conduit element adjacent to the cold water injection element from above. Geyser eruptions were described in the model as a time-dependent sink with a rate Q (unknown estimated parameter), active during a 5 min time interval [75 min + (k − 1) × IBE, 80 min + (k − 1) × IBE], where k is a cycle number (k = 1,2,3 ... 10).

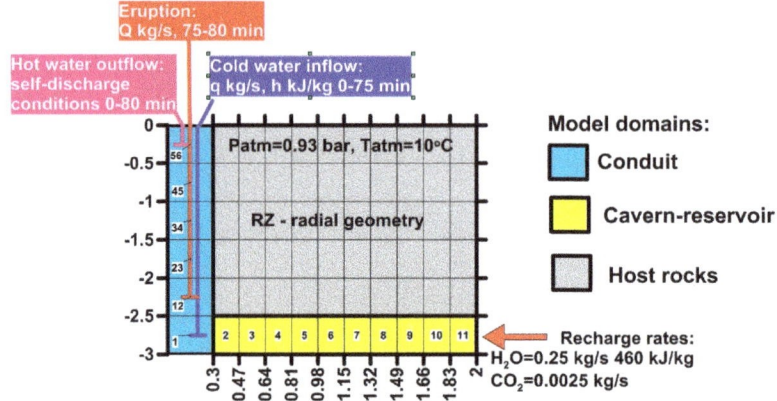

Figure 7. Cross-section of the Shaman geyser RZ-model showing the numerical grid, model zonation, and boundary conditions. Numbers are the names of selected model elements.

Table 3. Material properties assigned in the model, linear relative permeabilities with zero residual saturations were assigned in all model domains.

Model Domain	Rock Min. Density kg/m^3	Porosity	Permeability m^2	Wet Heat Conductivity W/m °C	Specific Heat J/kg °C
GEYSER CONDUIT	2600	0.99	1 × 10^{-8}	2	1000
CAVERN-RESERVOIR	2600	0.5	1 × 10^{-8}	2	1000
HOST ROCK	2600	0.1	1 × 10^{-17}	2	1000

6.2. Modeling Results

6.2.1. Steady-State Mode

Natural state (steady-state) modeling was run to define flow-gravitational equilibrium upon the boundary conditions defined above. During natural-state, run time-dependent sources and sinks were disabled in the model, hence this shows a constant rate of CO$_2$-hot spring steady-state conditions. Figure 8 shows the P–T–Sg–P$_{CO2}$–X$_{CO2liq}$ (pressure, temperature, gas saturation, partial CO$_2$ pressure, and mass fraction of CO$_2$ in liquid phase, respectively) conditions along streamlines from the point of reservoir recharge (element 11 of the model, where hot water inflow rate was assigned) to the point of discharge at the top of a vertical conduit (element 56 of the model, where self-discharge well on deliverability conditions were assigned) (see Figure 7 for numeration of the model elements).

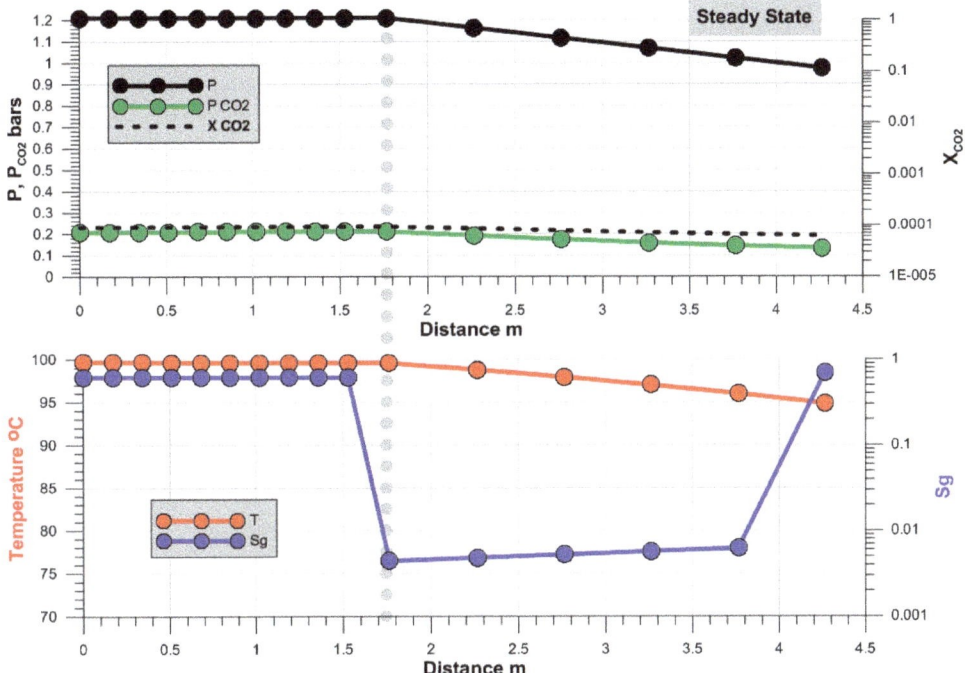

Figure 8. Steady-state model (#8_A3_NS) parameter distributions (temperature in °C, gas saturation S_g, pressure and partial CO_2 pressure P_{CO2} in bars, mass fraction of CO_2 in a liquid phase) along a streamline from hot water recharge element #11 to discharge element #56 of the model (see Figure 7 for model element names references). Vertical dotted line corresponds to the position of flow direction change from radial-horizontal to ascending-vertical.

The first part of the streamline goes horizontally in the reservoir (distance from 0 to 1.77 m), then goes up vertically in the conduit (distance from 1.77 to 4.77 m). Two-phase flow conditions were observed throughout the entire model. S_g remained constant at 0.62 and temperatures were at an average of 99.7 °C in the reservoir, then S_g decreased to 0.004 at the bottom, slowly rising to 0.006 in the middle part, then sharply increasing to 0.70 in the shallow part of the conduit. Simultaneously, the temperature slowly declined in the conduit from 99.6 to 94.9 °C. Note that the temperature was significantly lower (from 4 to 5 °C) than the steam saturation temperature at a given pressure due to the lowering effect of P_{CO2} (from 0.21 to 0.14 bars). Mass fraction CO_2 in a liquid phase varied from 9.21×10^{-5} to 6.21×10^{-5}. Pressure drops along an upflow segment of the model (distance along a streamline from 1.77 to 4.27 m) from 1.21 (bottom) to 0.97 bars (top).

It is worth noting that total CO_2 and H_2O mass (in gas and liquid phases) stored in the system were 0.767 kg and 1842 kg, respectively (Table 4). This factor yielded a significantly lower (24 times) mass fraction of CO_2 in the system (that is 0.00042) compared to the mass fraction of CO_2 injected in the source (which was 0.01). This is an effect of the CO_2 redistribution between phases. $CO_2 + H_2O$ goes mostly in the gas phase in this case, as the liquid phase with low CO_2 concentration was not in use, but accounted for bulk concentration estimates.

6.2.2. Geyser Cycling Mode

Once cycling model parameters appear in the model, then significant model cycling begins. We started from the steady-state discussed above, then applied 10 cycles. The model solution by cycle 10 was not sensitive to the initial conditions, which is was what we were looking for.

The cold water injection and geyser discharge rates in the model were adjusted to fit the in situ recorded temperature, when fountain activity was observed in the Shaman geyser (July 2019). The following estimates were obtained (using trial and error method) to obtain a match between a transient model and the observed temperatures: (1) Cold water injection q in the bottom of the geyser conduit (0.15 kg/s) with enthalpy h of 126 kJ/kg (water 30 °C) during a 75 min injection time; and (2) discharge rate Q = 3 kg/s during 5 min eruption time (Figure 9).

Figure 10 shows the geyser pre-eruption conditions, which follow 75 min of cold water injection in the bottom of the geyser conduit (0.15 kg/s, 126 kJ/kg or 30 °C) as a distribution of P–T–Sg–P_{CO2}–X_{CO2liq} conditions along streamlines from the recharge (model element 11) to discharge point (model element 56) of the geyser system. Although the pressure profile remains similar to steady-state conditions, a significant difference from the steady-state in other model parameters was observed: (1) Temperature dropped from 100 °C to 72–75 °C in the geyser conduit; (2) Sg dropped to 0.001 in the lower and mid part of the conduit, the maximum was 0.201 in the top; (3) P_{CO2} rose to 0.63–0.82 bars in the conduit; and (4) Mass fraction CO_2 in a liquid phase significantly rose to 4.6×10^{-4}–3.7×10^{-4}. We also noted that there was an increase in total CO_2 and H_2O mass (in gas and liquid phases) stored in the system: 1.081 kg and 2102.3 kg, correspondingly. Thus, total CO_2 mass stored in the system increase was equal to 0.314 kg, compared to the steady state.

The geyser after eruption conditions are shown in Figure 11. This is what happens in a geyser system after 5 min of discharge with a rate of 3 kg/s as specified in the model. Sg rose to 1.0 in the middle and top parts of the geyser conduit, which showed that it was almost completely emptied from the liquid phase. Pressure dropped to 0.99 bar in the reservoir and to 0.95 bars in the conduit. CO_2 mass fraction in a liquid phase varied from 0.18×10^{-4} to 4.5×10^{-4}. Total CO_2 and H_2O mass (in gas and liquid phases) stored in the system was 0.650 kg and 1271.9 kg, correspondingly. Thus, CO_2 mass release in a geyser model system during a geyser eruption was equal to 0.431 kg (corresponding to a volume of 219.4 L of CO_2 at normal conditions) (Table 4).

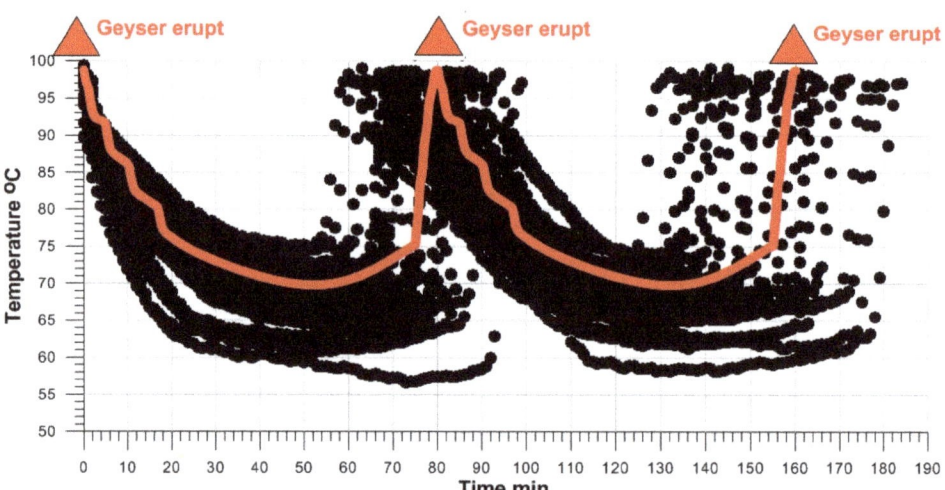

Figure 9. Modeling (run #9B, unknown parameters estimations are: q = 0.15 kg/s, h = 126 kJ/kg (water 30 °C), Q = 3 kg/s) of the transient bottom temperature (red line) in the Shaman geyser vs. the bottom temperature records (dots) obtained during the time period of observations from 24.07.2019 09:00 to 28.07.2019 15:00.

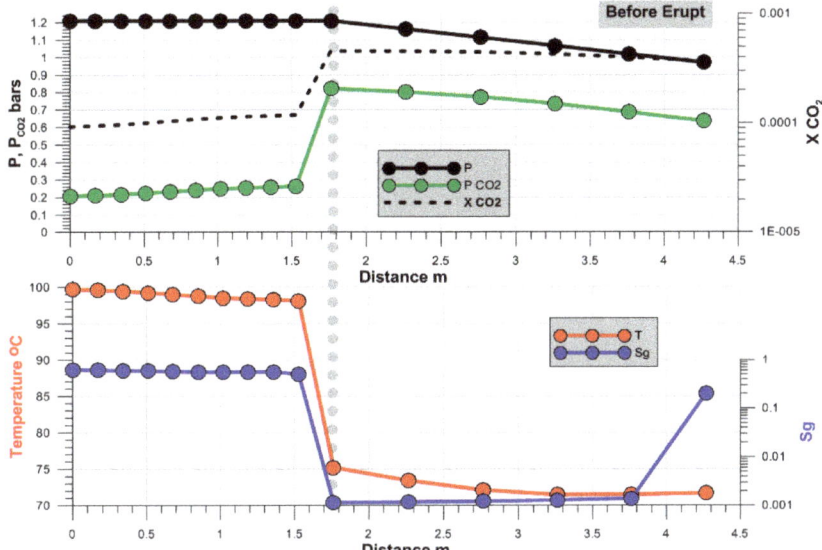

Figure 10. State before eruption (run #9B): model parameter distributions (temperature in °C, gas saturation Sg, pressure and partial CO_2 pressure P_{CO2} in bars, mass fraction of CO_2 in a liquid phase) along a streamline from hot water recharge element #11 to discharge element #56 of the model (see Figure 7 for model element name references). Vertical dotted line corresponds to the position of flow direction change from radial-horizontal to ascending-vertical.

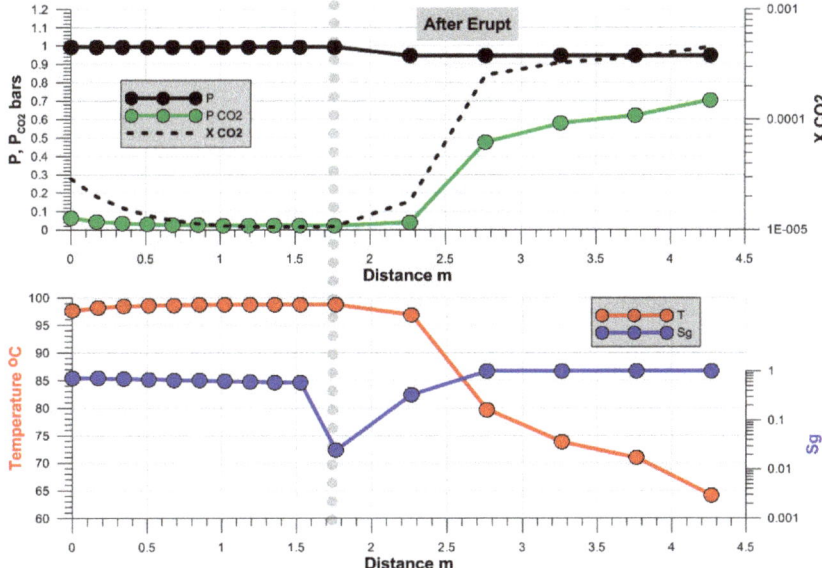

Figure 11. State after eruption (run #9B): model parameter distributions (temperature in °C, gas saturation Sg, pressure and partial CO_2 pressure P_{CO2} in bars, mass fraction of CO_2 in a liquid phase) along a streamline from hot water recharge element #11 to discharge element #56 of the model (see Figure 7 for model element name references). Vertical dotted line corresponds to the position of flow direction change from radial-horizontal to ascending-vertical.

6.2.3. Cycling Flowrate Mode

Flowrate cycling mode was modeled in the same way as a geyser cycling mode. Cold water injection rate and hot spring discharge rate model parameters were adjusted to fit the transient temperature records during the time when no fountain activity was observed in the Shaman geyser (June 2019). The following estimates were obtained: (1) Cold water injection in the bottom of the conduit 0.15 kg/s (q), enthalpy 84 kJ/kg (h) (water 20 °C) during 75 min injection time; and (2) Discharge rate 2 kg/s (Q) during 5 min eruption time (Figure 12).

Figure 13 shows the P-T-Sg-P_{CO2}-X_{CO2liq} conditions along the streamline from the recharge (model element 11) to discharge point (model element 56) of the hot spring system at the following hot spring pre-eruption conditions: (1) Pressure profile remains similar to steady-state conditions; (2) Temperature dropped from 100 °C to 64 °C in a geyser conduit; (2) Sg dropped to 0.001 in the lower and mid part of the conduit, where the maximum was 0.124 in the top; (3) P_{CO2} rose to 0.77–0.94 bars in the conduit; and (4) Mass fraction CO_2 in a liquid phase varied from 6.3×10^{-4} to 5.2×10^{-4} in the conduit. Total CO_2 and H_2O mass (in gas and liquid phases) stored in the system estimates: 1.199 kg and 2084.6 kg, respectively.

Hot spring after eruption conditions are shown in Figure 14. This is what happens in the studied hot spring system after 5 min of discharge with a rate of 2 kg/s specified in the model. Sg rose to 1.0 in a top part of the geyser conduit. Pressure dropped to 1.05 bar in the reservoir and to 0.95 bars in the conduit. CO_2 mass fraction in a liquid phase varied from 0.74×10^{-4} to 6.2×10^{-4}. Total CO_2 and H_2O mass (in gas and liquid phases) stored in the system was 0.885 kg and 1551.7 kg, correspondingly. Thus, the CO_2 mass release in a geyser model system during geyser eruption was equal to 0.314 kg (which corresponds to a volume of 159.9 L of CO_2 at normal conditions) (Table 4).

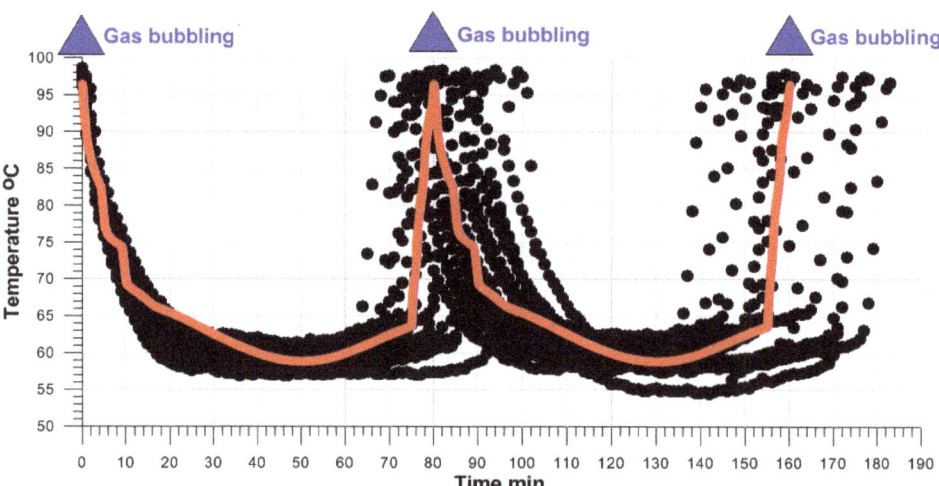

Figure 12. Modeling (run #9A2, unknown parameter estimations were: q = 0.15 kg/s, h = 84 kJ/kg (water 20 °C), Q = 2 kg/s)) of the transient bottom temperature in the Shaman geyser vs. the bottom temperature records obtained during the time period of observations from 27.06.2019 18:00 to 29.06.2019 18:00 (Figure 5).

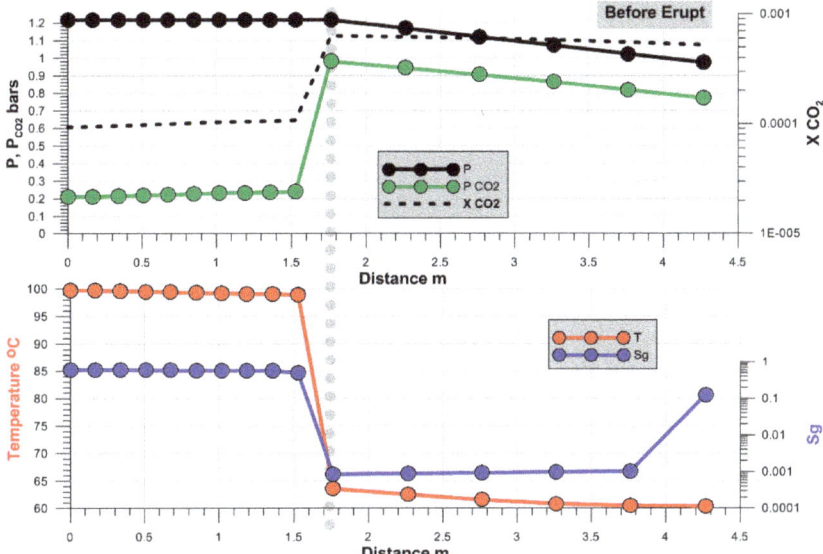

Figure 13. State before eruption (run #9A2): model parameter distributions (temperature in °C, gas saturation Sg, pressure and partial CO_2 pressure P_{CO2} in bars, mass fraction of CO_2 in a liquid phase) along a streamline from hot water recharge element #11 to discharge element #56 of the model (see Figure 7 for model element names references). Vertical dotted line corresponds to the position of flow direction change from radial-horizontal to ascending-vertical.

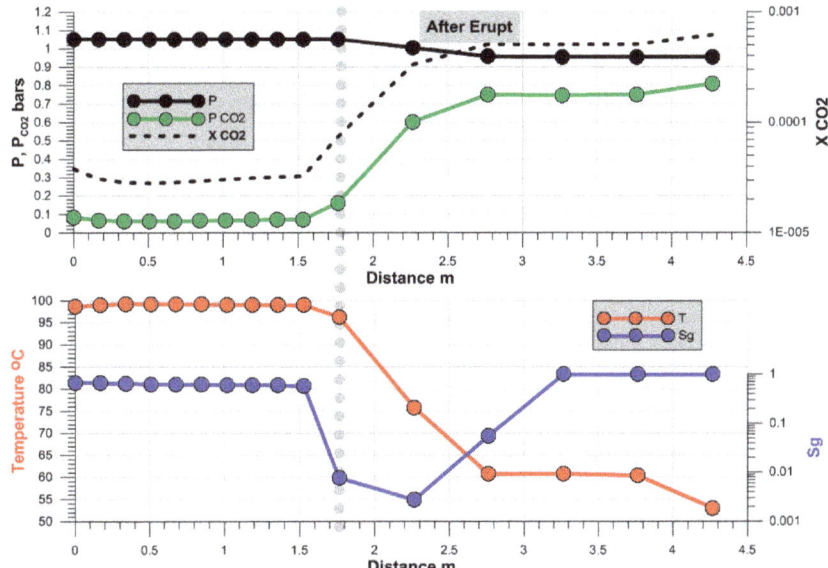

Figure 14. State after eruption (run #9A2): model parameter distributions (temperature in °C, gas saturation Sg, pressure and partial CO_2 pressure P_{CO2} in bars, mass fraction of CO_2 in a liquid phase) along a streamline from hot water recharge element #11 to discharge element #56 of the model (see Figure 7 for model element names references). Vertical dotted line corresponds to the position of flow direction change from radial-horizontal to ascending-vertical.

7. Discussion and Conclusions

The Shaman geyser was formed in the Uzon caldera, Kamchatka in autumn 2008 in the place of a former hot spring. The initial eruptions with an IBE from 12 to 20 min from a conduit with a top diameter of 30 cm and depth of 2.3 m, enlarged the geyser conduit size to a top diameter of 60 cm and depth of 2.9 m. In a few years, the geyser achieved rather stable parameters of eruptions maintained in 2012–2017: an IBE from 129 to 144 min and fountain heights increased up to 4 m. In 2019, the geyser's pool boiling temperature dropped to 80 °C, that is, a significantly lower value compared to the 97.6 °C of pure water under the local atmospheric pressure of 0.93 bar. That pointed to CO_2-gas lift driven conditions during geyser eruptions [19]. CO_2 dominance was proved by gas sampling and chemical analysis of non condensable gases NCG. Simultaneously, geyser pool subsidence took place, which caused the re-direction of the adjacent stream into the geyser conduit. As a sequence of cold water inflow, chemical component dilution was observed and IBE decreased to 80 min by 2019.

In June 2019, the Shaman geyser eruptions temporarily stopped, while cyclic discharge continued. Nevertheless, the Shaman geyser renewed its eruption mode in July 2019, when the cold water inflow rate declined and the inflow temperature rose due to seasonal change. Temperature records in the geyser bottom-hole conditions (depth of 2.9 m) were obtained in June and July 2019 in cyclic discharge and geyser mode correspondingly. Geyser conduit geometry, volume of geyser system, and CO_2 discharge/recharge rate estimates were used to design a conceptual model of the Shaman geyser, relevant to conditions of 2019.

The TOUGH2-EOS2 numerical model was applied to reproduce the transient temperature records obtained in both modes (cycling hot spring and geyser). A simple R-Z model grid represents the geyser conduit joint to the cavern-reservoir at a depth of 3 m and host rocks. Matches of modeling results with temperature records yielded the following estimates of cold inflow rate and enthalpy and eruption rate, which were used as estimated model parameters. The geyser mode of cycling was realized in the model at an inflow rate of 0.15 kg/s and enthalpy of 126 kJ/kg (water 30 °C), and geyser eruption rate of 3 kg/s. Hot spring mode of cycling was realized in the model at an inflow rate of 0.15 kg/s and enthalpy of 84 kJ/kg (water 20 °C), and the geyser eruption rate of 2 kg/s.

Modeling showed that the geyser system CO_2 storage capacity was significantly less (24 times), compared to the CO_2 inflow concentration. Modeling also yielded estimates of net CO_2 release during the geyser eruption cycle equal to 0.431 kg (corresponding to a volume of 219.4 L of CO_2 at normal conditions), while net CO_2 mass release during hot spring discharge cycle was equal to 0.314 kg (corresponding to a volume of 159.9 L of CO_2 at normal conditions) (Table 4). Thus, 60 L of CO_2 makes a difference for the Shaman geyser to be or not to be a "geyser".

Table 4. Mass in place (kg) during the three Shaman modes.

Mode	Before Discharge		After Discharge		CO_2 Release in kg	CO_2 Release in L
	CO_2	H_2O	CO_2	H_2O		
Steady State	0.767	1842	0.767	1842	0	0
Geyser erupt	1.081	2102.3	0.650	1271.9	0.431	219.4
Cycling Flowrate	1.199	2084.6	0.885	1551.7	0.314	159.9

The following are concluding remarks on the uncertainties of the Uzon geyser TOUGH2-based model: (1) The geyser conduit was assumed to be a cylindrical domain with 99% porosity, while this is rather an empty nonregular-conically-shaped hole; (2) The geyser adjacent reservoir was assumed to be a horizontal layer with specified radius, while this was more likely to be a discrete fracture network system; (3) The geyser-reservoir system geometry was limited to a 3 m depth and 2 m in the radial direction, and a constant hot water and CO_2 recharge was assumed; (4) Absence of the data of direct pressure measurements in a geyser conduit significantly restricted the inverse modeling capabilities; (5) Multiple fountain eruption was substituted by the piecewise constant discharge function; (6) Model grid sensitivity may be tested; and (7) Hot water and gas upflow recharge measurements using a probe with a pumping device atop may yield an overestimate of flow, while the tracer method may underestimate an active volume geyser system.

Nevertheless, one of the possible TOUGH2-model based solutions, which reasonably explained the geyser bottom hole transient temperature records, was obtained. We also used iTOUGH2-EOS2 inverse modeling capabilities [20] to verify the parameter estimates obtained above. This solution also explains the switch from cycling flowrate mode to geyser cycling mode in terms of maximizing the CO_2 release on eruption stage.

Based on the above, measures to re-direct the cold stream from the geyser conduit to maintain Shaman geyser activity are recommended. We also recommend online monitoring of the Shaman geyser activity so that it might be a precursor of a new diatrem event (Figure 2).

Author Contributions: Conceptualization, A.V.K.; methodology, A.V.K.; software, A.V.K.; validation, A.V.K., G.K.; formal analysis, A.V.K.; investigation, A.V.K., G.K.; resources, A.V.K.; data curation, A.V.K., G.K.; writing—original draft preparation, A.V.K.; writing—review and editing, A.V.K.; visualization, A.V.K., G.K.; supervision, A.V.K.; project administration, A.V.K. All authors have read and agreed to the published version of the manuscript.

Funding: This work was supported by RFBR project #18-05-00052-20.

Acknowledgments: Significant logistic support during the field survey was given by Elena Subbotina. The authors also express their gratitude to G. Kroshkin, P. Shpilenok, D. Panicheva, A. Sergeyeva, N. Zhuravlev, T. Rychkova, A. Polyakov, E. Chernykh, P. Voronin, E. Kartasheva, V. Lavrushin, V. Grigoriyev, and N. Ostrik for transportation support, fruitful discussions, useful comments, participation in field surveys, sampling support and analysis, and manuscript preparation. We also appreciate the comments of Dr. Atsuko Namiki and four unknown reviewers, whose helpful advice was used to improve this manuscript. Thorough editing of the manuscript was performed by Prof. J.C. Eichelberger and is highly appreciated.

Conflicts of Interest: The funders had no role in the design of the study; in the collection, analyses, or interpretation of data; in the writing of the manuscript, or in the decision to publish the results.

Appendix A

Figure A1. Shaman conduit view in 2010 (photo by G. Karpov, south view).

Figure A2. Shaman geyser eruption in 2010 (photo by G. Karpov, south view).

Figure A3. Shaman geyser area view, SH—Shaman geyser, SH2—adjacent CO_2 hot spring, SH3—hot spring in the upstream of creek, partially diverted into Shaman geyser conduit (photo by G. Karpov, 2015, northwest view).

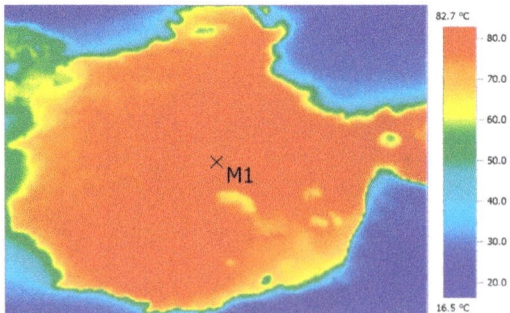

Figure A4. Shaman geyser pool pre-eruption infrared (IR) (testo-865) view. Photo by A. Kiryukhin, 24 August 2019.

Figure A5. Shaman geyser conduit top view, grid scale 20 cm. Photo by A. Kiryukhin, 24 August 2019.

Figure A6. Shaman geyser pool view, in front is partially diverted to the Shaman conduit creek, coming from SH3 (see Figure 2). Photo by A. Kiryukhin, 25 August 2019.

Figure A7. Shaman geyser eruption events on 24 August 2019 20:01. Times (negative): −07:42—discharge; −05:36—discharge increase; −05:15—eruption #1; −05:03—some backflow; −04:35—discharge increase; −04:28—eruption #2; −04:19—some backflow; −03:31—discharge increase; −03:27—eruption #3; −03:15—some backflow; −02:16—discharge increase, but no eruption commenced; −00:30—backflow recharge start; −00:15, −00:08; 00:00—cold water inflow in geyser conduit starts and continues, water level in conduit is 1 m in depth. Time stamp format minutes:seconds. Photo by A. Kiryukhin, 24 August 2019.

Figure A8. Shaman geyser eruptions on 06.09.2014, 21.04.2019, and 12.09.2019 and the conduit top view in 2019. Time stamp format in a left-bottom minutes:seconds. (photo by A. Kiryukhin (06.09.2014, 21.04.2019) and N. Ostrik (12.09.2019)).

Figure A9. Bannoye Lake/Maar. Photo by A. Kiryukhin, 24 August 2019.

Figure A10. Dalneye Lake/Maar. Photo by A. Kiryukhin, 24 August 2019.

Figure A11. Shaman-2 CO_2-hot spring located adjacent to the Shaman geyser (next candidate to be a geyser in the Uzon caldera). Photo by A. Kiryukhin, 25 August 2019.

References

1. Hurwitz, S.; Manga, M. The fascinating and complex dynamics of geyser eruptions. *Annu. Rev. Earth Planet. Sci.* **2017**, *45*, 31–59. [CrossRef]
2. Hurwitz, S.; Clor, L.E.; McCleskey, R.B.; Nordstrom, D.K.; Hunt, A.G.; Evans, W.C. Dissolved gases in hydrothermal (phreatic) and geyser eruptions at Yellowstone National Park. *USA Geol.* **2016**, *44*, 235–238. [CrossRef]
3. Han, W.S.; Lu, M.; MCPherson, B.J.; Keating, E.H.; Moore, J.; Park, E.; Watson, Z.T.; Jung, N.-H. Characteristics of CO_2-driven cold-water geyser, Crystal Geyser in Utah: Experimental observation and mechanism analyses. *Geofluids* **2013**, *13*, 283–297. [CrossRef]
4. Walter, T.R.; Jousset, P.; Allahbakhshi, M.; Witt, T.; Gudmundsson, M.T.; PállHersir, G. Underwater and drone based photogrammetry reveals structural control at Geysir geothermal field in Iceland. *J. Volcanol. Geotherm. Res.* **2018**, *391*. [CrossRef]
5. Namiki, A.; Ueno, Y.; Hurwitz, S.; Manga, M.; Munoz-Saez, C.; Murphy, F. An experimental study of the role of subsurface plumbing on geothermal discharge, Geochem. *Geophys. Geosyst.* **2016**, *17*. [CrossRef]
6. Kiryukhin, A.V.; Rychkova, T.V.; Dubrovskaya, I.K. Hydrothermal system in Geysers Valley (Kamchatka) and triggers of the Giant landslide. *Appl. Geochem.* **2012**, *27*, 1753–1766. [CrossRef]
7. Kiryukhin, A.V.; Rychkova, T.V.; Sergeeva, A.V. Modeling conditions of permeable geyser conduit formations in silicic volcanism areas. *J. Volcanol. Seismol.* **2020**, *2*, 3–16.
8. Kiryukhin, A. Modeling and observations of geyser activity in relation to catastrophic landslides–mudflows (Kronotsky nature reserve, Kamchatka, Russia). *J. Volcanol. Geotherm. Res.* **2016**, *323*, 129–147. [CrossRef]
9. Kiryukhin, A.; Sugrobov, V.; Sonnenthal, E. Geysers Valley CO_2 Cycling geological Engine (Kamchatka, Russia). *Geofluids J.* **2018**, *17*. [CrossRef]
10. Leonov, V.L. Geological Structure and History of Geysers Valley. In *The Valley of Geysers—The Pearl of Kamchatka (Scientific Guide)*; Petropavlovsk-Kamchatsky: Kamchatpress, Russia, 2009; pp. 45–51.
11. Lundgren, P.; Lu, Z. Inflation model of Uzon caldera, Kamchatka, constrained by satellite radar interferometry observations. *Geophys.Res. Lett.* **2006**, *33*, L06301. [CrossRef]
12. Karpov, G.A. Sulphur Melt in a Bottom of Thermal Lake in Uzon Caldera (Kamchatka). *J. Volcanol. Seismol.* **1996**, *2*, 34–47. (In Russian)
13. Droznin, V.A. New Geyser in Uzon Caldera. *Vestnik KRAUNC* **2009**, *2*, 10–12. (In Russian)
14. Karpov, G. New Geyser in Kronotsky Reserve. Characteristics and Reasons of Formation Proc. IVS FEB RAS Conf. Vulcan. Relat. Process. 2010. Available online: http://www.kscnet.ru/ivs/publication/volc_day/2010/art12.pdf (accessed on 14 May 2020).
15. Karpov, G. Evolution of Regime and Physical-Chemical Characteristics of the New Formed Geyser in Caldera Uzon (Kamchatka). *Volcanol. Seismol.* **2012**, *3*, 3–14.
16. Karpov, G.A.; Schroeder, P.A. and Nikolaeva, A.G. Geochemistry of rare-earth elements in thermal waters of the Uzon-Geyzernaya hydrothermal system (Kamchatka). *Geol. Geophys.* **2018**, *59*, 1152–1163.
17. Etiope, G.; Lollar, B.S. Abiotic methane on Earth. *Rev. Geophys.* **2013**, *51*, 276–299. [CrossRef]
18. Pruess, K.; Oldenburg, C.; Moridis, G. *TOUGH2 User's Guide, Version 2.0. Rep. LBNL-43134*; Lawrence Berkeley Natl. Lab.: Berkeley, CA, USA, 1999; 198p.
19. Rychkova, T.V. Hydrogeological Analysis of the Geysers Formation and Functionality (Kamchatka Hydrothermal Systems Examples). Ph.D. Thesis, F.O.N., OOO, Petropavlovsk-Kamchatsky, Russia, 2020; 20p.
20. Finsterle, S. iTOUGH2 V7.0 Command Reference. In *Rep. LBNL-40041 rev., 2014*; Lawrence Berkeley Natl. Lab.: Berkeley, CA, USA; 130p.

© 2020 by the authors. Licensee MDPI, Basel, Switzerland. This article is an open access article distributed under the terms and conditions of the Creative Commons Attribution (CC BY) license (http://creativecommons.org/licenses/by/4.0/).

Article

Magma Fracking Beneath Active Volcanoes Based on Seismic Data and Hydrothermal Activity Observations

Alexey Kiryukhin *, Evgenia Chernykh, Andrey Polyakov and Alexey Solomatin

Institute of Volcanology & Seismology FEB RAS, Petropavlovsk-Kamchatsky 683006, Russia; jenia.chev@yandex.ru (E.C.); pol@kscnet.ru (A.P.); alf110111@gmail.com (A.S.)
* Correspondence: avkiryukhin2@mail.ru

Received: 28 November 2019; Accepted: 26 January 2020; Published: 29 January 2020

Abstract: Active volcanoes are associated with microearthquake (MEQ) hypocenters that form plane-oriented cluster distributions. These are faults delineating a magma injection system of dykes and sills. The Frac-Digger program was used to track fracking faults in the Kamchatka active volcanic belt and fore-arc region of Russia. In the case of magma laterally injected from volcanoes into adjacent structures, high-temperature hydrothermal systems arise, for example at Mutnovsky and Koryaksky volcanoes. Thermal features adjacent to these active volcanoes respond to magma injection events by degassing CO_2 and by transient temperature changes. Geysers created by CO_2-gaslift activity in silicic volcanism areas also flag magma and CO_2 recharge and redistributions, for example at the Uzon-Geyserny, Kamchatka, Russia and Yellowstone, USA magma hydrothermal systems. Seismogenic faults in the Kamchatka fore-arc region are indicators of geofluid fracking; those faults can be traced down to 250 km depth, which is within the subduction slab below primary magma sources.

Keywords: magma; hydrothermal; fracking; volcanoes; Kamchatka

1. Introduction

Volcanoes and crustal magma chambers are the products of magma injection from primary magma chambers at depths of 150–200 km below the Kuril-Kamchatka volcanic zone (Figure 1). These magma injections create new hydraulic fractures and reactivate existing faults. The formation of shear cracks in adjacent to the main aseismic opening triggers microseismicity [1]. Earthquake magnitudes (M) for shear deformations with amplitudes of 0.1 mm–1 cm and fracture lengths of a few hundred meters are estimated to range from 1 to 2 (Ks from 3.5 to 5.5), corresponding to the sensitivity of local seismic networks operating in the areas of the Klyuchevskoy, Koryaksky-Avachinsky, and Mutnovsky-Gorely volcano groups. We propose that the planes defined by clusters of microearthquakes correspond to magmatic intrusions. For a search of plane-oriented clusters of earthquakes, we use our original programs Frac-Digger (RU reg. ## 2016616880), and for production feed-zones, we use Frac-Digger2 (RU reg. ## 2017618050) (these applications discrete fracture systems using seismic and geological data). In these programs, the following criteria are used to select clusters of earthquake hypocenters or production feed zones: (1) proximity in time δt (Frac-Digger only); (2) proximity within the horizontal plane δR; (3) proximity to plane orientation δZ (distance between the object and plane); and (4) minimum number of elements in cluster N. This method has already been tested on the example of the 2012 eruption of Tolbachik volcano [2], the 2000–2016 MEQ data of Koryaksky-Avachinsky volcanic cluster [3–5], Paratunsky graben (adjacent to Vilyuchinsky volcano) [6], the Mutnovsky-Gorely volcanic cluster [7], and the Klyuchevkoy group of volcanoes [8]. Note that the methodology of dyke tracking using the Frac-Digger program is presented in detail in Section 3.2.1, "Seismic Data and Method of Plane-Oriented Clusters Identification" [5].

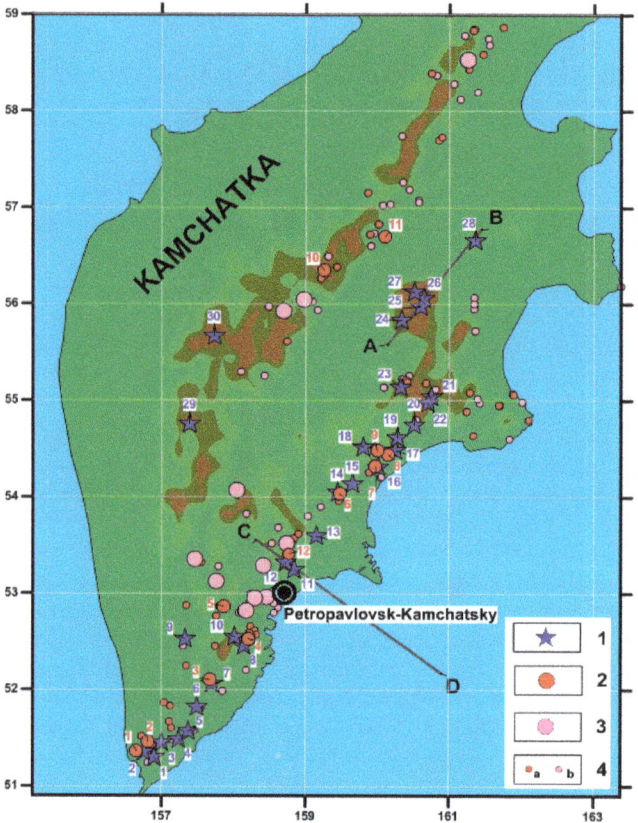

Figure 1. Location map of the volcanoes and hydrothermal systems of Kamchatka. 1—Active volcanoes (1, Kambalny; 2, Koshelevsky; 3, Diky Greben; 4, Ilynsky; 5, Zheltovsky; 6, Ksudach; 7, Asachinsky; 8, Mutnovsky; 9, Opala; 10, Gorely; 11, Avachinsky; 12, Koryaksky; 13, Zhupanovsky; 14, Karymsky; 15, Maly Semyachik; 16, Bolshoy Semyachik; 17, Kihpinych; 18, Taunshits; 19, Krasheninnikova; 20, Kronotsky; 21, Komarova; 22, Gamchen; 23, Kizimen; 24, Pl. Tolbachik; 25, Bezymyanny; 26, Kluchevskoy; 27, Ushkovsky; 28, Shiveluch; 29, Khangar; 30, Ichinsky); 2—High-temperature hydrothermal systems (1, Koshelevsky; 2, Pauzhetsky; 3, Hodutkinky; 4, Mutnovsky; 5, B-Banny; 6, Karymsky; 7, Semyachiksky; 8, Geyserny; 9, Uzonsky; 10, Apapelsky; 11, Kireunsky; 12, North-Koryaksky); 3—Hydrothermal systems with temperatures below 150 °C; 4—Groups of thermal springs: (a) temperature from 50 to 100 °C, (b) temperature from 20 to 50 °C. Notes: The AB and CD lines delineates the cross-sections shown below.

By coupling magma fracking observations beneath volcanoes with the monitoring of thermal features in adjacent hydrothermal systems, we can develop appropriate geofiltrational and thermal–hydrodynamic TOUGH2-models that allow an assessment of recoverable geothermal resources and explain the mechanisms of the anomalous hydrothermal perturbations. A recently revealed mechanism [9] of geysers activity due to cyclic magmatic CO_2-gaslift pointed to the possibility of understanding the mechanisms of relationship between anomalous hydrothermal perturbations, volcano activity, and strong earthquakes. Continuous measurements of CO_2 partial pressures in the Koryaksky Narzan thermal mineral springs (northern sector of Koryaksky volcano, 2017–2019) and in the Mutnovsky geothermal Power Plant condenser (2019) provide evidence of this process.

2. Active Faults and Magma Fracks Based on Seismic Data Analysis

2.1. Koryaksky-Avachinsky Volcanoes

Plane-oriented clusters of 5160 seismic events from data of the Kamchatka Branch Federal Research Center United Geophysical Survey Russia Academy of Sciences (KB FRC UGS RAS) of 01.2000–07.2016) below Koryaksky and Avachinsky volcanoes were used to identify magma injection zones [4,5]. Cluster identification was carried out using our Frac-Digger program. The previously defined criteria were used to include a new event in the cluster: $\delta t \leq 1$ day, $\delta R \leq 6$ km, $\delta Z \leq 0.2$ km, $N \geq 6$.

It was found that 30% of the total number of 5160 earthquakes formed 204 plane-oriented clusters. A careful analysis of the orientations of dykes and sills (using stereograms and histograms) showed the following: (1) the dykes beneath Koryaksky Volcano were mostly emplaced in the depth range from −5000 to 0 m masl; they mostly strike north–south (75% of dykes have strike azimuths between 320° and 40°) and have dip angles over 50° (70% of the dykes), and (2) the dykes beneath Avacha Volcano were mostly in the depth range 1000–2000 m masl. Their strikes are distributed over all possible directions, but there is a local maximum of dykes striking nearly north–south with 25% of dykes having azimuths between 350° and 10°. Steeply dipping dykes are not obviously dominant (61% have dip angles over 45°).

Additionally: (1) A shallow crustal magma chamber, likely as a combination of sills and dykes, lies below the southwestern base of Koryaksky Volcano at depths ranging from −2 to −5 km masl and a width of 2.5 km; (2) dyke accumulation in a nearly north–south zone (7.5 by 2.5 km, with absolute depths ranging from −2 to −5 km occurs beneath the northern sector of Koryaksky Volcano; and (3) a shallow magma chamber in the cone of Avachinsky Volcano lies at absolute depths ranging from 1 to 2 km and has a width of 1 km.

Thus, we propose the geomechanical state according to [10]: (1) extension is recorded beneath Koryaksky Volcano producing normal faulting (NF conditions), the vertical stress Sv is the maximum stress, the horizontal principal stresses are coincident with the north–south (SHmax) and the east–west (Shmin) directions, (2) the cone of Avacha Volcano is found to be under conditions of radial extension.

The relationship $M = 0.67 \log(V) − 0.82$ between the upper bound of the induced seismic event M and magmatic injection volume V, suggested by [11], may be used for recorded dykes volume estimates beneath Koryaksky in a range from 220 m^3 to 1.988×10^6 m^3, and the total recorded volume of the injected magma is estimated at 5.862×10^6 m^3 during 2008–2011. Avachinsky volcano characterized by a range from 111 m^3 to 0.064×10^6 m^3, and the total recorded volume of the injected magma is estimated at 0.169×10^6 m^3 during 2008–2011.

It is also worth noting that seismogenic planes and microearthquake (MEQ) mechanisms planes (19 mechanisms estimated were used) are intersected with the angle of 51° (on average) [12]. This may have the following geomechanical interpretation: seismogenic planes mark opening mode fractures (hydrofracking-type fractures) formed orthogonal to the least effective stress during magma injection, whereas MEQ mechanisms planes are mark shear faults, which initiated from the main opening mode fault (dyke) using pre-existing fractures systems.

There were six magma injections below the northern foothills of Koryaksky volcano detected in 2011–2019 based on MEQ analysis data using Frac-Digger criteria ($\delta t = 1$ day, $\delta R = 6$ km, $\delta Z = 0.2$ km, $N = 6$): one dyke on 02.08.2011, one dyke on 28.02.2016, one dyke on 22.08.2017, two dykes on 14.11.2017, and one dyke on 22.04.2019. All of these dykes were injected just below the Koryaksky Narzan and Isotovsky thermal mineral springs area. The depth of magma injections was estimated from seismic data from −7000 to +1000 masl, with M ranges from 0.85 to 2.95 (Figure 2, Table 1).

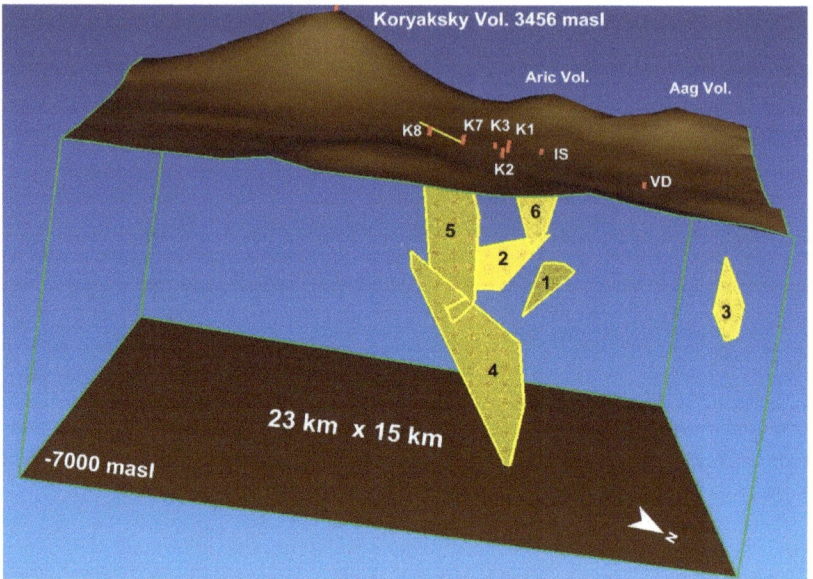

Figure 2. Magma injections (dykes) below the northern foothills of Koryaksky volcano during the time period from 2011 to 2019. Koryaksky Narzans are marked as K1, K2, K3, K7, and K8, Isotovsky hot spring—IS, Vodopadny hot spring—VD. Frac-Digger parameters are: $\delta t = 1$ day, $\delta R = 6$ km, $\delta Z = 0.2$ km, and $N = 6$. Dyke geometries and magnitudes are given in the Table 1 below. Note: Dykes patches in 3D view are convex polygons, with vertices in the points of projections of plane-oriented clusters of earthquakes hypocenters on an approximation plane.

Table 1. Dyke geometries and corresponding microearthquake (MEQ) magnitudes of the Koryaksky volcano for the period from 02.08.2011 to 22.04.2019.

##	Data	Dip (°)	Dip Direction (°)	Mmax	N	Area km²
1	02.08.11	39.3	115.1	2.95	7	1.4
2	28.02.16	70.2	64	2.45	7	1.7
3	22.08.17	69	246.3	1.30	6	1.0
4	14.11.17	78	291.2	1.25	17	11.5
5	14.11.17	63.7	283.2	0.85	8	10.0
6	22.04.19	73	266.5	1.95	6	2.2

If a less severe Frac-Digger criterion in time, $\delta t \leq 30$ day, is used, the number of detected dykes significantly increases. By using this approach, it was found that as a result, 91% of the total number of 5160 earthquakes 452 plane-oriented clusters were formed. Figure 3 shows that 68 of these features were interpreted as dykes from 2011 to 2019 beneath Koryaksky volcano. Most of them were injected in a northeast sector of the volcano structure, where thermal springs activity occurs.

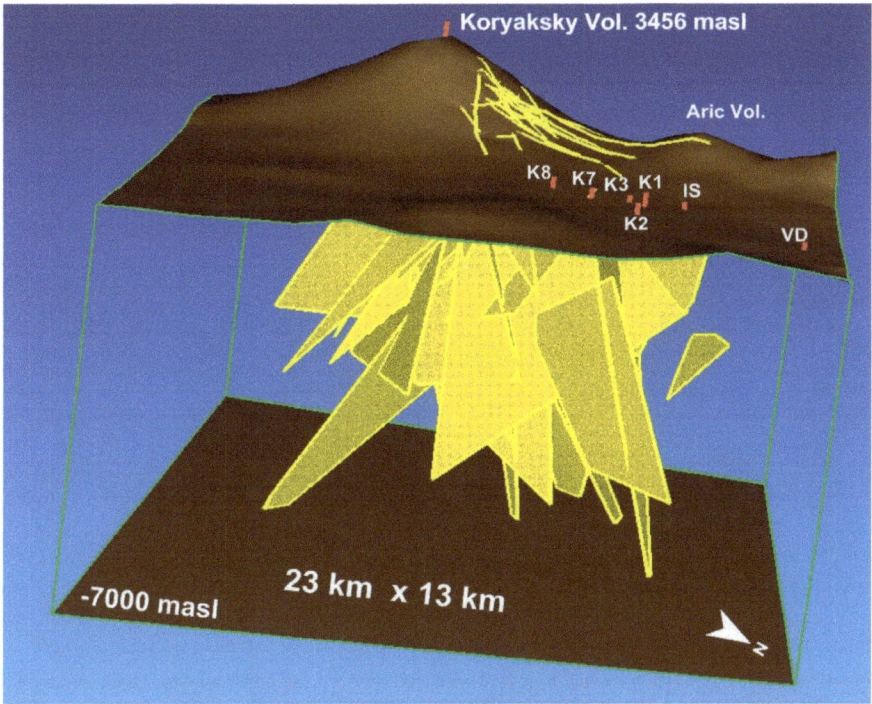

Figure 3. Magma injections (dykes) below northern foothills of Koryaksky volcano during the time period from 2011 to 2019. Koryaksky Narzans are marked as K1, K2, K3, K7, and K8, Isotovsky hot spring—IS, Vodopadny hot spring—VD. Frac-Digger criteria are: (δt = 30 day, δR = 6 km, δZ = 0.2 km, N = 6).

2.2. Mutnovsky and Gorely Volcanoes

We already used plane-oriented earthquake clusters (KB FRC UGS RAS, catalogs 2009–2017) to track the dykes injected beneath and around the Mutnovsky volcano by using the Frac-Digger program [5]. Thus, magmatic injection zones (shape-like dykes and sills, [13,14]) defined in this manner were treated as pathways and heat sources for ascending high-temperature upflows that feed adjacent hydrogeological reservoirs and surface feature (hot springs and fumaroles) discharge.

Four seismic stations record seismicity in the Mutnovsky–Gorely volcanic cluster. A total of 1336 earthquakes were recorded by the KB FRC UGS RAS in the edifices and basement of the Mutnovsky and Gorely volcanoes between January 2009 and February 2017. Cluster identification was carried out using our Frac-Digger program. The same criteria as above were used to include a new event in the cluster. Our treatment of these data used the method previously outlined, which yielded 973 earthquakes (72.8%) that make up 74 plane-oriented clusters between November 2013 and October 2016 (most of them associated with the Mutnovsky volcano) (Figure 4).

The analysis of seismic activity at the Mutnovsky volcano revealed the following geomechanical features. (1) Most of the dykes were injected below and in the northeastern sector of the Mutnovsky volcano in an area 2 km × 10 km. (2) Most of the dykes beneath the Mutnovsky volcano have a dip angle ranging from 20° to 40°. (3) Most of the dykes were injected at a depth ranging from −4.0 to −2.0 km masl and at strikes in a NE–NNE (20–50°) direction. (4) Six dykes were injected close to the SE boundary of the Mutnovsky production geothermal reservoir at a depth ranging from −6.0 to −4.0 km masl. (5) Seismic events of magnitude M during dyke injection ranged from 1.0 to 2.8. (6) There is a trend of dyke dip angle increase moving from the Mutnovsky volcano to the north (1°/km).

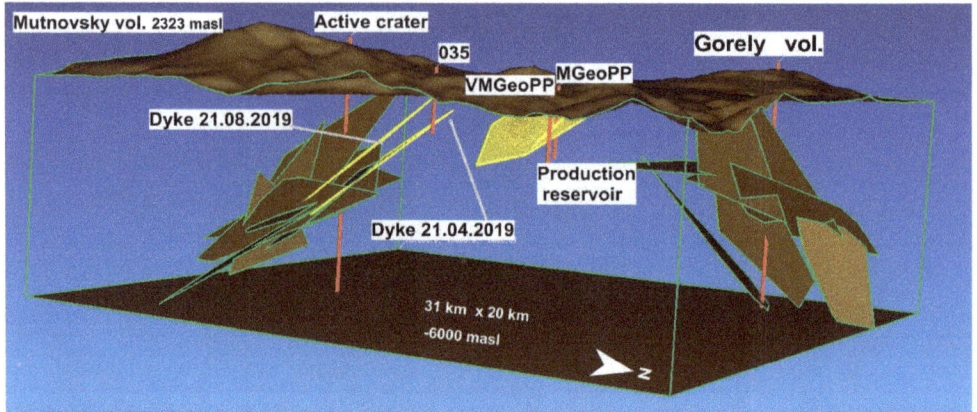

Figure 4. Mutnovsky–Gorely area magma fracking geometry vs. known production zone geometry. Dykes 21.04.2019 and 21.08.2019 injected close to well 035 are shown, too. This figure also gives a new approach for geothermal production targets in Mutnovsky area.

Thus, dyke geometry indicated reverse fault (RF) geomechanic conditions [10] in the vicinity of the Mutnovsky volcano (SHmax > Shmin > Sv, SHmax striking to NE), while there is a trend to normal fault (NF) conditions 10 km away in the region where the production geothermal reservoir formed. The relationship M = 0.67 log (V) − 0.82 between the upper bound of the induced seismic event M and magmatic injection volume V, suggested by [11], may be used for recorded dyke volume estimates in a range from 520 m^3 to 0.252×10^6 m^3, and the total recorded volume of the injected magma is estimated at 0.67×10^6 m^3 from 2009–2016.

There were additional 22 magma injections identified during 02.2017 to 10.2019 in the Mutnovsky–Gorely geothermal area. Most of the dykes beneath the Mutnovsky volcano have a dip angle ranging from 20° to 40° and were injected at the depth ranging from −4.0 to −2.0 km masl and dipping in the SEE–SE (120°–130°) direction. Most of the dykes beneath the Gorely volcano have a dip angle ranging from 40° to 70° and were injected at the depth ranging from −2.0 to −1.5 km masl and dipping in the NWW (280°–320°) direction. It is worth noting two sub-parallel dykes (dip angles from 39° to 43°, dip azimuth from 149° to 153°, depth range from −2.6 to −1.9 km masl, M from 1.25 to 1.7) injected in the vicinity of well 035 in 2019 (Figure 4).

2.3. Northern Group of Volcanoes

To recover the sequence of magma injections in the area of the Klyuchevskoy Volcanic Cluster, we used a catalog containing data of seismic monitoring at 19 seismic stations operated by the KB FRC UGS RAS for the 2000–2017 period (the total number of earthquakes that have been recorded between January 1, 2000 and August 23, 2017 is 122,451). The method that we used to identify plane-oriented earthquake clusters and fitting planes using the Frac-Digger software was described in [2–5]. The relevant calculated plane patches are interpreted as zones of magma emplacement in the form of dykes and sills. The calculations assumed the following parameters for the identification of plane-oriented clusters: $\delta t \leq 1$ month, $\delta R \leq 6$ km, $\delta Z \leq 1$ km, $N \geq 6$. The calculations yielded 1788 clusters that contained 117,412 earthquakes (96% of the total number of recorded events).

A vertical profile along the axis of the Klyuchevskoy Volcanic Cluster that extends from the Tolbachik volcanoes to Shiveluch (Figure 5) shows that when extrapolated downward, the dykes beneath Tolbachik and Shiveluch intersect at a point that lies between absolute depths of −165 and −205 km beneath Klyuchevskoy; it is at this location that the region of primary magma melting is thought to reside, or there is a primary magma chamber to provide magma for all active volcanoes in the Klyuchevskoy Cluster (Klyuchevskoy, Plosky Tolbachik, Bezymyanny, and Shiveluch).

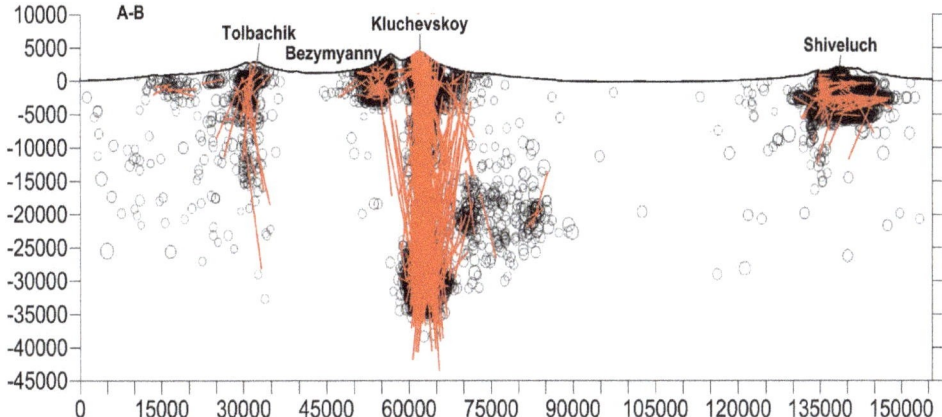

Figure 5. Magma fracking system beneath the Northern group of volcanoes. The AB cross-section line was shown in Figure 1.

The 2000–2017 magma injections directly beneath Klyuchevskoy were concentrated in the depth ranges between −31 and −28 km masl (35%) and between −1 and +2 km masl (20%), where we hypothesize the existence of a crustal chamber (C2) and a peripheral chamber (C1), respectively (we mean chamber as a plexus of sills and dykes here). The crustal magma chamber (C2) was found to contain dykes that dip at angles of 70–80° (30%) and sills that dip at 15–20° (8%); one also finds an increasing number of nearly east–west striking dykes that supply magma to the nearby Bezymianny and Krestovsky volcanoes. Nine episodes of intensive dyke emplacement and seven episodes of intensive sill emplacement were recorded in the C2 chamber during the period of interest (2000–2017); the dyke episodes preceded the sill episodes by an advance time of one year (two cases), the episodes occurred simultaneously (five cases), or they did not terminate in sill emplacement (one case), which provides evidence of a change in the geomechanical condition around the C2 chamber in the range from horizontal extension to radial compression.

The dip angles and azimuths of dykes and sills in the peripheral chamber (C1) are distributed rather uniformly; one notes some increase in the number of dykes that dip at angles of 35° to 50° and 65° to 80°. There is also a set of dykes that dip at northeast azimuths (i.e., striking along the Klyuchevskoy-Krestovsky line).

Tolbachik Volcano is characterized by injections of magma in the depth range between −8 and −1 km masl, with most dykes dipping more steeply than 60° (67%). In addition, we identified a set of dykes that dip at azimuths of 180–220° (26%) and strike along the Ostry Tolbachik to Plosky Tolbachik to the Bol'shaya Udina line.

Shiveluch is characterized by magma injections in the depth ranges between −4 and −2 km masl (37%) and between 0 and +2 km (47%), where two peripheral magma chambers are hypothesized to reside. The injections mostly occurred in the form of sills that dip at angles below 5°, with some trend of prevailing western dip azimuths.

Bezymianny Volcano receives magma from the crustal magma chamber beneath Klyuchevskoy Volcano (C2), which lies at a depth between −31 and −28 km masl. The injections beneath Bezymianny occurred in the depth range between −2 and +2 km (93%), where a peripheral magma chamber is hypothesized to exist. The injections mostly occurred in the form of sills that dip at angles below 15° (27%) and low-angle dykes that dip at 50° or less (58%); the dip azimuths are distributed uniformly over all directions.

The volumes of magma injection from magma chambers can be estimated using empirical relationships that relate the maximum magnitude of the triggered seismicity to the volume of injected magma [11]: $M = 0.67 \log(V) − 0.82$, where M is the maximum magnitude of triggered seismic events and V is the injection volume. The results from this assessment of dyke and sill injections volumes for the volcanoes in the Klyuchevskoy Cluster during 2000–2017 are Klyuchevskoy (peripheral chamber C1)—0.6×10^6 m^3, Klyuchevskoy (crustal chamber C2) —2.4×10^6 m^3, Bezymyanny—4.7×10^6 m^3, Tolbachik—13.2×10^6 m^3, Shiveluch—22×10^6 m^3. Assuming the enthalpy of magma at a temperature of 1200 °C to be 1000 kJ/kg and the density of magma to be 2800 kg/m^3, one can estimate the mass discharge and the corresponding thermal power due to magma injections for the time elapsed from January 2000 to August 2017.

The thermal power values for magma injections in the volcanic plumbing systems considered here and the output estimates for the respective volcanoes from [15] yields the following ratios between the discharge of magmas stored beneath volcanoes and that of magma ejected onto the ground surface (volume of intruded magma)/(volume of erupted magma): 0.8% for Klyuchevskoy, 14.9% for Bezymyannyy, 23.2% for Tolbachik, and 72.9% for Shiveluch.

Thus, Kluchevskoy volcano magma injections (dykes) took place in "normal fault" (NF) conditions, forming a permeable reservoir down to −35 km msl and two magma chambers within it; Shiveluch volcano magma injections (sills) occurs in "reverse fault" (RF) conditions at shallow levels from −4 km masl to −2 km masl, forming sills in an area 15 km across.

2.4. Kamchatka East Volcanic Belt and Adjacent Shelf Area

Critically stressed faults are key players in creation of the productive reservoirs and generation of strong earthquakes [10]. We performed a Frac-Digger analysis of the Kamchatka's regional seismicity to identify the network of such active faults, which are identified as plane-oriented clusters of hypocenters in Frac-Digger terms [16].

We used the regional catalog of earthquakes from KB FRC UGS RAS, which includes 5972 earthquakes with M values above 4.25 during the time period from 01.1980 to 02.2016. The following criteria for seismogenic planes selection were used: $N \geq 6$, $\delta t \leq \infty$, $\delta Z \leq 10$ km, $\delta R \leq 100$ km. These criteria correspond to the assumption of the existing network of continuously active regional faults.

Based on the above, we found 156 plane-oriented clusters of earthquakes, which are interpreted as active seismogenic faults. Most of them are located beneath Kamchatka's eastern shelf, between the shore line and ocean trench (Figure 6). Seventeen faults are found to be the most active (with more than 100 earthquakes each). Most of the active faults are characterized by dip angles from 50° to 70° and a dip azimuth NWW from 300° to 310°, striking subparallel to the ocean trench. This points to the extension in the NWW direction and NF geomechanical conditions as a whole.

Seismogenic fault #110 includes the hypocenter of earthquake M = 7.3 on 24 Nov 1971, which was the strongest felt in Petropavlovsk-Kamchatsky since 1959. Faults with other directions (especially in shallow conditions at elevations above −10 km masl) are found too, which reflects local geomechanical features. One of such faults is #4 (175 earthquakes included, dip angle 53°, dip azimuth 217°) striking in the direction of Petropavlovsk-Kamchatsky.

Seismogenic fault #35 includes the hypocenter of earthquake M = 6.9 on 17 Aug. 1983, which was accompanied with the ground deformations described by [17]. Seismogenic fault #35 has a dip angle of 60.3°, dip azimuth of 281.5°, and includes 77 hypocenters of earthquakes.

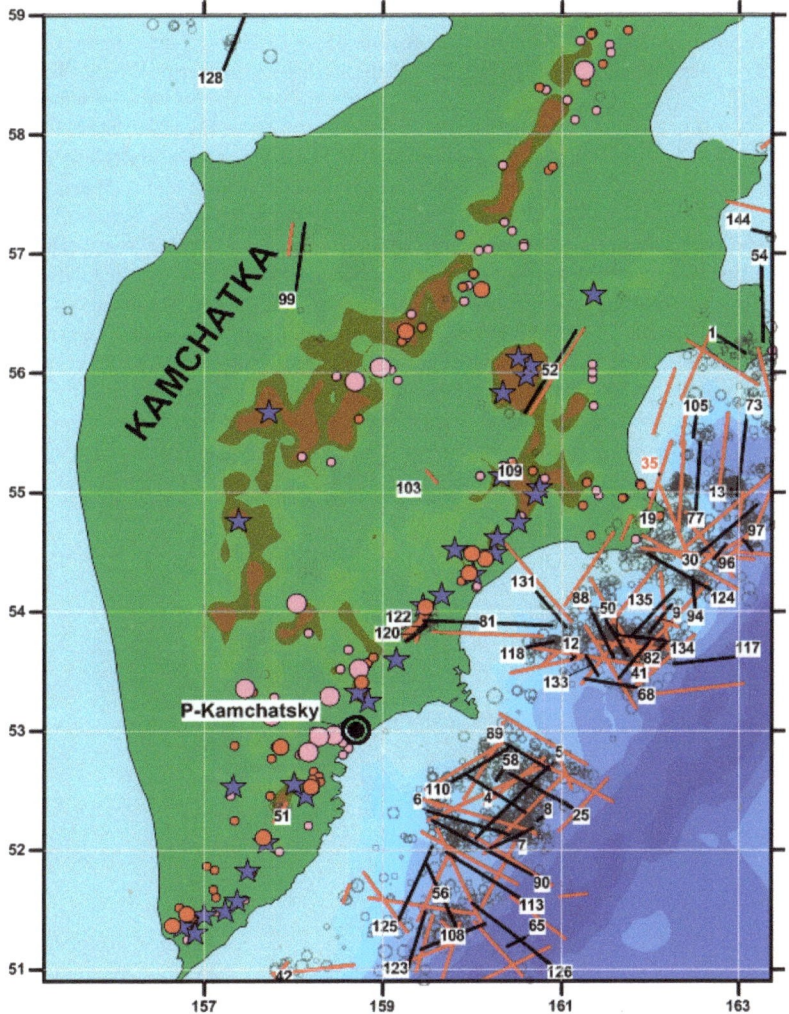

Figure 6. Kamchatka shelf active seismogenic faults. Traces of seismically active faults (black lines with numbers at elevation of −5 km masl, red lines at elevations of −20 km masl), circles corresponds to regional earthquakes epicenters in a range of depth above −30 km asl (size of circle proportional to earthquakes magnitudes). For other symbols, see the legend in Figure 1.

Seismogenic faults distributions point to a lower limit of the hydrofrack propagation into the subduction plate (down to −250 km masl), where this fluid can act as the key ingredient for magma melting to recharge the primary chambers of active volcanoes (Figure 7).

It is also worth noting that 98% of all earthquakes formed a plane-oriented clusters network, which characterizes the geomechanical conditions of collided plates and suggests hydrofracking mechanisms due to fluids (possible phases are: water, oil, gas) generation there.

In case of water, the empirical relationship $M = 0.67 \log (V) + 1.42$ between the upper bound of the induced seismic event M and water injection volume V, suggested by [11], may be used for water volumes recharge estimates into the active faults network of the Kamchatka shelf. This yields in a range from 0.00007 to 0.3 km^3, and the total volume of the recharged water is estimated at 1.87 km^3 during the time interval from 1980 to 2016 (over approximately 750 km of arc length).

In case of gas, the empirical relationship is slightly modified (supercritical CO_2 injection) to M = 0.67 log (V) − 0.30 [11]; then, this may be used for rough estimates of gas volumes generated from deep hydrocarbon sources to be injected into the active faults network of the Kamchatka shelf. This yields in a range from 0.024 to 111 km^3, and the total volume of the injected gas is estimated at 691 km^3 during the time interval from 1980 to 2016. In relation to this, it is worth noting that methane hydrates and methane submarine discharges are widely distributed along the east coast of Kamchatka.

Then, we perform an additional analysis of the 3D distribution of the 200 strongest earthquakes (M> 5.65) from the above-mentioned catalog of seismic events of KB FRC UGS RAS. The following selection criteria were used in Frac-Digger2: δZ = 4 km, δR = 100 km, N = 5. The output results show that 102 of the 200 strongest earthquakes form 11 plane-oriented clusters of hypocenters. A comparative analysis of seismogenic faults orientation with the mechanisms of corresponding earthquakes (estimated at http://www.globalcmt.org/CMTsearch.html) shows an average intersection angle of 33°. This may have the following geomechanical interpretation: seismogenic planes are marked as opening mode fractures (hydrofractures), which formed orthogonal to the least effective stress during magma injection, while the EQ's mechanisms planes are marked as shear faults that initiated from the main opening mode fault (dyke) using pre-existing fractures systems.

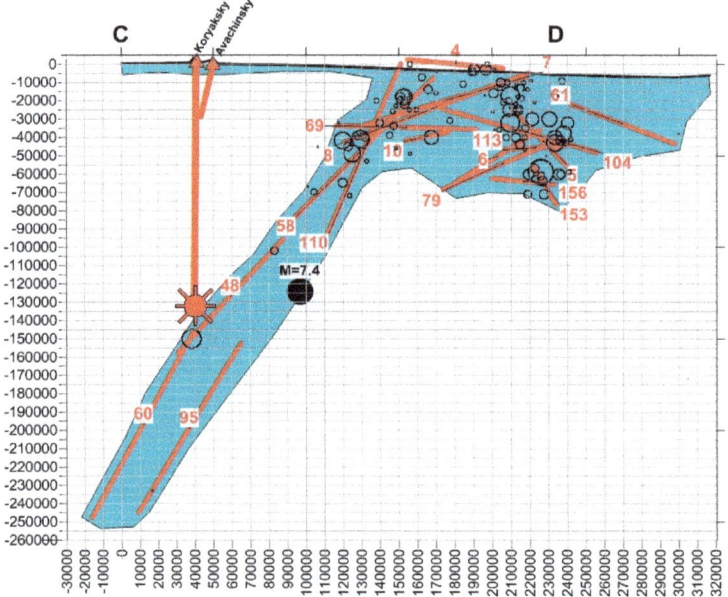

Figure 7. Cross-section of Kamchatka along the CD line (see Figure 1), showing traces of active seismogenic faults. The blue color covers the area where plane-oriented seismogenic faults are possible, which points to hydrofrack or brittle rock conditions. Numbers correspond to the fault numbers shown in Figure 6. The star denotes suggested active volcanoes' primary magma chamber positions.

3. Hydrothermal Response to Magma Fracking

3.1. Uzon-Geysernaya Caldera Geysers

Geysers are examples of cyclically erupted boiling hot springs. It has been shown that the driving mechanism of geysers cycling is gas-lift assisted eruptions, where non-condensable gases (mainly CO_2) are important players [9,18] due to CO_2 significantly drops boiling temperatures.

There is evidence from gas sampling at the Velikan Geyser that at the time of full activity before Jan. 3, 2014, gas composition was dominated by CO_2. Gas sampling from September 2013 showed that the gas component of the hydrothermal reservoir that fed the Velikan Geyser was dominated by carbon dioxide (CO_2, 61.5%) and nitrogen (N_2, 32.1%), along with a significant amount of methane (CH_4, 5.8%) and hydrogen (H_2, 0.45%) [9]. Gas composition in 2014–2019 in the Velikan and Bolshoy geysers reveals nitrogen becoming a dominant gas, while CO_2 declined to less than 1% [19]. In contrast, a new geyser, Shaman (Mutny), developed in 2008 in a channel of the former hot springs of Uzon caldera, just 12 km apart [20]. The gas composition of the newly formed Shaman Geyser was characterized by CO_2 domination according to our gas sampling in 2018 and 2019.

We interpret this as a redistribution of the magmatic gas recharge from magma plumbing systems of the Uzon-Geysernaya caldera in the following way: the Valley of Geysers' hydrothermal system magmatic CO_2 recharge was reduced, while that of the Uzon geothermal reservoir CO_2 recharge increased.

3.2. Koryaksky Narzan Thermal Springs

The Koryaksky Narzan CO_2-reach thermal springs (12–14 °C) and Isotovsky thermal springs (40–50 °C) are located in a north sector of Koryaksky volcano, where magma injections have taken place since 2008 [4,5]. We have been doing continuous temperature monitoring in Isotovsky spring (IS) since 2010 and in Koryaksky Narzan springs (KN1 and KN2) since 2017.

We already reported transient temperature anomalies recorded in Isotovsky in 2012–2013 and in 2015–2016, which were apparently related to dyke injections in adjacent areas that occurred on 2.08.2011 and 28.02.2016 correspondingly. The latest dyke injections on 14.11.2017 and 22.04.2019 are also associated with thermal tales in Isotovsky spring (Figure 8).

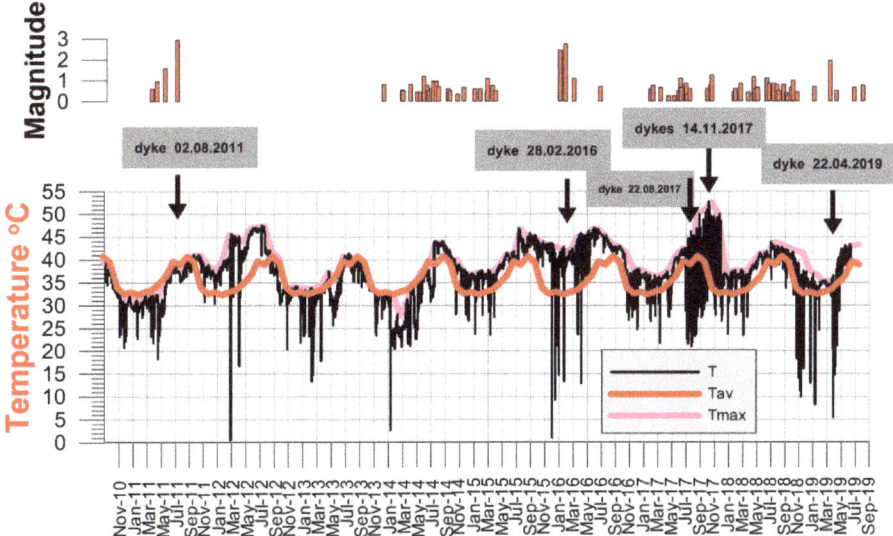

Figure 8. Temperatures recorded in Isotovsky hot spring and suggested times of dyke injections in adjacent areas. T—observational temperature; Tav—annual maximum monthly temperature (in a referenced year 2010/2011); Tmax—maximum monthly temperature (in a current year). Arrows above correspond to the dyke events shown in Figure 2, bars on a plot above correspond to the dykes' events magnitudes shown in Figure 3.

Koryaksky Narzan-2 (K2 in Figures 2 and 3) transient temperature records 2017–2019 look very stable with a few temperature drops from 0.1 to 0.4 °C (Figure 9). The most significant drop observed is 0.4 °C, which may be caused by a H_2O–CO_2 boiling temperature drop in a spring pool due to the magmatic CO_2 release associated with dykes injections in the adjacent area.

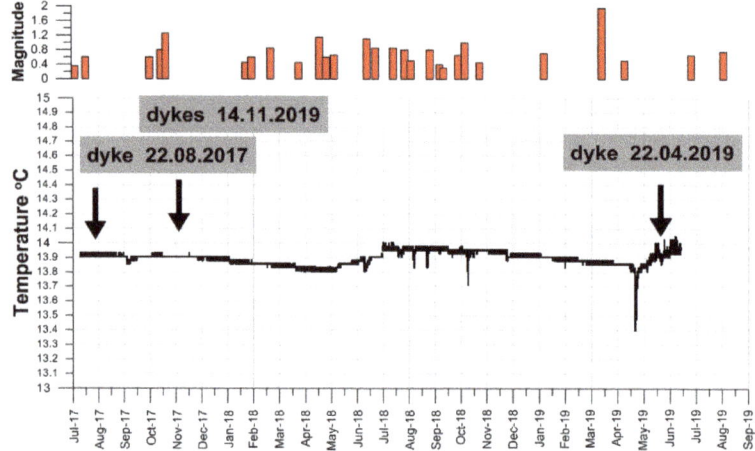

Figure 9. Koryaksky Narzan-2 transient temperature records 2017–2019. The significant temperature drop on 0.4 °C may be caused by boiling temperature decline due to additional magmatic gas recharge. The arrows above corresponds to the dyke events shown in Figure 2, while the bars on the plot above correspond to the dyke event magnitudes shown on Figure 3.

3.3. Mutnovsky Production Reservoir

The Mutnovsky (Dachny) geothermal reservoir includes at least two production faults geometrically and hydraulically connected to the magma fracking system of Mutnovsky volcano [7] (see also Figure 4).

Production fault #1 or the Main production zone (dip angle 58°, dip azimuth 110°) includes 20 production feed-zones as follows: wells O29W, A3-1, 4-E, O27, O19, O8, O45, O1, O14, A2-1, A2-2, O16, O13, 1, A3-2, A4, 26, 24, and 8 and the Dachny thermal feature. This nearly coincides with the "Main production zone" defined in [21,22] that has a dip angle of 60° and a dip azimuth of 106°.

Production fault #2 or the North-East production zone (dip angle 57°, dip azimuth 143°) includes 17 production feed zones as follows: wells A2-2, O29W, A3-2, A4, 26, A2-1, O16, O13, 1, A3-1, O8, O37, O42, O48, O53N, and O55 (10 of them also belong to production zone #1) and the Verkhne-Mutnovsky thermal feature.

We performed 30-day continuous monitoring of the non-condensable gases (most of gas content is CO_2) partial pressure in the turbine condenser of the geothermal power plant. To estimate PCO_2, we performed simultaneous measurements of the steam condensate pressure Pc and temperature Tc; then, PCO_2 was calculated as a difference between Pc and saturation pressure corresponding to temperature Tc.

Figure 10 shows the transient PCO_2 change during the observational period of time from 25.08.2019 to 25.09.2019. It is clearly seen that at least 14 maxima of PCO_2 synchronized with 14 minimums of Tc, which detect non-condensable gas arrivals into the turbine from the production geothermal reservoir. Some of these PCO_2 peaks may be related to magmatic gas recharge impulses, followed by the magma fracking processes described above (see Section 2.2 of this paper).

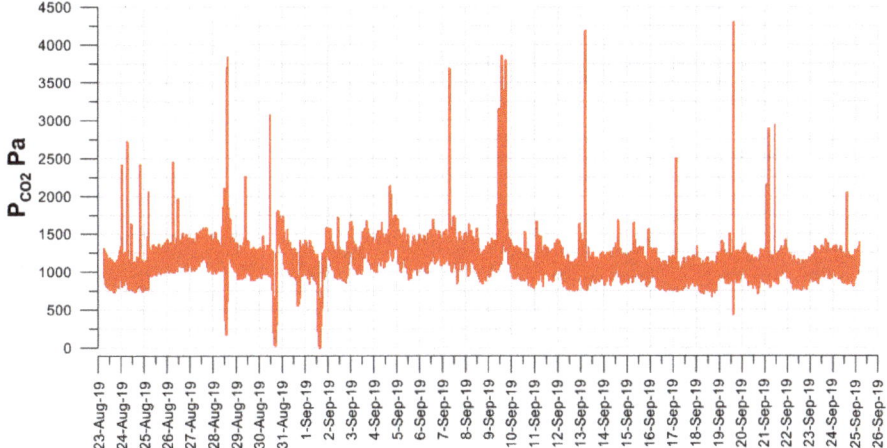

Figure 10. Estimated partial pressure of CO_2 in the condenser at the outlet of the turbine of the geothermal power plant.

4. Discussion

The key question in this paper is: How can we verify that plane-nested earthquakes are tracking magma injections in the shapes of dykes or sills beneath active volcanoes? The answer is in the form of a question, too: what other fluid except magma can demonstrate such hydrofracking capabilities? Superheated water and non-condensable gases (possibly CO_2) are other candidates for such working fluid duties. However, superheated water in shallow permeable fracture reservoirs conditions is very sensitive to host rock temperatures and is easily converted into high-compressible two-phase conditions, forming geothermal production fields. Another fluid is CO_2, especially magmatic CO_2 having less compressibility as compared to superheated steam, and it may act as a working fluid coupling with magma. If so, then we should extend our term of magma fracking beneath active volcanoes to the term magma+CO_2 fracking beneath active volcanoes. CO_2 impact is also useful to explain traces of fracks in a slope of Koryaksky volcano (Figure 3), with no associated magma discharge on the surface at the same time. In this connection, it is worth noting that magmatic CO_2 redistribution appears to be the key reason to explain the transfer of geysers activity from Geysers Valley to Uzon in Kamchatka and recent (2018) reactivation of Steamboat Geyser in Yellowstone.

Another important point is the accuracy of seismic hypocenter data and uniqueness of the Frac-Digger method for plane-oriented shapes definition. There is no uniqueness, since we found more fracks using less severe criteria of selections in Frac-Digger (see Section 2.1). Thus, we suggest that 3D distributions of magma fracks be considered a plausible scenario of magma fracking beneath active volcanoes, but these may include more or less magma+CO_2 dykes, depending on the Frac-Digger selection parameters. Nevertheless, some Frac-Digger selections pointed to 90%–95% of earthquake hypocenters belonging to plane-oriented clusters, meaning that fracking is a dominant process there.

High-temperature (HT) geothermal system formation due to magma fracking is also well explained in terms of the magma thermal-hydrodynamic modeling.

Conceptual 2D iTOUGH2-EOS1sc thermal hydrodynamic modeling of the Mutnovsky magmatic–hydrothermal system [7] reasonably explains its evolution over the most recent 1500–5000 years in terms of heat recharge (supplied by injected dykes from the active funnel Mutnovsky-4) and mass recharge (water injected through the dormant volcanic funnels Mutnovsky-3 and possibly Mutnovsky-2) conditions. We emphasize that the magmatic injection rate is approximately equivalent to the heat discharge rate (455 MWt). This is equivalent to approximately 455 kg/s of magma, which is in turn equivalent to from 15.3 to 25.6 km^3 of magma over 3000–5000 years. This volume value is comparable to the available space for dyke accommodation under regional horizontal extension conditions [23].

Conceptual TOUGH2 modeling was used to understand and explain the mechanism of the formation of the hydrothermal system beneath northern slope of Koryaksky Volcano [4,5]. For this purpose, the following terms were found to be crucial in this model: (1) heat sources of 20 MW/km^3 (340 MW total in 17 km^3 of reservoir rocks) and gas (CO_2) sources of 10 g/s/km^3 acting during 7000 years in the zones of magma injections; and (2) cold water recharge of 580 kg/s through the volcanic funnel to the deep dyke injection area. The modeling results reasonably match the Na–K geotemperature estimates of geothermal reservoirs (300 °C), the isotopic values (δD, $\delta^{18}O$) of high-elevation meteoric water recharge, the concentrations of magmatic CO_2 (up to 4 g/kg) in the hot springs on the northern slope of Koryaksky Volcano, and the thermal reactions to the dyke injections recorded in Isotovsky and Koryaksky Narzan hot springs. This modeling also indicates that a hidden high-temperature geothermal reservoir is present also beneath the southern slope of Koryaksky Volcano (at an elevation of −1 km), which may become a subject of future drilling explorations.

Another interesting related issue is where magma came from to recharge volcanoes for magma fracking and consequent eruptions. The possible answer is that water release from the subduction plate (due to hydro-fracking in the plate itself) forms upflows into the mantle edge, which creates melting conditions for primary magma chambers at the depth of 150 km (Figure 7). This closed loop of water–magma–water circulation in the subduction zone may be responsible for strong earthquakes, when hydrofracking drives the opening of shear faults. A mini-loop of such water–magma interplay may be Karymsky Volcano, which has been working almost continuously since 1996 due to water recharge fed from Karymsky Lake into dip seismogenic faults #120 and #122 and from there into the magma chambers of the volcano (Figure 6).

5. Conclusions

(1) Active volcanoes are injectors of magma and water into adjacent structures, which creates high-temperature production reservoirs.

(2) Seismic data reveal magma hydrofracking and stress conditions around active volcanoes.

(3) Magma fracking reservoirs may reside within production geothermal reservoirs.

(4) Seismogenic faults on the Kamchatka shelf are indicators of geofluid generation and water propagation to 250 km depth.

(5) Geysers result from CO_2-gaslift in active silicic volcanic areas. Continuously performed measurements of CO_2 partial pressures in hydrothermal features are a possible key to understanding the mechanisms of the relationship between anomalous hydrothermal perturbations [24], volcanic activity, and strong earthquakes.

Author Contributions: Conceptualization, methodology, software, validation, data curation, writing—original draft preparation: A.K. Koryaksky and Avachinsky volcanoes MEQ's data curation and analysis: E.C.; Mutnovsky MEQ's data curation and analysis, field works organization: A.P.; EQ mechanisms analysis: A.S. All authors have read and agreed to the published version of the manuscript.

Funding: This work was supported by RFBR project # 18-05-00052-19.

Acknowledgments: Authors express gratitude to P.A., P.O. Voronin, N.B. Zhuravlev, M.V. Lemzikov, O.O. Usacheva, T.V. Rychkova, M.A. Maguskin, A.I. Kozhurin and Sergeeva A.V. for their help in field survey and data analysis. Special thanks for constructive comments to A.A. Lyubin. Thorough editing of the manuscript performed by J.C. Eichelberger highly appreciated.

Conflicts of Interest: The authors declare no conflict of interest. The funders had no role in the design of the study; in the collection, analyses, or interpretation of data; in the writing of the manuscript, or in the decision to publish the results.

References

1. Sigmundsson, F.; Hooper, A.; Hreinsdóttir, S.; Vogfjörd, K.S.; Ófeigsson, B.G.; Heimisson, E.R.; Drouin, V. Segmented lateral dyke growth in a rifting event at Bárðarbunga volcanic system, Iceland. *Nature* **2015**, *517*, 191–195. [CrossRef] [PubMed]
2. Kiryukhin, A.V.; Fedotov, S.A.; Kiryukhin, P.A. A geomechanical interpretation of the local seismicity related to eruptions and renewed activity on Tolbachik, Koryaksky, and Avacha Volcanoes, Kamchatka, in 2008–2012. *J. Volcanol. Seismol.* **2016**, *10*, 275–291. [CrossRef]
3. Kiryukhin, A. Analysis of magma injection beneath an active volcano using a hydromechanical numerical model. *Horiz. Wells* **2017**, *1*, 1–5.
4. Kiryukhin, A.V.; Fedotov, S.A.; Chernykh, E.V. Magmatic plumbing systems of the Koryakskii–Avacha Volcanic Cluster as inferred from observations of local seismicity and from the regime of adjacent thermal springs. *J. Volcanol. Seismol.* **2017**, *11*, 321–334. [CrossRef]
5. Kiryukhin, A.; Lavrushin, V.; Kiryukhin, P.; Voronin, P. Geofluid systems of Koryaksky-Avachinsky volcanoes (Kamchatka, Russia). *Geofluids* **2017**, *2017*, 1–21. [CrossRef]
6. Kiryukhin, A.V.; Vorozheikina, L.A.; Voronin, P.O.; Kiryukhin, P.A. Thermal-Permeability structure and recharge conditions of the low temperature Paratunsky geothermal reservoirs, Kamchatka, Russia. *Geothermics* **2017**, *70*, 47–61. [CrossRef]
7. Kiryukhin, A.; Polyakov, A.; Usacheva, O.; Kiryukhin, P. Thermal-permeability structure and recharge conditions of the Mutnovsky high-temperature geothermal field (Kamchatka, Russia). *J. Volcanol. Geotherm. Res.* **2018**, *356*, 36–55. [CrossRef]
8. Kiryukhin, A.V.; Fedotov, S.A.; Kiryukhin, P.A. Magmatic systems and the conditions for hydrothermal circulation at depth in the klyuchevskoy volcanic cluster as inferred from observations of local seismicity and thermo-hydrodynamic simulation. *J. Volcanol. Seismol.* **2018**, *12*, 231–241. [CrossRef]
9. Kiryukhin, A. Modeling and observations of geyser activity in relation to catastrophic landslides–mudflows (Kronotsky nature reserve, Kamchatka, Russia). *J. Volcanol. Geotherm. Res.* **2016**, *323*, 129–147. [CrossRef]
10. Zoback, M.D. *Reservoir Geomechanics*; Cambridge University Press: Cambridge, UK, 2010; p. 448.
11. Fujii, Y.; Kodama, J.; Fukuda, D. Upper bounds of seismic events in induced seismicity. *J. Open Geosci.* **2010**. submitted for publication.
12. Chernykh, E.V.; Kiryukhin, A.V. Comparison of the geometry of seismogenic plane-sets and the mechanisms of the foci of earthquakes of the Koryaksky Volcano in 2008–2009. In Proceedings of the Geothermal Volcanology Workshop, Petropavlovsk Kamchatsky, Russia, 4–9 September 2019; pp. 58–63.
13. Selyangin, O.B. Mutnovsky Volcano, Kamchatka: New evidence on structure, evolution, and future activity. *J. Volcanol. Seismol.* **1993**, *15*, 17–38.
14. Selyangin, O.B. *Wonderful World of Mutnovsky and Gorely Volcanoes: Volcanologic and Traveller's Guide*; Novaya Kniga: Petropavlovsk-Kamchatsky, Russia, 2009; p. 108.
15. Polyak, B.G.; Melekestsev, I.V. On the output of volcanoes. *J. Volcanol. Seismol.* **1981**, *5*, 22–37.
16. Kiryukhin, A.; Fedotov, S.; Solomatin, A.; Kiryukhin, P. Geomechanical interpretation of seismicity on Kamchatka Shelf: Applications for seismic forecast and hydrocarbon exploration. In Proceedings of the 20th Conference on Oil and Gas Geological Exploration and Development, Gelendzhik, Russia, 10–14 September 2018.
17. Fedotov, S.A.; Maguskin, M.A.; Kirienko, A.P.; Zharinov, N.A. Vertical ground movements in the coast of the Kamchatka Gulf: their specific features in the epicentral zone on August 17, 1983 Earthquake M= 6.9, before and after. *Tectonophysics* **1992**, *202*, 157–162. [CrossRef]
18. Hurwitz, S.; Clor, L.E.; McCleskey, R.B.; Nordstrom, D.K.; Hunt, A.G.; Evans, W.C. Dissolved gases in hydrothermal (phreatic) and geyser eruptions at Yellowstone National Park, USA. *Geology* **2016**, *44*, 235–238. [CrossRef]
19. Kiryukhin, A.; Sugrobov, V.; Sonnenthal, E. Geysers VALLEY CO2 cycling geological engine (Kamchatka, Russia). *Geofluids* **2018**, *2018*, 1–16. [CrossRef]
20. Karpov, G. Evolution of regime and physical–chemical characteristics of the new formed Geyser in Caldera Uzon (Kamchatka). *J. Volcanol. Seismol.* **2012**, *3*, 3–14.
21. Kiryukhin, A.V. High temperature fluid flows in the Mutnovsky hydrothermal system, Kamchatka. *Geothermics* **1993**, *23*, 49–64. [CrossRef]

22. Kiryukhin, A.V.; Takahashi, M.; Poliakov, A.Y.; Lesnykh, M.D.; Bataeva, O.P. Origin of water in the Mutnovsky geothermal field an oxygen (δ 18O) and hydrogen (δD) study. *J. Volcanol. Seismol.* **1999**, *20*, 441–450.
23. Basmanov, O.L.; Kiryukhin, A.V.; Maguskin, M.A.; Dvigalo, V.N.; Rutqvist, J. Thermo-hydrogeomechanical modeling of vertical ground deformation during the operation of the Mutnovskii Geothermal Field. *J. Volcanol. Seismol.* **2016**, *10*, 138–149. [CrossRef]
24. Kiryukhin, A.V.; Korneev, V.A.; Polyakov, A.Y. On possibility of relationship between strong earthquakes and anomalous pressure variations in two-phase geothermal reservoir. *J. Volcanol. Seismol.* **2006**, *6*, 3–11.

© 2020 by the authors. Licensee MDPI, Basel, Switzerland. This article is an open access article distributed under the terms and conditions of the Creative Commons Attribution (CC BY) license (http://creativecommons.org/licenses/by/4.0/).

Article

Transport and Evolution of Supercritical Fluids During the Formation of the Erdenet Cu–Mo Deposit, Mongolia

Geri Agroli, Atsushi Okamoto, Masaoki Uno and Noriyoshi Tsuchiya *

Graduated School of Environmental Studies, Tohoku University, Sendai 980-8579, Japan; geri@geo.kankyo.tohoku.ac.jp (G.A.); atsushi.okamoto.d4@tohoku.ac.jp (A.O.); uno@geo.kankyo.tohoku.ac.jp (M.U.)
* Correspondence: Noriyoshi.Tsuchiya.e6@tohoku.ac.jp

Received: 4 March 2020; Accepted: 20 May 2020; Published: 25 May 2020

Abstract: Petrological and fluid inclusion data were used to characterize multiple generations of veins within the Erdenet Cu–Mo deposit, Mongolia, and constrain the evolution of fluids within the magmatic–hydrothermal system. Three types of veins are present (from early to late): quartz–molybdenite, quartz–pyrite, and quartz. The host rock was emplaced at temperatures of 700–750 °C, the first quartz was precipitated from magma-derived supercritical fluids at 650–700 °C, quartz–molybdenite and quartz–pyrite veins were formed at ~600 °C, and the quartz veins were precipitated in response to retrograde silica solubility caused by decreasing temperatures at <500 °C. We infer that over-pressured fluid beneath the cupola caused localized fluid injection, or that accumulated stress caused ruptures and earthquakes related to sector collapse; these events disrupted impermeable layers and allowed fluids to percolate through weakened zones.

Keywords: Erdenet Cu–Mo deposit; cathodoluminescence; supercritical fluid; transient fluid pressure; magmatic-hydrothermal system; fluid inclusion

1. Introduction

Studies of intrusive magmatism can provide insights into the fluid activity beneath volcanoes. Fluids derived from melts are important distributors of mass and energy in the upper crust. Such fluids form a link between magmatic and hydrothermal systems and provide valuable evidence of transport processes within the crust [1,2]. Slip events, or other external events that reduce principal stresses, induce transient pressure phenomena [3] (e.g., earthquakes, sector collapse) that cause the instantaneous precipitation of ore minerals. Over-pressured fluids can release energy to an overlying hydrostatic regime and generate a series of earthquakes within a relatively short period of time, known as swarm phenomena [4].

Studies of natural [5] and experimental [6,7] systems, including geophysical observations of hypocenter migration [8,9], confirm that transient pressure shifts are common occurrences in volcanic regions. This fluid behavior is also relevant to mineral and energy exploration because transient events such as ruptures can cause ore precipitation [10], and energy can be transported from the magma by high-enthalpy supercritical fluids to be utilized as enhanced geothermal systems [11–13].

However, there is limited access to active magmatic–hydrothermal systems, so we used a porphyry copper system as a natural analogue to constrain the evolution of magmatic–hydrothermal fluids. The Erdenet Cu–Mo porphyry copper deposit, Mongolia, become our case study of magmatic–hydrothermal processes within a supercritical geothermal system. Our interpretations of petrological and fluid inclusion data provide insights into the formation of the different generations of veins and their relationship to fluid activity within the Erdenet system.

2. Geological Setting

The Erdenet Cu–Mo deposit is located in northern Mongolia within the Permian–Triassic Selenge–Orkhon Trough, a volcanic–plutonic belt formed by the collision of the Siberian Craton in the north with the Central Mongolian Block in the south [14]. The volcanic activity was associated with voluminous intrusive magmatism, including the production of the Selenge Complex (SC) and the Erdenet porphyry complex.

The Erdenet porphyry complex consists of the Erdenet intrusive suites, referred to here as the porphyry association (PA), which is contained within the SC (Figure 1a). The SC consists of three phases—gabbroids, granitoids, and subalkaline granites and syenites—that were emplaced at 253–221 Ma [15]. The PA consists mainly of diorite and granodiorite porphyry and comprises five ore-related stages (Figure 1b). The first stage is syn-mineralization activity, followed by explosive events and brecciation. Field observations have documented two large explosion pipes of up to 250 m in diameter that are connected to the surface. The second stage porphyries are represented by granodiorite and granite porphyries. The granite porphyries are pink-grey and the granodiorite porphyries of the second stage are grey massive rocks with around 40% phenocrysts of plagioclase, hornblende, chloritized biotite, quartz, and very rare K feldspar, set in a fine-grained micropoikilitic and micrographic groundmass of K feldspar, plagioclase, quartz, and ore minerals. These porphyries crosscut both granodiorite porphyries and dacites of the first stage and quartz-sericite alteration. The third stage porphyries are represented by biotite plagioclase, plagiogranite, and granodiorite porphyries, which cut both first and second stages porphyries. The fourth stage consists of leucocratic porphyries and rhyodacite, truncating the last three stage porphyries as very rare dykes as well as the fifth stage. The fifth stage is described as diorite porphyries, andesite (amphibole-plagioclase), and granodiorite porphyry, which is associated with propylitic alteration [15].

K-Ar dating of the PA has yielded ages of 259–243 Ma [16], U-Pb zircon yielded ages of 245.9 ± 3.3–235.6 ± 4.4 Ma [17], and the latest obtained $^{40}Ar/^{39}Ar$ dates of 239.7 ± 1.6 and 240 ± 2 Ma for the emplacement of the intrusive sequences [18]. The ore formation at the Erdenet deposit occurred at ~240 ± 0.8–235.9 ± 1.9 Ma, based on molybdenum Re–Os [19] and sericite $^{40}Ar/^{39}Ar$ dating [16]. These ages indicate that mineralization is related to the intrusion of PA stocks and dikes, and that intrusions modified the wall rocks prior to mineralization. Three stages of alteration and mineralization are recognized at the Erdenet deposit. These stages, from the deep central parts of the deposit towards the shallower and outer parts, are (1) quartz–sericite, (2) chlorite–sericite alteration at the periphery of the ore-body, and (3) silica-rich and propylitic alteration, which is characterized by chloritized biotite [20,21].

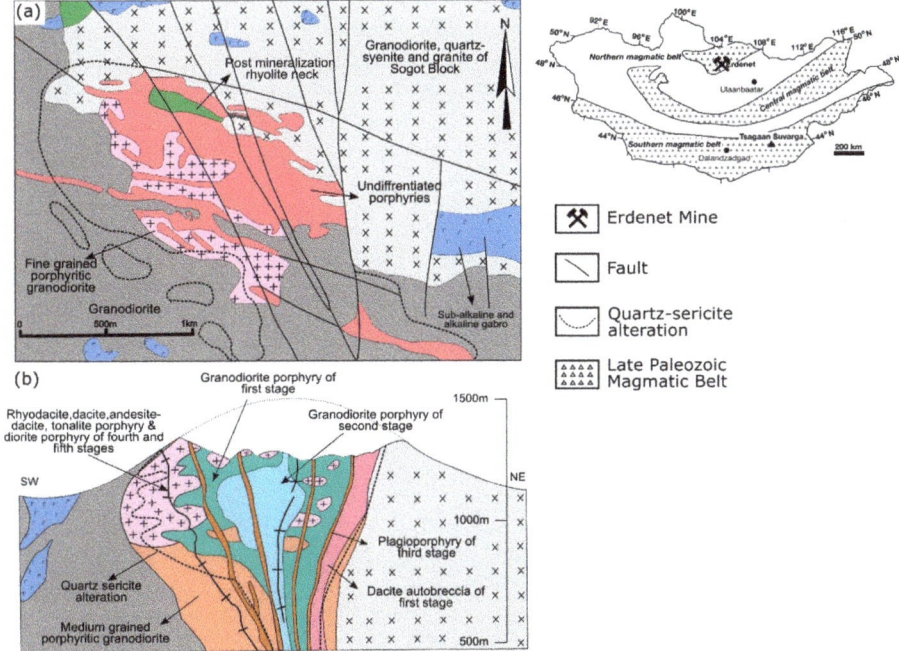

Figure 1. Geology and structure of the Erdenet deposit, modified after [20]. Insert shows the location of Paleozoic magmatic belts and the study area within the outline of Mongolia. (**a**) Geological map. (**b**) Schematic cross-section through the deposit, showing multiple stages of intrusion.

3. Sampling and Analytical Methods

A total of 57 samples of granodiorite and associated porphyries were taken from a hole drilled by the Erdenet Mining Corporation and from the Erdenet open pit. Twenty-five representative samples were selected for study on the basis of hand specimen observations. Polished thin sections of these samples were examined in detail using an optical microscope. The veins in two samples from 567 m and 646 m depths show crosscutting relationships (Figure 2b); these veins represent different hydrothermal events [5]. Double-polished thick sections, ~100 μm thick, were prepared for fluid inclusion microthermometry. The homogenization temperatures (T_h) were measured using a Linkam THMS600 heating stage, but the inclusions were too small (1–3 μm) to determine salinity from the ice-melting temperature ($T_{m\text{-ice}}$). The microthermometry data were combined with the Raman spectroscopy data to constrain the fluid composition of each vein type.

The textures of the quartz veins were characterized using a Hitachi-S3400N scanning electron microscope (SEM) equipped with an Oxford cathodoluminescence (CL) detector and photomultiplier at the Graduate School of Science, Tohoku University, Japan. Standard polished thin sections were analyzed at an accelerating voltage of 25 kV and a beam current of 90 μA. The textural characteristics of the different generations of veins are distinctive in SEM–CL, and qualitative differences in luminescence were used to classify the veins as CL-gray, CL-dark, and CL-bright [22].

The chemical compositions of minerals in the granodiorite and associated veins were analyzed by electron probe microanalysis (EPMA) on a JEOL JXA 8200 instrument at the Graduate School of Environmental Studies, Tohoku University, Japan. For most of the elements, the accelerating voltage was set to 15 kV, the beam current was set to 12 nA, and the counting time for each element was 20 s. The data were corrected using a ZAF correction method. Trace elements, including the Ti in quartz veins, were measured with an accelerating voltage of 20 kV, a beam current of 120 nA, and a beam diameter of 5 µm, to minimize specimen damage. The count times were 300 s on the relevant peak and 150 s for the high and low background measurements. These conditions corresponded to detection limits of 7 ppm and 22 ppm for Ti and Al, respectively. Element mapping for Ti, Al, and Fe was performed by EPMA on selected areas where SEM–CL data were acquired, using an accelerating voltage of 20 kV, a beam current of 120 nA, a focused beam, and dwell times of 1 s per pixel [23].

4. Petrography and Microstructure

4.1. Host Rocks

The drill core was brecciated locally and cut by a distinctive stockwork of veinlets that is typical of porphyry deposits (Figure 2). The rocks were divided into a gray-colored quartz porphyry (Figure 2a) and green-gray tonalite/granodiorite porphyries (Figure 2b); these rocks are referred to as the host rocks. The quartz porphyry occurred from the surface to the middle of the core (0–500 m), and the tonalite/granodiorite porphyries occurred at a >500 m depth.

Figure 2. Simplified core log and representative hand specimens, with sampling depths shown. (**a**) Quartz porphyry. (**b**) Tonalite/granodiorite porphyry.

The quartz porphyry was pervasively sericitized with a sericite content of 40–60%; the quartz occurred as phenocrysts within the sericite groundmass (Figure 3a). Ore minerals, such as pyrite, chalcopyrite, and molybdenite, were disseminated within the host rock (Figure 3b), and ore minerals and other sulfides also occurred within the veins. The sericite-bearing samples contained pseudomorphs of plagioclase and chlorite, but the original texture of these minerals was commonly overprinted by white mica and sericite in the more intensely altered samples. The accessory minerals included rutile, anhydrite, and apatite.

The tonalite/granodiorite porphyry consisted mainly of plagioclase (~40 vol%), quartz (~30 vol%), minor chlorite, and rare biotite. These minerals occurred as phenocrysts within a groundmass of microcrystalline quartz, K-feldspar, and plagioclase (Figure 3c). The groundmass formed ~20% of the rock by volume and filled between the phenocrysts. The plagioclase showed cryptic zoning, had a dusty/cloudy appearance, and was partially altered to white mica (Figure 3d). The quartz grains were embayed, some were rounded, and CL-dark fractures were common. Biotite is considered indicative of the PA [15,18]. It occurred as relatively large crystals of up to ~2 mm diameter; included rutile, calcite, and quartz; and was partly altered to chlorite. Chlorite occurs as an alteration after mafic minerals (e.g., amphibole); chlorite rims were commonly altered to white mica/sericite and contained inclusions of apatite, quartz, calcite, and rare anhydrite.

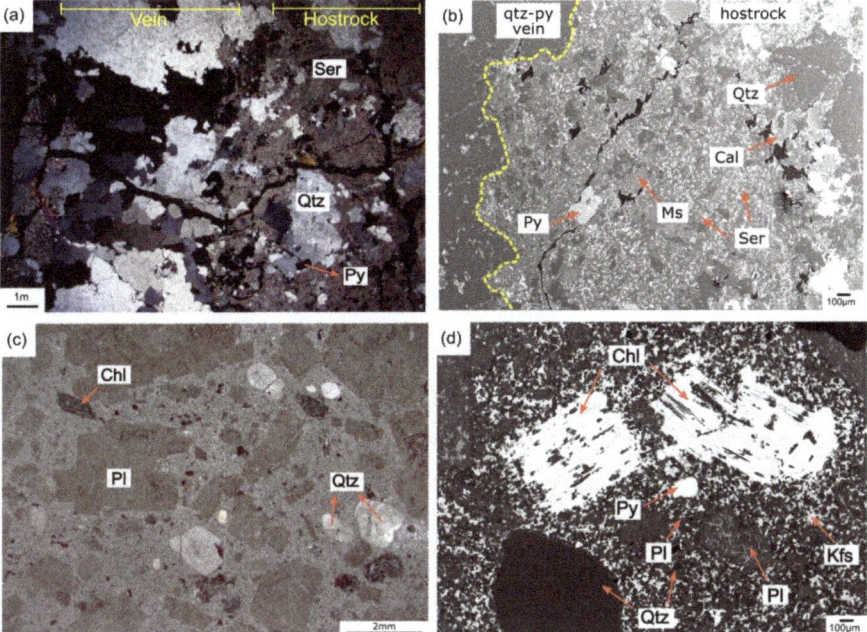

Figure 3. (**a**) Photomicrograph of sericitized quartz porphyry (221 m depth), cross-polarized light. (**b**) Backscattered electron (BSE) image of quartz porphyry and vein. (**c**) Tonalite/granodiorite porphyry (567 m depth), plane-polarized light. (**d**) Representative BSE image of the sample shown in (**c**). Mineral abbreviations are after [24].

4.2. Veins

4.2.1. Vein Types

Veins occur throughout the Erdenet deposit as a stockwork within the quartz and tonalite/granodiorite porphyries. The veins are millimeters to centimeters wide and contain sulfide minerals such as molybdenite, pyrite, chalcopyrite, and bornite, with minor calcite and rutile. At least three types of vein were recognized on the basis of crosscutting relationships in the samples Er-22 (567 m) and Er-24 (646 m). The different vein generations recorded progressive stages of fluid movement and hydrothermal activity. From early to late, they were as follows (Figure 4d):

- Quartz–Molybdenite ± Calcite veins.
- Quartz–molybdenite ± calcite (qtz–mol) veins occurred within the tonalite/granodiorite and quartz porphyries. The veins were 10–15 mm wide in the shallow parts of the core and <5 mm wide in the deeper parts of the core. Molybdenite with a rectangular and platy shape occurred in the vein margins and vein walls (Figure 4a).
- Quartz–Pyrite ± Calcite veins.
- Pyrite was concentrated within the center of the quartz–pyrite ± calcite (qtz–py) veins and also occurred as euhedral grains that were disseminated in the host rock of quartz and tonalite/granodiorite porphyry. In the deeper part of the core, pyrite occurred in the walls of the veins and was disseminated throughout the vein. The pyrite-bearing veins were 20–25 mm

wide. A veinlet of sericitized material was observed parallel to the vein wall. Most directions of elongation of the quartz grains were oriented perpendicular to the vein wall, but some quartz was oriented parallel to the vein wall (Figure 4b).

- Quartz ± Calcite.
- Quartz (qtz) veins cut the earlier vein types and represented the final stage of vein formation within the Erdenet deposit. This vein type was rare in all samples. The veins were 15–20 mm wide and occurred within the tonalite/granodiorite and quartz porphyry. These veins had a distinctive bright appearance, which was attributed to the quartz (Figure 4c)

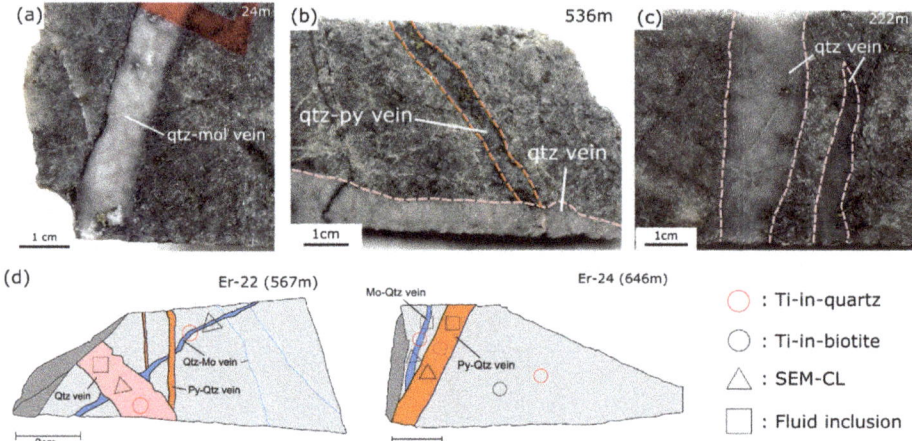

Figure 4. Photographs of the three vein types of the Erdenet deposit. (**a**) Quartz–molybdenite vein cut by later veins, (**b**) Quartz–pyrite vein cut by the later qtz vein. (**c**) Quartz vein within the quartz porphyry. (**d**) Sketch of samples showing analytical spots and crosscutting relationships among vein generations on tonalite/granodiorite porphyry.

4.2.2. Vein Textures

SEM–CL images reveal multiple generations of complex qtz veins that show textures that are not visible under an optical microscope and which record diverse processes (e.g., the dissolution of quartz grain cores, recrystallization, and fracturing). All three stages of veins are characterized by primary oscillatory zoning formed during initial quartz precipitation and the physicochemical changes of hydrothermal fluid cause precipitation of a secondary texture—for instance, recrystallization, fracturing, and the dissolution of quartz. These features crosscut or overgrow on the primary textures and can be distinguished from the primary quartz in the SEM–CL images [22].

The qtz–mol veins are the earliest generation of veins. The constituent grains have oscillatory zoning and are CL-gray in the SEM–CL images. The crystals grew from the vein wall towards the center of the vein, and calcite is common in the vein centers (Figure 5a,b). These features indicate that the quartz growth was syntaxial and that fractures provided voids for quartz precipitation from a silica-saturated fluid within the fluid conduit [25]. The zoned quartz grains are cut by interconnected CL-dark fractures knows as cobweb texture [22]; this texture records dissolution after the formation of the qtz–mol veins.

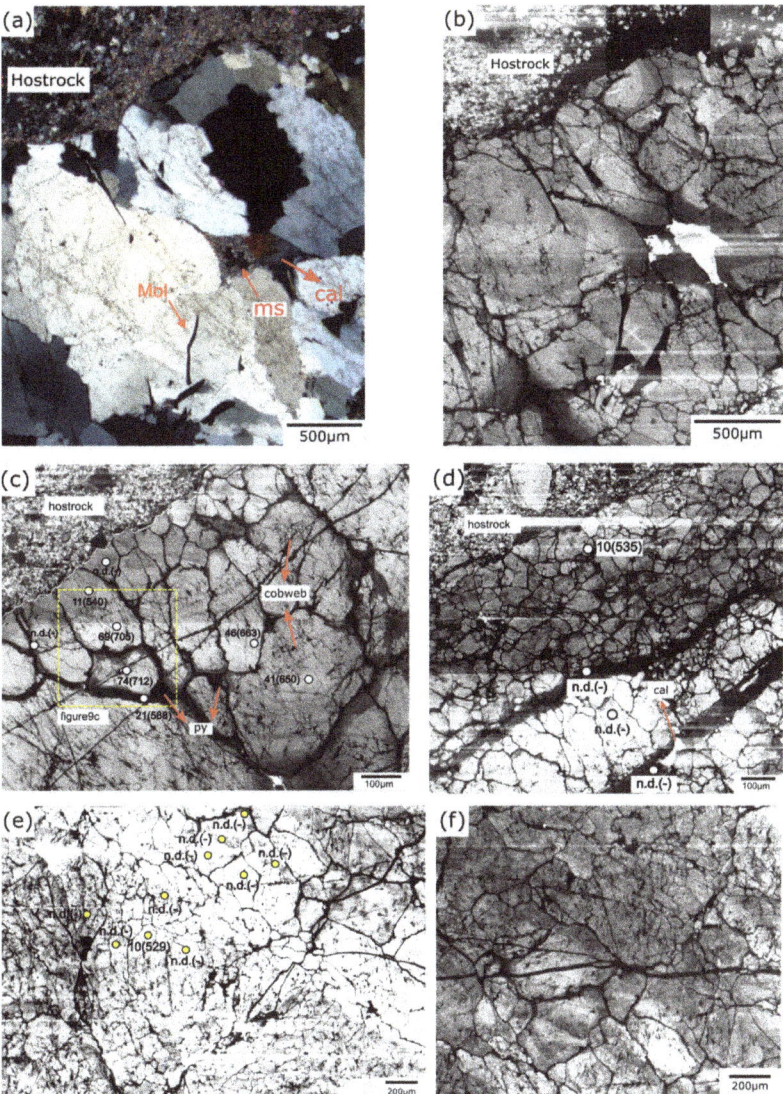

Figure 5. Images of veins. (**a**) Crystallographic orientation in cross-polarized light of the qtz–mol vein. (**b**) SEM–cathodoluminescence (CL) image of the qtz–mol vein shown in (**a**). (**c**,**d**) SEM–CL images of the qtz–py veins showing Ti analyses and temperatures. (**e**,**f**) SEM–CL images of the qtz veins showing Ti analyses and temperatures.

The qtz–py veins are associated with a dense fracture network and developed during the second stage of vein formation. Small euhedral crystals with vaguely defined oscillatory zoning occur on the walls of the veins. CL-dark quartz records a fracturing event after the formation of the initial qtz–py veins. Sulfides and calcite within the fractures and on the grain boundaries are associated with the CL-dark bands (Figure 5c,d). These veins are inferred to be syntaxial, based on the oscillatory zoning, crystallographic orientation, and presence of calcite.

Quartz veins, which represent the last stages of fluid activity on this system, are characterized by less CL-dark compared to the prior quartz veins. These veins contain a granular CL-bright quartz that contains a cobweb-like pattern of fractures filled with CL-dark quartz (Figure 5e,f). Primary textures, such as oscillatory zoning, are absent within this type of vein.

5. Estimation of Formation Temperatures

5.1. Application of Ti-in-Quartz and Ti-in-Biotite Geothermometers to the Host Rock

The crystallization temperature of the host rock was estimated using the Ti-in-quartz [26,27] and Ti-in-biotite [28] geothermometers (Figure 6). The biotite-bearing tonalite/granodiorite is part of the PA, which is associated with mineralization [15].

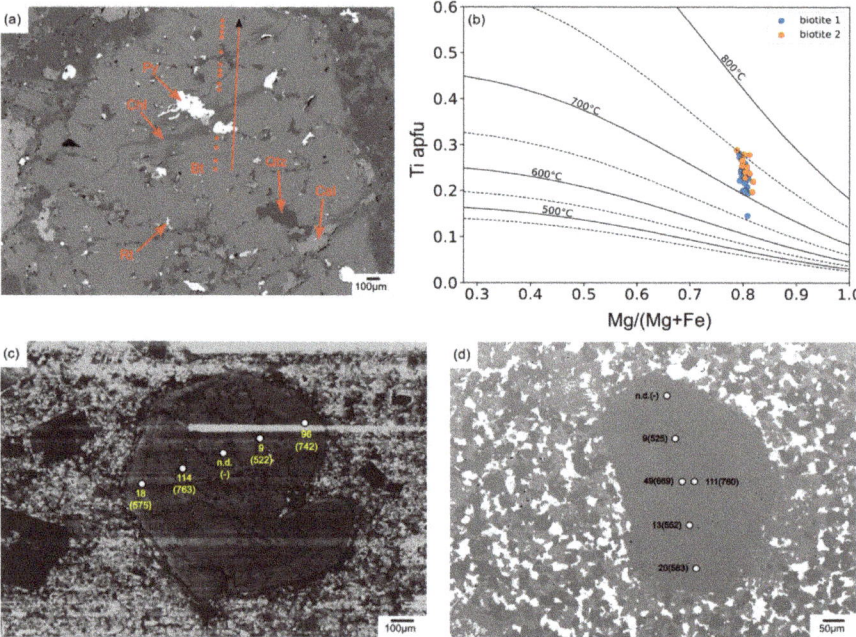

Figure 6. (**a**,**b**) Temperatures estimated using the Ti-in-biotite geothermometer of [26] (apfu = atom per formula unit). The host rock records temperatures of 700–750 °C. (**c**,**d**) Temperatures estimated using the Ti-in-quartz geothermometer. The quartz phenocrysts record temperatures of 400–750 °C.

Application of the Ti-in-quartz geothermometer to two quartz phenocrysts from the granodiorite porphyry yielded temperatures of 520–760 °C (Table 1). The Ti content was measured for a profile across the quartz grains (rim to rim) to determine if there was a temperature gradient (e.g., high and low temperatures recorded by the core and rim, respectively). Quartz grain (QtzHr1) (Figure 6c) does not show a systematic pattern, and the measured temperatures range from 550 to 750 °C. The core and rim of grain QtzHr2 (Figure 6d) record higher and lower temperatures, respectively, and a wide range of temperatures. The temperature recorded by two biotite grains was estimated using the Ti-in-biotite thermometer, and this provided more consistent results than those of the Ti-in-quartz thermometer (Figure 6b). Measurements were taken at points located on a profile from the core to the rim of the biotite grains (Figure 6a). The calculated temperatures ranged from 700 to 750 °C, based upon the biotite compositions provided in Table 2. The maximum temperature of quartz phenocryst calculated using the Ti-in-quartz geothermometer is similar to the temperatures calculated using the Ti-in-biotite thermometer which represent initial quartz formation or nucleation. Therefore, we infer that the host rock was emplaced at a temperature of 700–750 °C.

Table 1. Compositions of quartz phenocrysts in the host granodiorite porphyry (number in parentheses next to each analysis represents 1σ and given in the term of the least unit cited; n.d. = not detected).

	Sample No.	SiO$_2$ (wt%)	Al (ppm)	Ti (ppm)	Ti-in-Quartz (°C)	
					Wark & Watson	Huang & Audétat
Er-24	qtzHr1-1	100.23	n.d.	96 (5)	742 (6)	702 (6)
	qtzHr1-2	99.77	829 (53)	9 (5)	522 (33)	489 (32)
	qtzHr1-3	100.23	n.d.	n.d.	-	-
	qtzHr1-4	98.72	95 (52)	114 (5)	763 (5)	722 (5)
	qtzHr1-5	100.53	n.d.	18 (5)	575 (21)	539 (21)
	qtzHr2-1	100.29	n.d.	n.d.	-	-
	qtzHr2-2	100.72	n.d.	9 (5)	525 (33)	492 (32)
	qtzHr2-3	100.72	n.d.	49 (5)	669 (10)	631 (10)
	qtzHr2-4	100.79	14 (50)	13 (5)	552 (26)	518 (25)
	qtzHr2-5	100.60	20 (51)	20 (5)	583 (20)	548 (19)

Table 2. Representative biotite compositions (wt%) used for Ti-in-biotite geothermometry [28]. Numbers in parentheses next to each analysis represent 1σ and given in terms of least unit cited; n.d. = not detected.

	Biotite_1-1	Biotite_1-2	Biotite_1-3	Biotite_1-4	Biotite_2-1	Biotite_2-2	Biotite_2-3	Biotite_2-4
SiO$_2$	39.09 (0.21)	39.23 (0.21)	39.72 (0.21)	39.09 (0.21)	39.30 (0.21)	38.77 (0.21)	39.40 (0.21)	39.27 (0.21)
TiO$_2$	2.04 (0.07)	2.26 (0.07)	1.93 (0.07)	1.80 (0.07)	2.45 (0.08)	2.18 (0.07)	2.05 (0.07)	2.26 (0.08)
Al$_2$O$_3$	17.14 (0.15)	16.29 (0.15)	17.28 (0.15)	17.46 (0.15)	16.45 (0.15)	17.15 (0.15)	17.86 (0.15)	17.20 (0.15)
FeO	7.87 (0.12)	7.97 (0.12)	7.63 (0.12)	7.69 (0.12)	8.08 (0.13)	7.47 (0.12)	7.16 (0.12)	7.77 (0.13)
MnO	0.54 (0.08)	0.50 (0.08)	0.44 (0.08)	0.39 (0.07)	0.33 (0.07)	0.39 (0.08)	0.47 (0.07)	0.53 (0.07)
MgO	17.85 (0.17)	17.65 (0.17)	17.96 (0.17)	18.25 (0.17)	17.88 (0.17)	18.16 (0.17)	18.20 (0.17)	18.05 (0.17)
CaO	0.01	n.d.	0.01	n.d.	n.d.	n.d.	n.d.	n.d.
Na$_2$O	0.18 (0.04)	0.16 (0.04)	0.20 (0.04)	0.12 (0.04)	0.11 (0.04)	0.12 (0.04)	0.04 (0.04)	0.17 (0.04)
K$_2$O	11.20 (0.13)	10.91 (0.13)	11.45 (0.13)	11.37 (0.13)	11.12 (0.13)	11.17 (0.13)	11.75 (0.13)	10.83 (0.13)
Total	95.91	95.00	96.64	96.25	95.71	95.44	96.91	96.08
Si	5.64	5.71	5.68	5.61	5.68	5.61	5.61	5.64
Ti (apfu)	0.22	0.25	0.21	0.19	0.27	0.24	0.22	0.24
Al	2.92	2.79	2.91	2.96	2.80	2.93	3.00	2.91
Fe^{2+}	0.95	0.97	0.91	0.92	0.98	0.90	0.85	0.93
Mn	0.07	0.06	0.05	0.05	0.04	0.05	0.06	0.06
Mg	3.84	3.83	3.82	3.91	3.85	3.92	3.86	3.86
Ca	n.d.	n.d.	n.d.	n.d.	n.d.	n.d.	n.d.	n.d.
Na	0.05	0.05	0.06	0.03	0.03	0.03	0.01	0.05
K	2.06	2.03	2.09	2.08	2.05	2.06	2.13	1.98
X_{mg}	0.80	0.80	0.81	0.81	0.80	0.81	0.82	0.81
T (°C)	716 (3)	729 (3)	710 (3)	701 (3)	739 (2)	730 (3)	723 (3)	738 (2)

5.2. Application of the Ti-in-Quartz Thermometer to Veins

Quartz is ubiquitous within the veins, so the Ti-in-quartz geothermometer from [26] was applied to constrain the vein formation temperatures and the pressure-dependent calibration of [27] was applied as a comparison. The pressure of ~1.6 kbar, at which the Diorite porphyry was (fifth stage) emplaced, was chosen to estimate the temperature [16]. The Ti activity (α) was assumed to be unity ($\alpha = 1$), which provides a minimum estimate of the temperature of vein formation (from hereafter referred as quartz formation temperature); if α were assumed to be 0.5, then the calculated temperatures would be ~65 °C higher. The detection limit for Titanium (Ti) by EPMA is 7 ppm, a concentration that corresponds to 500 °C, and the uncertainties are exponentially larger for Ti concentrations of <7 ppm [29]. Ti measurement points were selected within areas of different CL intensities to investigate the different physicochemical and fluid dynamic properties of fluids that percolated through fractures in the rocks.

5.2.1. Quartz ± Molybdenite Veins

The qtz–mol veins are associated with the early and middle stages of mineralization within the Erdenet deposit [15,18]; the Ti content of quartz and estimated temperatures for these veins are shown in Figure 7a. In general, quartz within the qtz–mol veins has Ti contents of 6–82 ppm, which correspond to temperatures of 500–750 °C (Figure 8). The calculated temperature for barren CL-gray quartz with oscillatory zoning is consistent at ~500–600 °C. The CL-bright areas yield temperatures as high as 700 °C, whereas the CL-dark quartz associated with calcite has no Ti contents, so the temperature is inferred to be <500 °C. Molybdenite has a platy shape and is associated with CL-dark fractures. The CL-gray which is associated with molybdenite truncates the oscillatory zoning and yields temperatures of ~600 °C. Molybdenite is also associated with an interconnected fracture which cuts the CL-bright quartz.

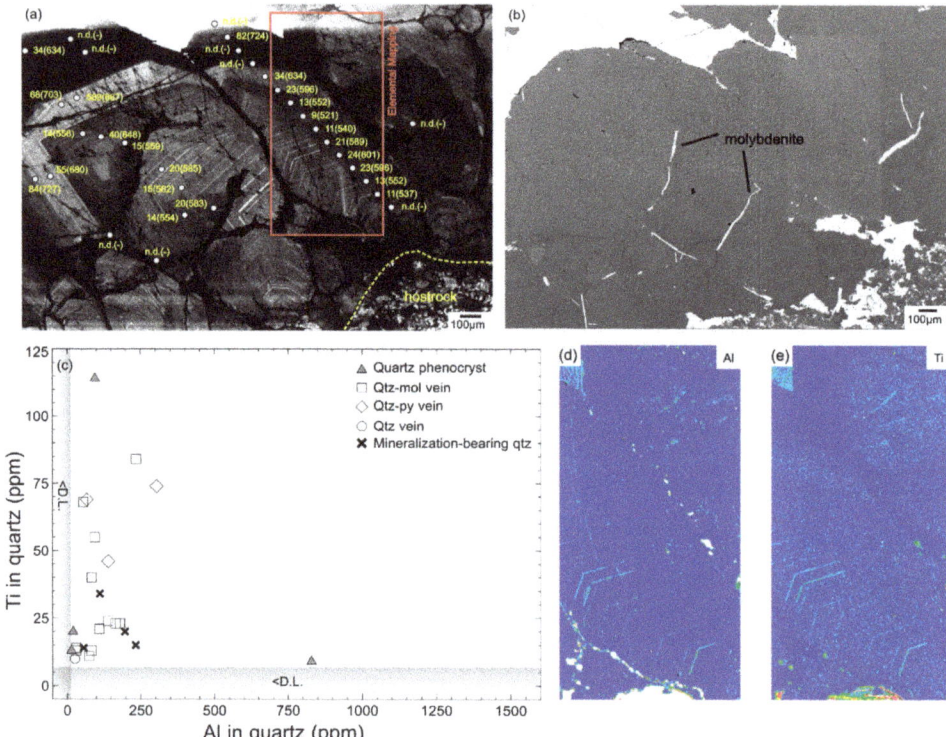

Figure 7. Representative qtz–mol vein. (**a**) SEM–CL image. (**b**) BSE image showing Ti analyses. The estimated temperature is shown in parentheses. (**c**) Ti and Al concentrations of quartz. (**d**,**e**) Trace element maps showing a positive relationship between CL-brightness, Ti, and Al.

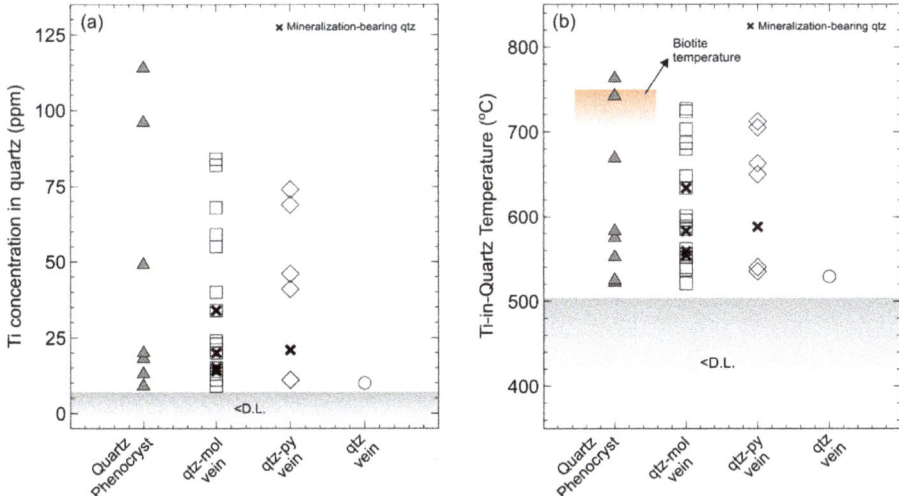

Figure 8. (a) Ti concentrations of quartz phenocrysts and vein quartz from the Erdenet deposit. (b) Minimum temperature of quartz vein formation calculated using the model of [28]. D.L. = detection limit.

Element maps show that the Ti and Al concentrations are proportional to the CL-brightness (Figure 7d,e), so it is assumed that these elements record the fluid conditions and diffusional states of Ti during quartz precipitation [30]. Therefore, the Ti concentration and temperature of CL-bright quartz are higher than those of CL-gray. Al content is positively correlated with Ti, and the Al concentrations are low (<250 ppm) relative to the average Al contents of other porphyry deposits (>2000 ppm), which show the negative correlations of Al and Ti contents with CL-brightness [23].

5.2.2. Quartz ± Pyrite Veins

The quartz–pyrite veins were formed during the early and middle stages of mineralization at the Erdenet deposit [15]. The middle stages of mineralization are also commonly associated with the formation of metasomatic sericite and low homogenization temperatures of fluid inclusions (<250 °C). Multiple measurements of the Ti content of primary quartz with oscillatory zoning and quartz in the CL-dark fractures associated with pyrite yielded Ti concentrations of 3–74 ppm, which correspond to temperatures of 450–700 °C (Figure 8). The relationship between the CL-brightness and Ti concentration is similar to that of the qtz–mol veins. The CL-gray quartz, which shows oscillatory zoning between the vein wall and vein center, yields temperatures of ~500–600 °C. The Ti concentration of CL-dark quartz have lies below the detection limit (Figure 5c,d), which suggests it occurs as low-temperature quartz. Pyrite occurs on margins of grains that record temperatures of ~600 °C and on fractures that connect to the calcite at the center of the vein.

The low formation temperatures inferred for CL-dark quartz, which is associated with sericite and calcite, indicate that the phyllic and propylitic alteration zones, which formed after the sericite alteration, also formed at low temperatures [31–33].

5.2.3. Quartz Veins

Only two points within the qtz veins yielded detectable Ti concentrations (Table 3, Figure 5e), which indicate temperatures of ~500 °C. However, most of the measured Ti concentrations were less than the detection limit. Thus, we suggest that th qtz veins formed at lower temperatures than those of the qtz–mol veins and qtz-py veins (Figure 8).

The moderate brightness of the CL signal indicates that the qtz veins formed under stable conditions and that the magmatic fluids mixed with lower-temperature fluids derived from meteoric water to produce a relatively low-temperature hydrothermal solution. This inference is supported by the fluid inclusion data (see below).

Table 3. Representative trace element contents of quartz veins and minimum temperatures calculated using Ti-in-quartz geothermometry (number in parentheses next to each analysis represents 1σ and given in the term of the least unit cited; n.d. = not detected).

	Sample No.	SiO_2 (wt%)	Al (ppm)	Ti (ppm)	Ti-in-Quartz (°C)	
					Wark & Watson	Huang & Audétat
			Qtz-mol vein			
Er-24	qtz-mol-2	99.11	n.d.	34 (5)	634 (12)	596 (12)
	qtz-mol-6	100.55	31 (49)	14 (5)	556 (25)	521 (24)
	qtz-mol-7	100.67	84 (48)	40 (5)	648 (11)	611 (11)
	qtz-mol-13	99.33	51 (49)	n.d.	-	-
	qtz-mol-14	99.33	1463 (56)	n.d.	-	-
	qtz-mol-17	100.03	97 (50)	6 (5)	493 (44)	460 (43)
	(qtz-mol) Line 5	100.45	140 (49)	24 (5)	601 (17)	564 (17)
	(qtz-mol) Line 9	100.45	30 (50)	13 (5)	552 (26)	517 (25)
	(qtz-mol) Line 14	99.05	n.d.	82 (5)	724 (7)	684 (7)
	(qtz-mol) Line 15	100.84	167 (51)	n.d.	-	-
			Qtz-py vein			
Er-24	qtz-py-1	100.14	67 (49)	69 (5)	705 (8)	666 (8)
	qtz-py-3	100.07	n.d.	11 (5)	540 (29)	506 (28)
	qtz-py-6	99.94	n.d.	41 (5)	650 (12)	612 (11)
	qtz-py-10	100.05	49 (49)	n.d.	-	-
	qtz-py-13	100.21	n.d.	21 (5)	588 (18)	553 (17)
			Qtz vein			
Er-22	qtz-1	99.90	26 (49)	10 (5)	529 (32)	495 (31)
	qtz-5	100.01	34 (49)	n.d.	-	-
	qtz-7	99.92	n.d.	n.d.	-	-
	qtz-9	100.15	56 (49)	n.d.	-	-
	qtz-11	99.91	654 (52)	n.d.	-	-

5.3. Fluid Inclusion Microthermometry and Compositions

5.3.1. Microthermometry

Fluid inclusion microthermometry was performed on all qtz vein stages. However, the vast majority of inclusions were too small for complete microthermometric analysis, so only the homogenization temperature (T_h) was measured [34]. All the examined samples show dense populations of fluid inclusion (cloudy area), so the FIA (fluid inclusion assemblages) could not be determined and it could not be distinguished whether the inclusions were primary or secondary. The size of inclusions ranges from 1 to 3 µm (mean value 1.5 ± 0.5 µm). Fluid inclusion microthermometry was performed on several fields of interest along the vein and focused on single inclusion with a moving bubble at room temperature up to 200–300 °C; at temperatures >350 °C, the moving bubble can no longer be seen. Most fluid inclusions show liquid and vapor phases with dark to colorless appearances. Inclusions have rounded-elongate shapes and sometimes occur as trails or solitary inclusions (Figure 9c).

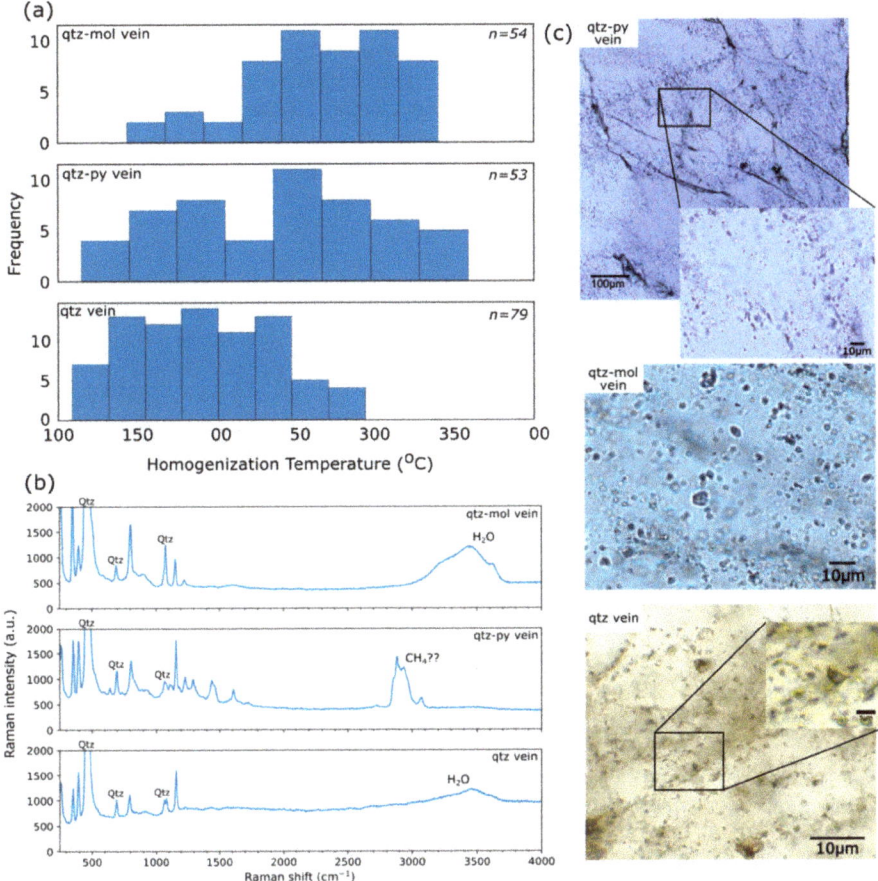

Figure 9. (a,b). Homogenization temperatures and fluid compositions of stage 1–3 veins. (c) Photomicrograph of representative fluid inclusions (plane-polarized light).

Fluid inclusions in the qtz–mol veins homogenize at 142–340 °C (mean value, 267 ± 46°C), and the T_h of qtz–py veins span from 115 to 359 °C (mean value, 242 ± 65). The homogenization temperatures of fluid inclusions in the qtz veins are in range 108–295 °C (mean value, 194 ± 45) (Figure 9a and summary of microthermometric data is provided in Supplementary). The fluid inclusions exhibit a bimodal population, consisting of a low mean value of the qtz vein and a relatively higher value of the qtz-mol and qtz-py veins. This indicates that two different processes occurred in the Erdenet system. Later, we use the homogenization temperature (T_h) value to estimate the minimum pressure condition of vein formation by using the isochore of T_h and the $T_{m\text{-ice}}$ of fluid inclusion in P-T space provided by [35,36].

5.3.2. Fluid Compositions

The compositions of the fluid inclusions within various veins were investigated by micro-Raman spectrometry using a Horiba XploRA PLUS instrument equipped with a 532 nm Ar laser and gratings set to 600 grooves mm^{-1} at Tohoku University, Japan. The qtz–mol and qtz veins are dominated by H2O, which might have occurred at the vapor or liquid phase, based on petrographic observations. Inclusions within the qtz–py veins that formed during the second stage might contain CH4 in the vapor phase (Figure 9b).

6. Discussion

6.1. Vein-Related Mineralization Processes

The vein formation temperatures and SEM–CL observations provide insights into the conditions that formed the quartz and ore/sulfide minerals of the Erdenet deposit. The qtz–mol veins were precipitated at high pressures and temperatures (500–700 °C) [37], and the relatively low Al content in the qtz-mol veins indicates that the veins formed at high temperatures [23]. Platy molybdenite occurs close to the CL-dark fractures. The sulfide mineralization was formed from fluids derived from intermediate–silicic magmas [38]; the CL-dark quartz associated with molybdenite formed at ~600 °C and crosscuts the oscillatory zoning. The molybdenite mineralization forms part of the first stage of mineralization at the Erdenet deposit. In contrast, the qtz–py veins show identical growing patterns, with the qtz-mol vein where the pyrite or chalcopyrite occur on the margins of the vein and grain boundary. The mineralization corresponds to CL-dark quartz that crosscuts primary quartz with oscillatory zoning and is oriented parallel or perpendicular to the vein wall. The formation temperature recorded by the pyrite-bearing quartz is ~600 °C. The qtz veins record the latest stage of quartz precipitation within this system and are relatively unmineralized, but some pyrite occurs within these veins in association with the CL-dark quartz. We infer that the barren quartz with oscillatory zoning formed at the beginning of the vein formation and that this quartz was precipitated on the vein wall. The molybdenite and pyrite mineralization formed after the initial quartz in association with the CL-dark fractures that cut the oscillatory zoning.

6.2. Estimated Fluid Pressures

The fluid pressures were estimated by combining isochores calculated from the homogenization and ice-melting of fluid inclusion [35,36] with temperatures from the Ti-in-qtz thermometry. The maximum value of T_h was used to construct the isochores to represent the minimum pressure condition. The temperature of ice melting could not be determined, but halite is not present within the inclusions, so the salinity was assumed to be 0–25 wt%. This assumption is consistent with previous reports of salinities of 5–25 wt% for fluid inclusions from surface samples [17]. The calculated pressure for the qtz–mol veins is 1.6–3.0 kbar, assuming a temperature of ~600 °C (Figure 10). A pressure of 1.1–2.5 kbar was estimated for the qtz–py veins, based on a Ti-in-quartz temperature of ~600 °C (Figure 10). A pressure of 2.1–2.8 kbar was calculated for the qtz veins, based on an assumed temperature of ~500 °C. However, the crosscutting relationships and the relatively lower Ti content as well as homogenization temperature (T_h) of the fluid inclusion compared to other vein types indicate that qtz veins record the later fluid activity at the Erdenet deposit and were formed at considerably lower pressures. Therefore, we suggest that the qtz veins must have formed at temperatures of <500 °C, based on their low Ti concentrations. In this case, the estimated pressure is 1–2 kbar assuming temperatures of 400–450 °C, although it could be lower (green b, Figure 10).

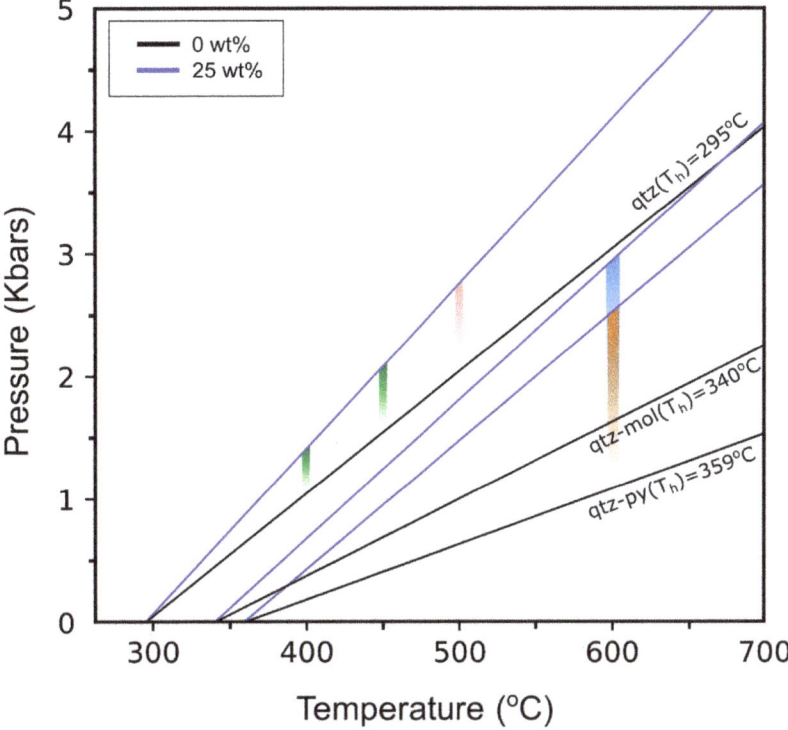

Figure 10. Estimated pressures of vein formation at the Erdenet deposit. Temperatures are derived from the Ti-in-quartz geothermometer, and the isochores correspond to salinities of 0–25 wt% [35].

6.3. Fluid Evolution at the Erdenet Porphyry Deposit

Porphyry copper deposits have distinctive characteristic veining and alteration patterns [31–33]. Systematic observations of the crosscutting relationships among veins the and textures within veins provide insights into fluid behavior, and quartz solubility models [39] have revealed fluid processes and evolution at individual porphyry deposits. At the Erdenet deposit, the magma chamber started to solidify and generate a crystal mush as the temperature decreased to ~700–800 °C after magma emplacement. Fluids released from crystallization processes (dehydration) moved upwards so that the cupola was saturated with fluid (Figure 11a). The initial pulses of magma injection caused fracturing in the host rock, and silica-saturated fluids percolated through these fractures to form the pre-mineralization quartz that grew from the vein walls (Figure 11b,e). This fluid might represent the magmatic fluid, based on the temperatures of 650–700 °C calculated for a high Ti activity (solid arrow in Figure 12).

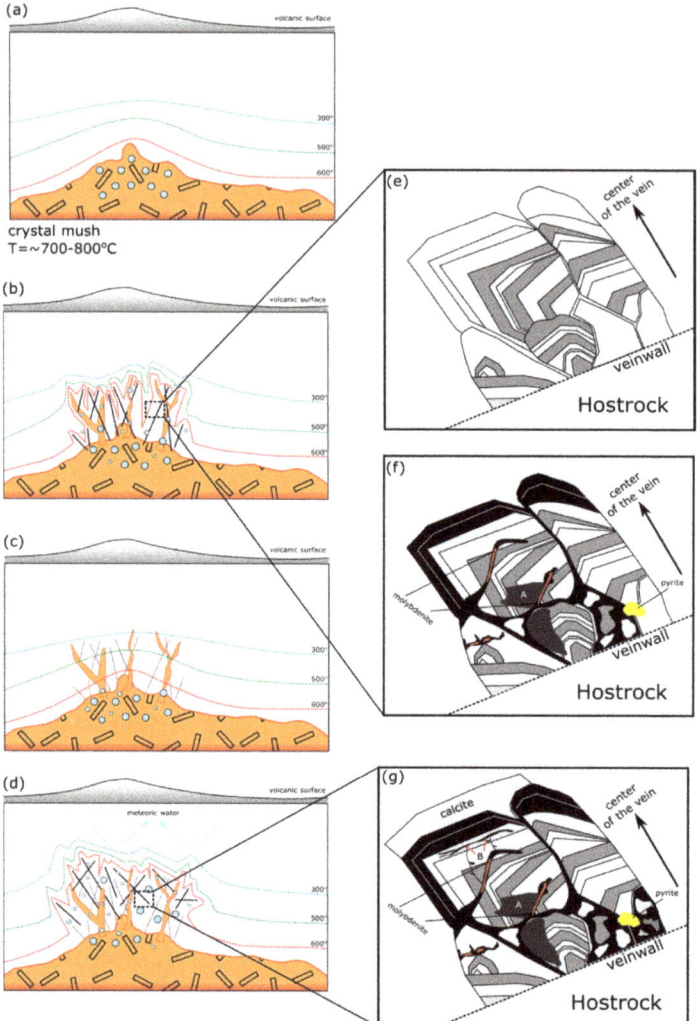

Figure 11. Schematic model of the evolution of magmatic–hydrothermal fluids at the Erdenet deposit and illustration of the crystallization of various quartz veins. Initial magmatic fluid accumulated at the cupola (**a**). Magmatic pulse (**b**) induced the precipitation of the initial euhedral zoning quartz, which has a temperature close to the host rock (**e**). Precipitation of the initial quartz sealed the fracture and increased the fluid pressure. Episodic transience occurred and the fluid was reinjected into the existing fracture, followed by the precipitation of mineralization-bearing quartz (**f**). In the final stages, the magmatic front descends (**c**,**d**) and the quartz veins precipitate and cut previous generations of veins with cooler temperatures (hydrothermal system) (**g**).

The molybdenite and pyrite mineralization were formed from later fluids by one of two possible mechanisms as follows: (1) The mineralization of molybdenite and pyrite formed by later fluid processes possibly occurred in two scenarios. The first is that both molybdenite and pyrite were precipitated together. This mechanism is suggested by the occurrences of molybdenite and pyrite relating to fractures and grain boundaries (represent by CL-dark). Additionally, the fractures cut the primary oscillatory zoning of qtz, which we assume as the initial fluid activity in Erdenet. Moreover, the temperature inferred for the mineralization related to the CL-dark quartz is close to 600 °C (Figure 11f). Mineral precipitation might have been induced by fluid decompression after the initial pulse of magma. Meanwhile, the fluid temperature decreased and fluids were released episodically into the existing fracture network, where they precipitated quartz and sulfides (dashed arrow Figure 12). (2) Molybdenite was precipitated first from the cooling fluid, followed by pyrite. However, regardless of the precise mechanism, we infer that these minerals were precipitated within a lithostatic pressure regime from supercritical fluids.

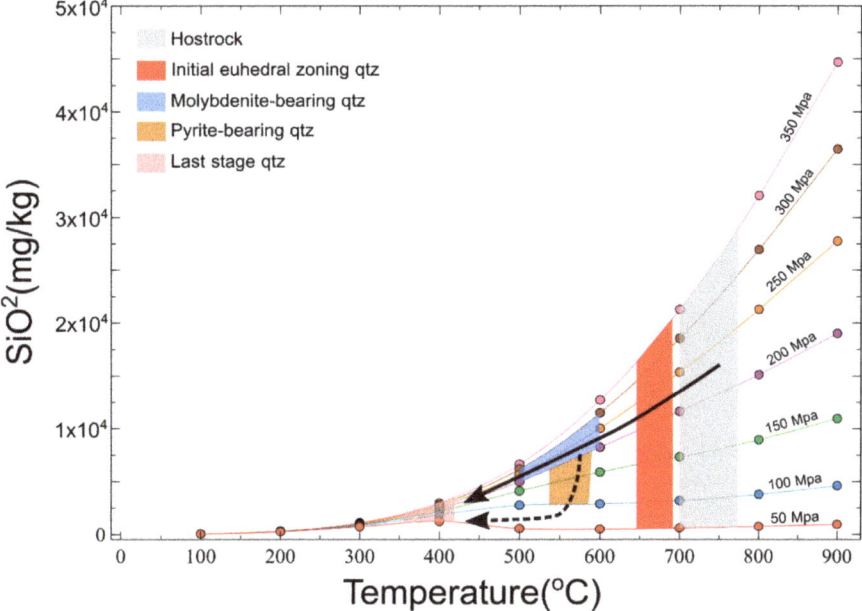

Figure 12. Evolution of supercritical fluid based on inferred pressures and temperatures (solid and dashed arrow are the inferred P-T path) and the model of quartz solubility in H_2O [40]. Two possible scenarios are presented here. The solid arrow represents the magmatic fluid upon gradual cooling that precipitated various quartz vein stages. The dashed arrow depicts a second scenario, where the decompression of the initial fluid takes place and induces the precipitation of mineralization-bearing quartz. Both mechanisms will tend to decrease at lower temperatures and precipitate the last stage of the qtz vein.

The final stages of fluid activity in this system involved the precipitation of qtz veins. The continued cooling and retreat of the magmatic front away from the position of initial emplacement to deeper parts of the system [3,40] are associated with a shift towards a hydrostatic pressure regime. The prograde quartz solubility caused the precipitation of qtz veins and related CL-dark quartz ± calcite (Figure 11g). The CL images show that the fluid generated intense fractures within the veins and intact rock at all stages. Fracturing might record infiltration by over-pressured fluid, whereby the build-up of pressure within the cooling pluton generated additional fluids that were transported into the cupola (Figure 11c). This process caused the injection of fluids into the surrounding rocks, which was controlled by the existing fractures [41]. This mechanism was associated with transient fluid pressure changes during the shift from a lithostatic to a hydrostatic fluid pressure regime (Figure 11d) [5].

7. Conclusions

We investigated the magmatic–hydrothermal system recorded by a porphyry Cu–Mo deposit in the Erdenet area of northern Mongolia. The characteristic features of the vein systems are as follows.

1. The veins of the Erdenet deposit are divided into three types: qtz–mol veins, qtz–py veins, and monomineralic qtz veins. Crosscutting relationships reveal that the order of formation was qtz–mol veins (stage 1), qtz–py veins (stage 2), and qtz veins (stage 3).
2. Euhedral quartz grains with oscillatory zoning grew from both side walls of the qtz–mol and qtz–py veins. There is no clear zoning within the monomineralic qtz veins. All the vein types are cut by a network of late CL-dark quartz veins.
3. Ti-in-quartz and Ti-in-biotite thermometry records temperatures of 700–750 °C for the host rock, 650–700 °C for the earliest quartz, and ~600 °C for the mineralization-bearing quartz. Fluid inclusion thermometry records homogenization temperatures of 180–325 °C for the qtz–mol veins, 147–360 °C for the qtz–py veins, and 108–245 °C for the monomineralic qtz veins. Assuming a salinity of 0–25%, the stage 1 veins are estimated to have formed at 2–3 kbar and the stage 2 veins at 1–2 kbar.
4. Fluids within the Erdenet deposit evolved from high-temperature fluids at moderate pressures to low-temperature fluids at lower pressures.
5. Transient fluid transport phenomena occurred within the Erdenet deposit as the fluid pressure regime shifted from lithostatic to near hydrostatic. The fluid injection was caused by the development of a fluid-saturated region under the cupola during the formation of the Erdenet deposit.

Supplementary Materials: Complete microprobe analysis used to calculate of Ti-in-quartz and Ti-in-biotite Geothermometers are available online at http://www.mdpi.com/2076-3263/10/5/201/s1.

Author Contributions: N.T. conducted the fieldwork and contributed to discussions during the manuscript preparation. G.A. prepared the samples and obtained petrological and geochemical data by EPMA, assisted by M.U., who also evaluated the data. A.O. assisted with the SEM–CL and fluid inclusion study. G.A. wrote the manuscript, supervised by N.T., A.O., and M.U. This paper is part of the first author's Masters project. All authors have read and agreed to the published version of the manuscript.

Funding: This research was funded by SATREPS project by JICA-JST.

Acknowledgments: We thank Gerel Orchir, Baatar Munkhtsengel, and Batkhishig Bayaraa of the Mongolian University of Science and Technology for kindly providing us with samples from Erdenet and the surrounding area and for assistance with geological fieldwork.

Conflicts of Interest: The authors declare that there is no conflict of interest that could have influenced the work and results of this paper.

References

1. Fournier, R.O. Hydrothermal processes related to movement of fluid from plastic into brittle rock in the magmatic-epithermal environment. *Econ. Geol.* **1999**, *94*, 1193–1211. [CrossRef]
2. Fournier, R.O. The transition from hydrostatic to greater than hydrostatic fluid pressure in presently ac-tive continental hydrothermal systems in crystalline rock. *Geophys. Res. Lett.* **1991**, *18*, 955–958. [CrossRef]
3. Burnham, C.W. Magmas and hydrothermal fluids. In *Geochemistry of Hydrothermal Ore Deposits*; Barnes, H.L., Ed.; John Wiley & Sons: New York, NY, USA, 1979; pp. 71–136.
4. Cox, S. Injection-Driven Swarm Seismicity and Permeability Enhancement: Implications for the Dynamics of Hydrothermal Ore Systems in High Fluid-Flux, Overpressured Faulting Regimes—An Invited Paper. *Econ. Geol.* **2016**, *111*, 559–587. [CrossRef]
5. Tsuchiya, N.; Yamada, R.; Uno, M. Supercritical geothermal reservoir revealed by a granite–porphyry sys-tem. *Geothermics* **2016**, *63*, 182–194. [CrossRef]
6. Weatherley, D.K.; Henley, R.W. Flash vaporization during earthquakes evidenced by gold deposits. *Nat. Geosci.* **2013**, *6*, 294–298. [CrossRef]

7. Amagai, T.; Okamoto, A.; Niibe, T.; Hirano, N.; Motomiya, K.; Tsuchiya, N. Silica nanoparticles produced by explosive flash vaporization during earthquakes. *Sci. Rep.* **2019**, *9*, 9738. [CrossRef] [PubMed]
8. Yoshida, K.; Hasegawa, A. Sendai-Okura earthquake swarm induced by the 2011 Tohoku-Oki earthquake in the stress shadow of NE Japan: Detailed fault structure and hypocenter migration. *Tectonophysics* **2018**, *733*, 132–147. [CrossRef]
9. Okada, T.; Matsuzawa, T.; Umino, N.; Yoshida, K.; Hasegawa, A.; Takahashi, H.; Yamada, T.; Kosuga, M.; Takeda, T.; Kato, A. Hypocenter migration and crustal seismic velocity distribution observed for the inland earthquake swarms induced by the 2011 Tohoku–Oki earthquake in NE Japan: Implications for crustal flu-id distribution and crustal permeability. *Geofluids* **2015**, *15*, 293–309. [CrossRef]
10. Sibson, R.H. Earthquake rupturing as a mineralizing agent in hydrothermal systems. *Geology* **1987**, *15*, 701. [CrossRef]
11. Manning, C.E.; Ingebritsen, S.E. Permeability of the continental crust: Implications of geothermal data and metamorphic systems. *Rev. Geophys.* **1999**, *37*, 127–150. [CrossRef]
12. Friðleifsson, G.Ó.; Pálsson, B.; Stefánsson, B.; Albertsson, A.; Gunnlaugsson, E.; Ketilsson, J.; Lamarche, R.; Andersen, P.E. Iceland Deep Drilling Project. The first IDDP drill hole drilled and completed in 2009. In Proceedings of the World Geothermal Congress 2010, Bali, Indonesia, 25–30 April 2010.
13. Weis, P.; Driesner, T.; Heinrich, C.A. Porphyry-Copper Ore Shells Form at Stable Pressure-Temperature Fronts Within Dynamic Fluid Plumes. *Science* **2012**, *338*, 1613–1616. [CrossRef] [PubMed]
14. Khasin, R.A.; Marinov, N.A.; Khurtz, C.; Yakimov, L.I. The copper–molybdenum deposit at Erdenetiin Ovoo in northern Mongolia. *Geol. Ore Depos.* **1977**, *6*, 3–15.
15. Gerel, O.; Munkhtsengel, B. *Erdenetiin Ovoo Porphyry Copper—Molybdenum Deposit in Northern Mongo-Lia. Geodynamics and Metallogeny of Mongolia with a Special Emphasis on Copper and Gold Deposits*; CERCAMS Natural History Museum: London, UK, 2005; Volume 85, p. 103.
16. Sotnikov, V.I.; Ponomarchuk, V.A.; Shevchenko, D.O.; Berzina, A.P. The Erdenetiyn-Ovoo porphyry Cu–Mo deposit, Northern Mongolia: Ar-40/Ar-39 geochronology and factors of large-scale mineralization. *Russ. Geol. Geophys.* **2005**, *46*, 620–631.
17. Munkhtsengel, B. Magmatic and Mineralization Processes of the Erdenetiin Ovoo Porphyry Copper-Molybdenum Deposit and Environmental Assessment, Northern Mongolia. Ph.D. Thesis, Tohoku University, Miyagi, Japan, 2007.
18. Kavalieris, I.; Khashgerel, B.-E.; Morgan, L.E.; Undrakhtamir, A.; Borohul, A. Characteristics and 40Ar/39Ar Geochronology of the Erdenet Cu-Mo Deposit, Mongolia. *Econ. Geol.* **2017**, *112*, 1033–1054. [CrossRef]
19. Watanabe, Y.; Stein, H.J. Re-Os ages for the Erdenet and Tsagaan Suvarga porphyry Cu–Mo deposits, Mongolia, and tectonic implications. *Econ. Geol.* **2000**, *95*, 1537–1542. [CrossRef]
20. Dejidmaa, G.; Naito, K. Previous studies on the Erdenetiin ovoo porphyry copper-molybdenum deposit, Mongolia. *Bull. Geol. Surv. Jpn.* **1998**, *49*, 299–308.
21. Berzina, A.P.; Sotnikov, V.I. Character of formation of the Erdenet-Ovoo porphyry Cu-Mo magmatic cen-ter (northern Mongolia) in the zone of influence of a Permo-Triassic plume. *Russ. Geol. Geophys.* **2007**, *48*, 141–156. [CrossRef]
22. Rusk, B.; Reed, M. Scanning electron microscope–cathodoluminescence analysis of quartz reveals complex growth histories in veins from the Butte porphyry copper deposit, Montana. *Geology* **2002**, *30*, 727. [CrossRef]
23. Rusk, B.G.; Lowers, H.A.; Reed, M.H. Trace elements in hydrothermal quartz: Relationships to cathodoluminescent textures and insights into vein formation. *Geology* **2008**, *36*, 547. [CrossRef]
24. Kretz, R. Symbols for rock-forming minerals. *Am. Mineral.* **1983**, *68*, 277–279.
25. Okamoto, A.; Sekine, K. Textures of syntaxial quartz veins synthesized by hydrothermal experiments. *J. Struct. Geol.* **2011**, *33*, 1764–1775. [CrossRef]
26. Wark, D.A.; Watson, E.B. TitaniQ: A titanium-in-quartz geothermometer. *Contrib. Mineral. Petrol.* **2006**, *152*, 743–754. [CrossRef]
27. Huang, R.; Audétat, A. The titanium-in-quartz (TitaniQ) thermobarometer: A critical examination and re-calibration. *Geochim. Cosmochim. Acta* **2012**, *84*, 75–89. [CrossRef]
28. Henry, D.J.; Guidotti, C.V.; Thomson, J.A. The Ti-saturation surface for low-to-medium pressure metape-litic biotites: Implications for geothermometry and Ti-substitution mechanisms. *Am. Mineral.* **2005**, *90*, 316–328. [CrossRef]
29. Mercer, C.N.; Reed, M.H. Porphyry Cu-Mo Stockwork Formation by Dynamic, Transient Hydrothermal Pulses: Mineralogic Insights from the Deposit at Butte, Montana. *Econ. Geol.* **2013**, *108*, 1347–1377. [CrossRef]

30. Penniston-Dorland, S.C. Illumination of vein quartz textures in a porphyry copper ore deposit using scanned cathodoluminescence: Grasberg Igneous Complex, Irian Jaya, Indonesia. *Am. Mineral.* **2001**, *86*, 652–666. [CrossRef]
31. Gustafson, L.B.; Hunt, J.P. The porphyry copper deposit at El Salvador, Chile. *Econ. Geol.* **1975**, *70*, 857–912. [CrossRef]
32. Lowell, J.D.; Guilbert, J.M. Lateral and vertical alteration-mineralization zoning in porphyry ore depos-its. *Econ. Geol.* **1970**, *65*, 373–408. [CrossRef]
33. Sillitoe, R.H. Porphyry Copper Systems. *Econ. Geol.* **2010**, *105*, 3–41. [CrossRef]
34. Goldstein, H.R.; Reynolds, T.J. *Systematics of Fluid Inclusions in Diagenetic Minerals*; SEPM Short Course; SEPM Society for Sedimentary Geology: Tulsa, OK, USA, 1994; pp. 31–99.
35. Bodnar, R.J.; Vityk, M.O. *Fluid Inclusions in Minerals: Methods and Applications*; Pontignano: Siena, Italy, 1994; pp. 117–130.
36. Goldstein, R.H.; Samson, I. Petrographic analysis of fluid inclusions. In *Fluid Inclusions: Analysis and InterPretation*; Mineralogical Asscociation of Canada: Québec, QC, Canada, 2003; pp. 9–53.
37. Wilkinson, J.J.; Johnston, J.D. Pressure fluctuations, phase separation, and gold precipitation during seismic fracture propagation. *Geology* **1996**, *24*, 395. [CrossRef]
38. Yardley, B.W.; Bodnar, R.J. Fluids in the Continental Crust. *Geochem. Perspect.* **2014**, *3*, 1–127. [CrossRef]
39. Akinfiev, N.N.; Diamond, L.W. A simple predictive model of quartz solubility in water–salt–CO_2 sys-tems at temperatures up to 1000 C and pressures up to 1000 MPa. *GCA* **2009**, *73*, 1597–1608.
40. Scott, S.; Driesner, T.; Weis, P. The thermal structure and temporal evolution of high-enthalpy geother-mal systems. *Geothermics* **2016**, *62*, 33–47. [CrossRef]
41. Ingebritsen, S.E.; Manning, C.E. Permeability of the continental crust: Dynamic variations inferred from seismicity and metamorphism. *Geofluids* **2010**, *10*, 193–205. [CrossRef]

© 2020 by the authors. Licensee MDPI, Basel, Switzerland. This article is an open access article distributed under the terms and conditions of the Creative Commons Attribution (CC BY) license (http://creativecommons.org/licenses/by/4.0/).

Article

Simultaneous Magmatic and Hydrothermal Regimes in Alta–Little Cottonwood Stocks, Utah, USA, Recorded Using Multiphase U-Pb Petrochronology

Michael A. Stearns [1],*, John M. Bartley [2], John R. Bowman [2], Clayton W. Forster [1], Carl J. Beno [2], Daniel D. Riddle [2], Samuel J. Callis [2] and Nicholas D. Udy [1]

1. Earth Science Department, Utah Valley University, Orem, UT 84058, USA; cforster2110@gmail.com (C.W.F.); nicholas.udy@uvu.edu (N.D.U.)
2. Department of Geology and Geophysics, University of Utah, Salt Lake City, Ut 84112, USA; john.bartley@utah.edu (J.M.B.); john.bowman@utah.edu (J.R.B.); carl.beno@utah.edu (C.J.B.); daniel.riddle@gmail.com (D.D.R.); samueljcallis@gmail.com (S.J.C.)
* Correspondence: mstearns@uvu.edu

Received: 15 February 2020; Accepted: 29 March 2020; Published: 2 April 2020

Abstract: Magmatic and hydrothermal systems are intimately linked, significantly overlapping through time but persisting in different parts of a system. New preliminary U-Pb and trace element petrochronology from zircon and titanite demonstrate the protracted and episodic record of magmatic and hydrothermal processes in the Alta stock–Little Cottonwood stock plutonic and volcanic system. This system spans the upper ~11.5 km of the crust and includes a large composite pluton (e.g., Little Cottonwood stock), dike-like conduit (e.g., Alta stock), and surficial volcanic edifices (East Traverse and Park City volcanic units). A temperature–time path for the system was constructed using U-Pb and tetravalent cation thermometry to establish a record of >10 Myr of pluton emplacement, magma transport, volcanic eruption, and coeval hydrothermal circulation. Zircons from the Alta and Little Cottonwood stocks recorded a single population of apparent temperatures of ~625 ± 35 °C, while titanite apparent temperatures formed two distinct populations interpreted as magmatic (~725 ± 50 °C) and hydrothermal (~575 ± 50 °C). The spatial and temporal variations required episodic magma input, which overlapped in time with hydrothermal fluid flow in the structurally higher portions of the system. The hydrothermal system was itself episodic and migrated within the margin of the Alta stock and its aureole through time, and eventually focused at the contact of the Alta stock. First-order estimates of magma flux in this system suggest that the volcanic flux was 2–5× higher than the intrusive magma accumulation rate throughout its lifespan, consistent with intrusive volcanic systems around the world.

Keywords: incremental pluton emplacement; contact metamorphism; petrochronology; titanite; zircon; U-Pb dating; thermometry; hydrothermal fluids

1. Introduction

Transport of magma within the Earth's crust has far-reaching effects: pluton emplacement at a variety of structural levels, development of associated contact and hydrothermal metamorphism in and around the magma conduits (plumbing systems), and commonly, but not always, volcanic eruptions. The relative timing of magma and/or fluid transport in the system is important for understanding mass and heat transfer, but also for volcanic–plutonic connections and volcanic hazard mitigation. The rock record at any one location depends on the interplay between the different processes and parts of the system. The magma emplacement rate and duration [1], tectonic forces [2], pre-existing and evolving permeability structure [3], and metamorphic mineral reactions [4] all serve to modify and cause feedbacks within the system. Simplifying assumptions, such as an instantaneous magma emplacement (e.g., Annen [5] and Reverdatto et al. [6]) or a lack of magmato-metamorphic processes (e.g., Paterson et al. [7]) exclude critical aspects of mass and heat transfer and obscure our understanding of the integrated system. The term "magmato-metamorphic" is used here to encompass a variety of near- and sub-solidus processes, such as melt and/or magmatic water flow, major and accessory phase (re)crystallization, and mineral reactions, that continue to modify the rock and obscure the record of super-solidus processes. More recently, authors have been applying sophisticated modeling that accounts for incremental pluton emplacement (e.g., Annen [8]) and pulsed conduit flow (e.g., Floess and Baumgartner [9]) to better understand the complex thermochemical feedback between magmas and wall rocks. This paper presents new preliminary petrochronology data from the Alta stock-Little Cottonwood stock system, which demonstrates the complex spatial and temporal patterns of magmatism, hydrothermal infiltration, and contact metamorphism over a crustal section from the surface to an ~11.5 km paleodepth [10].

The Wasatch Intrusive Belt comprises a series of Eocene–Oligocene plutons that intruded the thickened crust in northern Utah following the Sevier orogeny [11–13] (Figure 1A). The Wasatch Intrusive Belt magmas likely resulted from lower-crustal metasomatism and melting caused by the last stages of subduction of the Farallon plate [12,14]. The emplacement and space-making involved extension driven by a combination of far-field tectonic and gravitational forces acting on the thickened crust prior to the Basin and Range extension [15–17]. The plutons intrude a sequence of siliciclastic and carbonate rocks ranging in age from Proterozoic to Triassic. Following emplacement, Wasatch Intrusive Belt plutons were exhumed by range-bounding normal faults of the Oquirrh and Wasatch ranges (Figure 1B) beginning at ~18 Ma (17.6 ± 06 Ma K-Ar sericite date [18] and are still being actively exhumed (e.g., References [19–21]). Offset across the Wasatch fault has tilted the range by ~20° to the east along a roughly horizontal N–S rotation axis [22]. This rotation and exhumation exposed a crustal section from ~11.5 km at the southeast corner of the Salt Lake Valley [10] to the paleosurface (Figure 1C) east of Park City, Utah. From structurally deepest to shallowest, this section contains the Little Cottonwood & Ferguson stocks (included in the Little Cottonwood stock from here forward), Alta stock, Clayton Peak and other eastern stocks, and the Keetley volcanic deposits. Previous geochronology from the Wasatch Intrusive Belt includes a combination of multigrain thermal ionization mass spectrometry U-Pb zircon; K-Ar mica and amphibole; fission-track titanite and zircon; and (U-Th)/He and fission-track in titanite, zircon, and apatite dates [19,20,23–25]. These data suggested that pluton emplacement began at ~36 Ma with the eastern stocks, continued with the emplacement of the Alta stock at ~35 Ma, and ended with the Little Cottonwood stock at ~30 Ma. New petrochronology presented here establishes magmatism in the Little Cottonwood and Alta stocks from ~36–26 Ma, beginning and ending with different portions of the Little Cottonwood stock, thus indicating a different sequence and a much longer duration of pluton emplacement from that previously inferred.

The Alta stock is a structurally shallow (~5–5.5-km depth current exposure [22,26,27] dike-like intrusion with subvertical walls (Figure 1)). The two mappable intrusive phases that comprise the Alta stock, namely an equigranular border phase and a later porphyritic central phase, are distinguished by crystal size, microscopic rock texture, and locally recognizable cross-cutting relationships [28]. The major minerals of the Alta stock are plagioclase + K-feldspar + quartz + biotite + amphibole, with titanite + apatite + zircon + ilmenite ± magnetite accessory phases. The contact metamorphic aureole surrounding the Alta stock is especially well-studied [28–34]. Past and continuing work has largely been motivated by the apparent mismatch between the size of the intrusion (~2 km at its widest) and the surrounding contact aureole (≤1.25 km in width). Cook and Bowman [30,31] demonstrated that metamorphism of the carbonate wall rocks was largely driven by the infiltration of high-temperature (~625 °C), H_2O-rich fluids laterally outward from the border phase of the Alta stock, which exploited and modified the existing permeability structure surrounding the pluton (Figures 1 and 2). Numerical models assuming instantaneous emplacement of the pluton showed that the locations of prograde metamorphic isograds, isotherms based on calcite-dolomite thermometry, and the pattern of $^{18}O/^{16}O$ depletion across the aureole could be duplicated with ~5000 years of advective heat and fluid flow, followed by ~20,000–30,000 years of conductive heating to produce the outer aureole [33]. This numerical modeling indicates that the observed extent of the contact aureole could be matched with heat supplied solely from cooling of the Alta stock only if the duration of pluton emplacement was <5000 years.

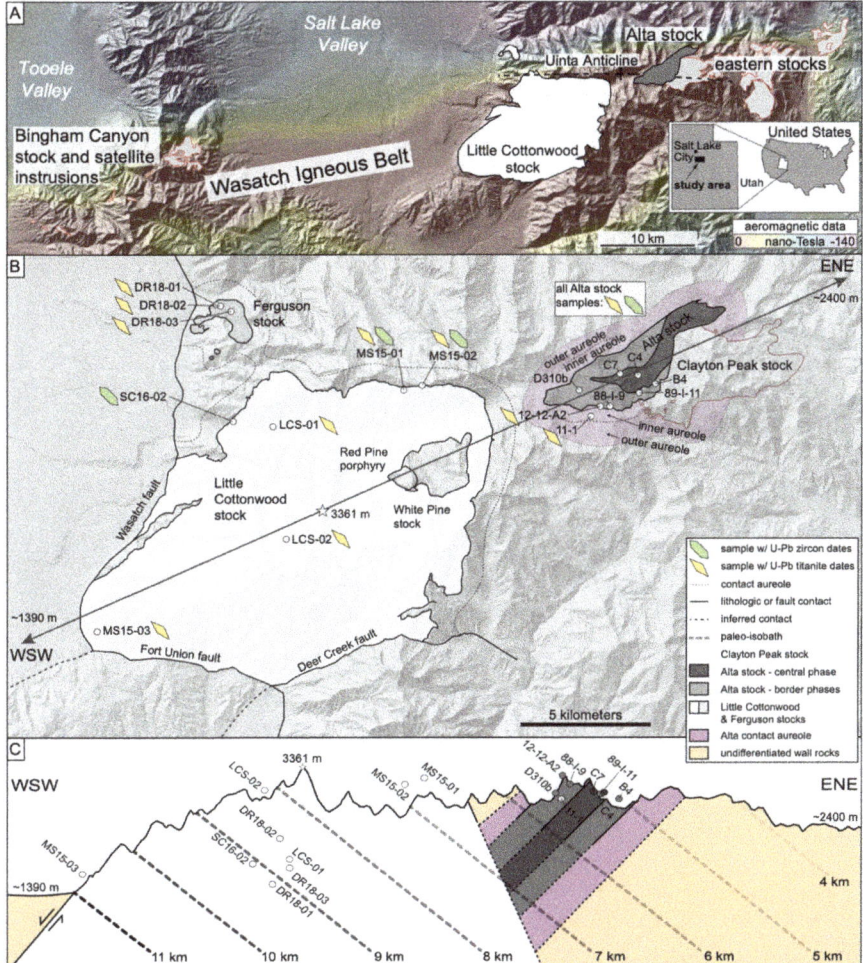

Figure 1. (**A**) A simplified geologic map (modified from Wohlers and Baumgartner [35]) shows the location of the Wasatch Intrusive Belt in northern Utah; the correlation of the Wasatch Intrusive Belt with the Uinta Anticline and the aeromagnetic anomaly (nano-Tesla scale bar) that likely represents a continuous body of intrusive rocks at depth [36]. (**B**) The more detailed geologic map overlaid on a shaded relief map and accompanying schematic cross-section (**C**) that illustrates the sample locations and the estimated paleodepth (isobaths) of the analyzed samples [10,22,37]. The cross-section is highly vertically exaggerated (see noted elevations on the map and cross-section) and was calculated using a ~15° rotation due to the differing azimuth from John's [22] ~20° rotation of the Wasatch footwall block. The western Alta and eastern Little Cottonwood contacts are ~75° to WSW and ~50° to ENE, respectively, based on three-point solutions.

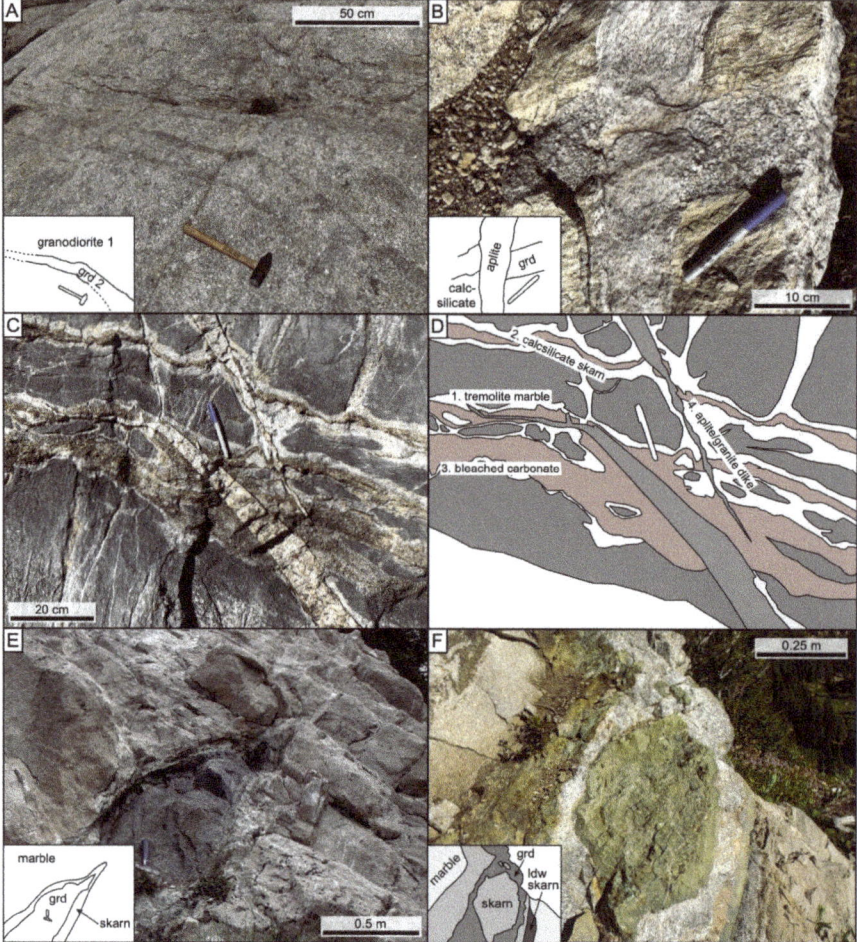

Figure 2. Photographs and interpretation schematics showing the highly composite nature of the magmatic and hydrothermal system. (**A**) A granodiorite (grd in schematics **A**, **B**, **E**, and **F**) dike with slightly more mafic margins cross-cutting nearly identical granodiorite in the Little Cottonwood stock. (**B**) A granodiorite dike that intruded the wall rock and was later crosscut by an aplite dike near the southeastern margin of the Little Cottonwood stock. (**C**) A photograph and (**D**) interpretation of a highly composite outcrop that reflects the injection of both magma and aqueous fluid into carbonate wall rock near the southeastern margin of the Little Cottonwood stock. (**E**,**F**) Calcite marbles intruded by granodiorite sills with calcsilicate skarns developed at the contacts. Calcsilicate skarns are also found adjacent to and included in the Alta stock. Ldw = ludwigite, resulting from boron metasomatism in the Alta aureole [69].

In contrast, if space for the Alta stock was made by horizontal dilation perpendicular to the long axis of the pluton [12] at a reasonable tectonic rate for a continental lithosphere (~1–2 mm·yr^{-1}), emplacement of the Alta stock must have taken ~1–2 Myr [38]. Thus, the preferred emplacement model based on field observations implies that the Alta stock grew too slowly by nearly three orders of magnitude to account for the thermal aureole that surrounds it. This extreme mismatch poses two important questions: What is the thermal history of the system, including the Alta conduit, aureole, and associated Little Cottonwood stock? What are the sources of heat and fluids, beyond the Alta stock, that drove metamorphism in the Alta aureole?

Because the contact aureole is centered on the Alta stock, it is unsurprising that previous studies have assumed that the Alta stock was the source of the heat and fluids that produced the aureole. However, a protracted emplacement of the Alta stock, coupled with the recognition that fluid infiltration is required to drive metamorphic reactions and $^{18}O/^{16}O$ depletion in the aureole, opens up the possibility that at least some of the hot fluid that emanated from the Alta stock originated from another source. The structurally deeper (current exposures correspond to a paleodepth range of ~6.5–11.5 km) and much larger Little Cottonwood stock is bounded by the Wasatch and Deer Creek faults to the west and south, respectively (Figure 1). Data from this study confirmed that the Ferguson stock, interpreted as a satellite to the main Little Cottonwood stock body, is cogenetic. In general, the Little Cottonwood stock is more felsic than the Alta stock, is structurally composite, and its modal mineralogy, crystal size distribution, and chemistry are highly variable [39,40] (Figure 2). The major minerals that make up the Little Cottonwood stock are plagioclase + K-feldspar + quartz + biotite ± amphibole, with titanite + apatite + zircon + ilmenite ± magnetite accessory phases (Figure 3). The Little Cottonwood stock intruded the Neoproterozoic Big Cottonwood Formation on the north and contains a ~1 km × ~200 m screen of these rocks near the western interior of the intrusion (Figure 1). The Little Cottonwood stock magmas intruded Cambrian siliciclastic rocks and Mississippian carbonates on its southeastern margin [41]. Emplacement of the Little Cottonwood stock contact metamorphosed and locally melted the surrounding wall rocks. The Big Cottonwood Formation is composed of interbedded quartzite and shale, and pelitic layers contain the peak contact metamorphic assemblage of biotite + cordierite + sillimanite + K-feldspar + melt [35].

The mineral textures (sensu lato) of intrusive igneous rocks are a topic of active conversation, and a growing body of both field (e.g., Johnson and Glazner [42]) and experimental (e.g., Lundstrom [43]) data suggest that both major and accessory phases in granitic rocks are susceptible to recrystallization at temperatures lower than traditional solidi of ~700 ± 50 °C [44]. For example, the coarsening of orthoclase to so-called megacrysts that are common in calc-alkaline intrusions, including the Little Cottonwood stock, may involve melt (e.g., Higgins [45]) or may occur completely in the solid state [46]. Cathodoluminescence imaging, titanium-in-quartz thermometry in the Alta stock [47], and numerical diffusion modeling of similar data from the Tuolumne Intrusive Suite [48] suggest that quartz in granitic plutons has commonly recrystallized at temperatures in the hydrothermal regime well below traditional granitic solidi. Accessory phases, such as titanite, have been shown to be reactive and recrystallized at similar thermal and fluid conditions [49–51]. Proper interpretation of such overprinting and continued modification using a range of processes spanning magmatic (e.g., Bartley et al. [52]) to hydrothermal (e.g., Smirnov [53]) requires a robust, multifaceted petrochronology approach to relate the chemistry and the time for the process to occur.

Figure 3. Plane polarized (top) and cross-polarized (bottom) photomicrographs that illustrate the range of titanite morphologies and phase relationships observed in both the Alta (**A**,**B**) and Little Cottonwood stocks (**C**). Titanites range from euhedral with few inclusions to anhedral rims on oxide phases (typically ilmenite). Zircons (not seen here) are typically inclusion free and included in a variety of early crystallizing phases, such as plagioclase and biotite. We interpreted these different groups to represent assemblages that grew via different processes. (**A**) Alta central phase sample C4, (**B**) Alta border phase sample D310b, and (**C**) Little Cottonwood stock sample MS15-03. All images were taken at 4× objective magnification.

Petrochronology exploits the accuracy and spatial context of in situ laser ablation split stream (LASS) inductively coupled plasma mass spectrometry (ICP-MS) analysis to simultaneously collect U-Pb isotopes and trace element contents in a variety of chronometer phases [54–56]. Measuring U-Pb and tetravalent cations in multiple phases, such as zircon and titanite from the same rock, gives more complete information about the thermal history of the rock due to the different reactions responsible for the paragenesis of each phase. Both zircon and titanite are common in calc-alkaline rocks like the Wasatch Intrusive Belt (WIB) (Figure 3) [57,58]. Zircon saturation depends on the Zr content and the network modifier/former ratio M of the magma [59,60]. In the Alta–Little Cottonwood system, whole-rock chemistry predicts the zircon saturation at ~725 ± 25 °C. Hanson [39] documented the occurrence of inherited zircon based on petrographic observations, which prolonged the Zr saturation in both the Little Cottonwood and Alta stocks. Prolonged zircon saturation suggests that the zircon saturation temperature (T_{Zr}) would be a maximum emplacement temperature for the pluton [39,61]. Titanite paragenesis in calc-alkaline rocks is more complex, and as mentioned above, titanite (re)crystallizes via multiple processes, such as de novo crystallization from melt [62], redox reactions involving Fe-Ti oxides [58], and/or (re)crystallization in the presence of fluids [63]. Titanite is stable in calc-silicate skarns [64–66] and often grows due to the breakdown of Ti-bearing clinopyroxene, and may (re)crystallize during the infiltration of fluorine and/or H_2O-rich fluids [50,67,68].

This paper presents titanite and zircon U-Pb dates, trace element concentrations, and tetravalent cation thermometry from 17 samples in the Alta–Little Cottonwood intrusive–hydrothermal system, which spans the upper ~11.5 km of the Eocene–Oligocene crust in what is now Utah. The new data indicate significant differences from previous conclusions regarding the timing, duration, and spatial extent of magma emplacement and hydrothermal activity within the Alta–Little Cottonwood system. The data together define a temperature–time path that spans >10 Myr from ~36–23 Ma. They further indicate that magmatic and hydrothermal processes were episodic at any given location but they were active in some portion of the system continuously throughout the >10 Myr duration.

2. Materials and Methods

Samples were taken from the freshest or least weathered outcrops and trimmed to remove any stained or visually altered rock before cutting thin section billets. As much as possible, titanite and zircon crystals were imaged and inspected for internal chemical zoning prior to analysis using backscattered electron and cathodoluminescence detectors, respectively. These images guided the laser spot placements to minimize the mechanical mixing of chemically and/or isotopically distinct domains and to ensure an analysis of the full range of textural populations in each sample. Many of the zircons analyzed, especially in the Alta stock samples, were chemically zoned at a scale smaller than the analytical beam (~20–25 μm diameter for zircon analyses), and thus the dates were certainly mechanically mixed during some analyses and represent minimum durations of zircon growth.

The majority of samples (except LCS-01,02 and MS15-01,02) were analyzed in thin sections to maintain the petrologic context of the U-Pb dates and trace-element analyses via the laser ablation split stream (LASS) technique; see the method and instrument details of References [49,54,70]. The crystals were ablated using a Photon Machines/Teledyne 193 nm Excimer laser equipped with a two-volume Helex® stage [71] and the ablated material was introduced via He carrier gas to the multi-collector inductively-coupled plasma (ICP-MS) and quadrupole (Q-MS) mass spectrometers to measure uranium and lead isotopes and trace elements, respectively. Isotopic ratios were measured on a Nu Plasma 3 and Thermo NeptunePlus multicollector ICP-MS and trace elements were measured on Agilent 7500ce and 8900 quadrupole mass spectrometers. A laser spot diameter of 20–40 μm for titanite and 20–25 μm for zircon, and a laser repetition rate of 4–6 Hz and fluence of ~2–3 J/cm^2 were used during the in situ analyses.

Isotopic analyses were bracketed using and standardized to a matrix-matched primary isotopic reference material (91500 zircon, Bear Lake Road and MKED-1 titanites [72–76]). Matrix-matched secondary reference materials were analyzed during all the zircon analyses (Plesovice zircon [77]) and titanite analyses from twelve samples (BLR titanite) to determine the accuracy and propagate external uncertainties. During the titanite analyses from six samples, both "Yates Mine" titanite [70] and Plesovice zircon (standardized to 91500 primary reference material) were used as a secondary reference material due to a lack of titanite standards at the time (see the Data Repository File).

Mass spectrometry data were reduced and interpreted using the VizualAge Data Reduction Scheme [78] for the Iolite plugin within IgorPro [79]. Homogeneous portions of the time-resolved data were selected to minimize the common Pb by monitoring the ^{204}Pb and ^{208}Pb channels and maximize the concordance. The isotopic dates were not used as a criteria for selecting portions of the data. The Iolite output data were further reduced using the Isoplot plugin for Excel [80] and the IsoplotR package for the R statistical program [81]. Analytical uncertainty, counting statistics, and extra error for the homogeneity and accuracy of the secondary reference materials were propagated in quadrature for the final reported uncertainty (see the Data Repository File), which was typically ~2–3% or ~0.5–1.0 Ma for these samples. The typically high variance and discordance of U-Pb in titanite data, and some in situ zircon data, require an interpretation of the raw dates using a partial Pb isochron method [82]. Initial ^{207}Pb/^{206}Pb values were chosen using a combination of linear regression, visual data fitting, calculation using Stacey–Kramer Pb growth curves, and comparison to Pb isotope data from K-feldspar. All of the initial Pb values used to calculate the ^{207}Pb-corrected dates fell within the expected range of 0.84–0.87, and the uncertainty in the initial Pb isotope ratio was propagated into the reported individual dates from each analysis point.

Ti-in-zircon and Zr-in-titanite apparent temperatures were calculated using the Ferry and Watson [83] and Hayden et al. [84] calibrations, respectively. Both calibrations depend on the activities of TiO_2 and SiO_2. The activity of TiO_2 was chosen to be $\alpha_{TiO2} = 0.5 \pm 0.2$ for both thermometers based on three observations: (1) titanite was present and rutile was absent in all samples except SC-02, which also lacked titanite ($\alpha_{TiO2} \geq 0.5$ except SC-02 [84]); (2) non-exsolved Fe-Ti oxide pairs from the cogenetic WIB extrusive rocks produced α_{TiO2} ~ 0.5–0.7 [39,85]; and (3) the rhyolite-MELTS model matching the modal abundance of major phases (for the Alta stock–Little Cottonwood stock (AS–LCS) range of whole-rock chemistry) also produced α_{TiO2} ~ 0.5–0.7 [86]. For the Ti-in-zircon thermometer, the reported uncertainty of ~±35 °C includes the analytical uncertainty (~10% 2SE), uncertainties in the activity terms (~0%–40%), and P dependence (~10%). The plotted apparent temperatures include the estimated pressure dependence of 35 °C/GPa. The Data Repository File includes temperatures with and without the pressure corrections calculated using both the lower [83] and the higher pressure dependence of Ferriss et al. [87] of 100 °C/GPa. The Zr-in-titanite thermometer is much more dependent on pressure and the typical uncertainty is ±40–50 °C.

3. Results

3.1. Petrologic Observations

The Alta and Little Cottonwood stocks are texturally [28,40] and compositionally variable [12]. Internal intrusive contacts were locally visible and suggest that the stocks are also highly composite. Granitic and aplite dikes were common both within and outside the intrusions (Figure 2A,B). Additionally, wall rocks adjacent to both intrusions recorded many generations of melt and fluid infiltration (Figure 2C–F). Where dikes intruded the wall rocks at the margins of the Alta and Little Cottonwood stocks, the dikes were commonly deformed and suggested syn-deformation emplacement (Figure 2C,D).

At the thin section scale, titanite in both intrusions ranged from euhedral with no inclusions to anhedral and overgrowing ilmenite (Figure 3). Titanite was commonly associated with both ilmenite and magnetite, and with amphibole, biotite, apatite, and sometimes plagioclase. This assemblage suggests a titanite-forming reaction similar to that seen in amphibolite. Titanite was commonly included within quartz, alkali feldspar, and amphibole (Figure 3), but titanite included within plagioclase was not observed. Titanites in both intrusions contained internal chemical zoning, including oscillatory zoning, sector zoning, and patchy zoning indicative of a range of processes from neocrystallization from a granitic melt, neocrystallization from pre-existing phases, and fluid-aided interface-coupled dissolution-reprecipitation recrystallization [51]. Zircons were typically euhedral and were observed to be included within most major minerals, including plagioclase (Figure 3A). Zircons ranged from nearly equant to acicular and oscillatory zoning was common. Zircons in the Little Cottonwood stock often contained distinct mantles and rims that were typically cathodoluminescent dark and bright, respectively. Resorbed, inherited cores were observed and are interpreted to have traveled from the melt source and served as nuclei for further zircon growth (Figure 4).

3.2. Petrochronology

3.2.1. U-Pb Dates

We presently have many more isotopic dates from titanite (n = 17 samples, 766 analyses) than from zircon (n = 9 samples, 254 analyses), particularly in the Little Cottonwood stock (titanite: n = 8 samples, 326 analyses; zircon: n = 3 samples, 129 analyses) (Figure S1 and Table S1). Isotopic dates are reported as ^{207}Pb-corrected dates projected to concordia from an initial ^{207}Pb/^{206}Pb ratio for each sample (Data Repository). For discussion of this method, see the Materials and Methods Section above. The mean square weighted deviations (MSWD [88]) of the Little Cottonwood stock zircon date populations ranged from ~6 to 21. Dates from three samples (MS15-01, 02, and SC16-02) ranged from 34–26 Ma, 35–25 Ma, and 33–27 Ma, respectively, with Proterozoic (~1000 Ma) inherited cores. Older Cenozoic dates (~35–34 Ma) came from distinct cores with zircon mantles that produced dates from ~33–30 Ma. The youngest dates, ~30–25 Ma, came from cathodoluminescent dark zircon rims, most notably in sample MS15-01 (Figure 4). MSWD values from LCS titanites ranged from ~1.2 to 7.2. Analyses from two of the eight samples, namely LCS-01 and 02, formed single populations (MSWD = 1.2). Titanites from these samples were concentrated by mineral separation techniques and analyzed in a 25-mm-diameter epoxy mount rather than with in vivo thin section analysis. It is possible (or likely) that mineral separation steps and grain picking biased the analyses toward a single titanite population. Little Cottonwood stock titanite grains from all eight samples produced dates from 37–27 Ma with modes at 34.5 and 31.5 Ma, and a different distribution for the zircon dates (Figure 5C). The oldest titanite dates ranged from ~37–32 Ma and came from structurally deeper western samples (MS15-03 and DR18-01, 02, and 03; Figure 5) near the Wasatch fault and from the satellite Ferguson stock. Samples from the center and structurally higher portions of the stock defined the middle and younger portions of the range from ~34–27 Ma. However, the incomplete spatial coverage of presently dated samples led us to postpone interpretation of an emplacement pattern pending collection of more spatially complete data.

Both zircon (125 analyses) and titanite (316 analyses) were analyzed in six samples from the Alta stock, including four samples from the border phases (D310b, B4, 89-I-11, and 88-I-9) and two samples from the central phase (C4 and C7). Zircons within the Alta stock range from an equant with a diameter of ~15 μm to elongate and ~100 μm in the long dimension. Uranium-lead dates from zircon in the AS border phases ranged from 35–31 Ma with a mode slightly younger than 34 Ma. Dates from zircons in the Alta central phase ranged from 34.5–32 Ma with a mode of 33.5 Ma. Modes of titanite dates from each Alta stock phase were younger than the zircon modes, and the ranges of titanite dates exceeded the ranges of zircon dates. The majority of titanite in the Alta border phases ranged from ~35–30 Ma. However, sample 88-I-9 from near the southern Alta stock contact contained titanites that spanned 35–25 Ma (excluding four analyses; Figure 4).

Titanites were also dated from a metasomatized Alta stock granitic rock (endoskarn sample 12-12-A2; 67 analyses) and from a wollastonite-bearing skarn (sample 11-1; 45 analyses) in the inner Alta contact aureole. Dates from the endoskarn sample ranged from 35–23 Ma (excluding three analyses) and had a complicated polymodal distribution (Figure 5A). The modes did not correlate with modes from sample 88-I-9 or in any of the Little Cottonwood stock distributions. The majority of titanite dates from the wollastonite skarn ranged from 37–29 Ma with modes at 34 Ma and 32 Ma and a single analysis at ~26.5 Ma. These modes did not correlate to the Alta stock zircon or titanite date modes, but did correlate both with modes of Little Cottonwood stock titanite dates and with the mode of Ferguson stock titanite dates (34 Ma; Figure 5).

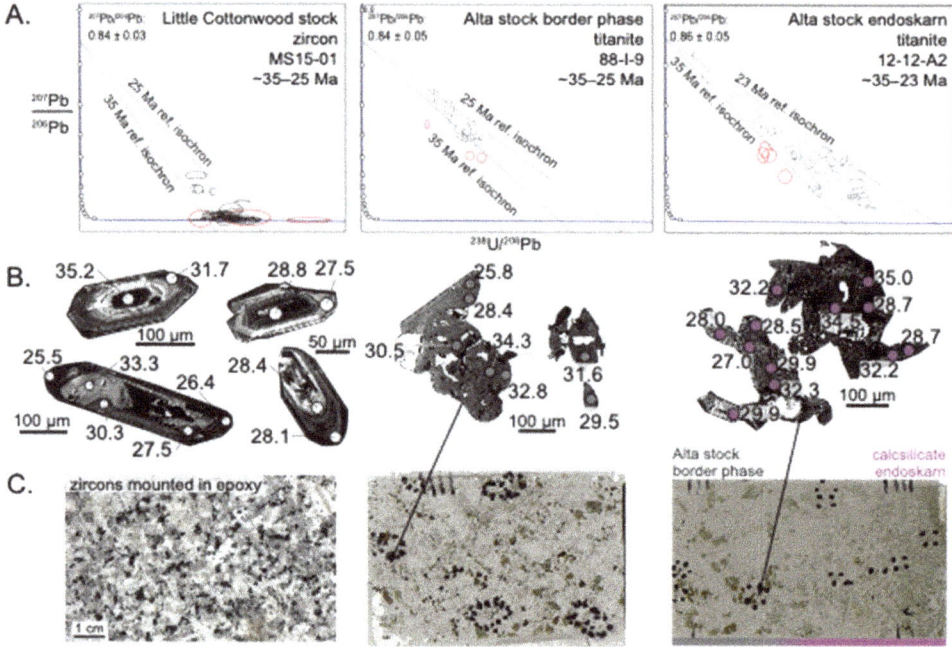

Figure 4. (**A**) The typical Tera–Wasserburg plots of U-Pb isotopes in zircon were nearly concordant with some outliers. The zircons in the Alta and Little Cottonwood stocks contained dates that reflected multiple generations of growth. In contrast, the Tera–Wasserburg plots of U-Pb isotopes in titanite have a large range of radiogenic and common Pb [89] and were interpreted to indicate an age range within these two samples from the Alta stock. The grey lines that bound the data were reference isochrons for the age range listed for each sample [82]. Red error ellipses were excluded based on either anomalously high ^{204}Pb or low Si and/or Ca concentrations, which indicated that a mineral other than titanite was sampled by the laser. (**B**) Representative cathodoluminescence (CL) and backscattered electron (BSE) images of zircon and titanite grains show the zoning and range of spot dates (colored by lithologic unit) observed in the Little Cottonwood stock (white data) and Alta stock (grey data) border phase and endoskarn samples. Laser sampling locations are labeled with the ^{207}Pb-corrected date and show that neocrystallized and reaction-rim grains were typically younger than recrystallized titanites. (**C**) Hand sample and petrographic thin section photographs show the unaltered macroscopic appearance of MS15-01, 88-I-9, and the gradational replacement of igneous rock by endoskarn (light purple data) in sample 12-12-A2.

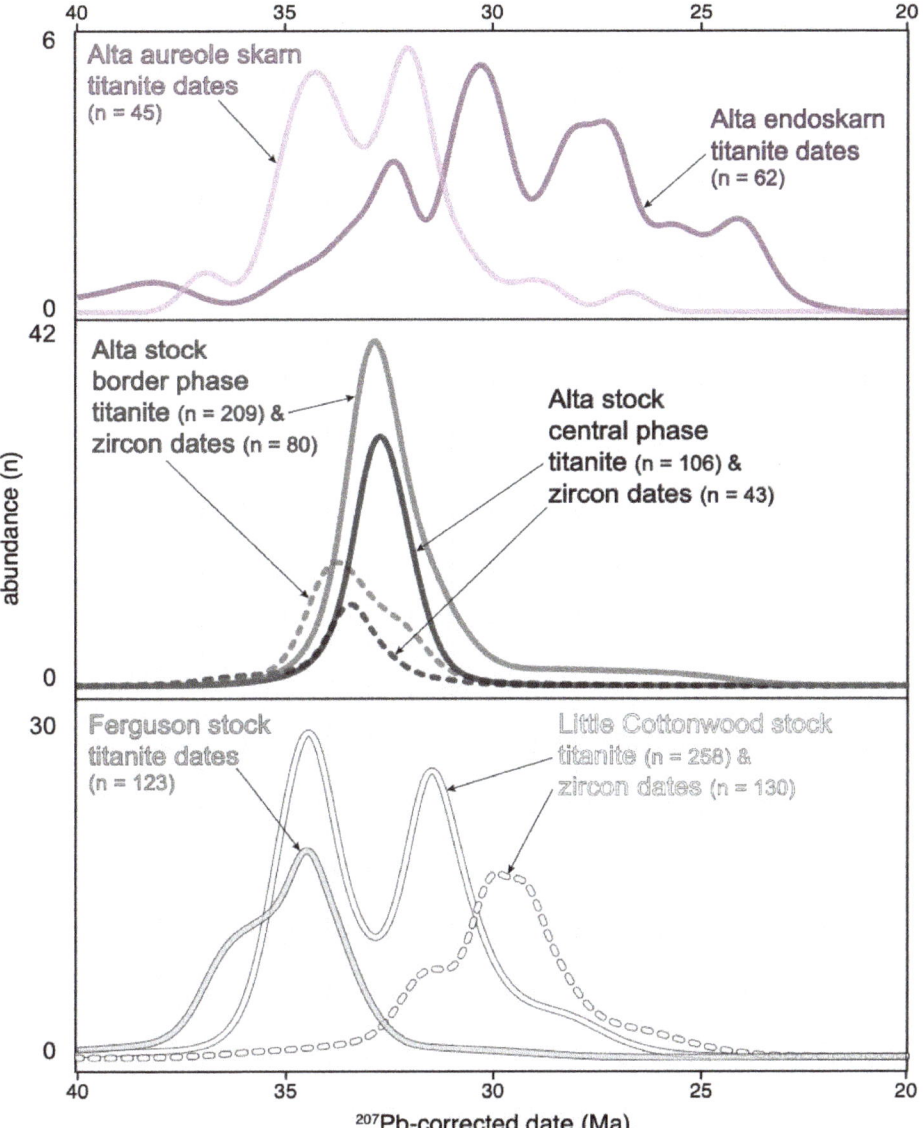

Figure 5. Kernel density estimates (KDEs) colored by lithologic unit of 40–20 Ma ^{207}Pb-corrected dates that were calculated and plotted using IsoplotR [81] for the (**top**) Alta aureole and endoskarn titanite dates (solid purple), (**middle**) Alta stock border and central phases titanite (dashed dark greys) and zircon dates (solid dark greys), and (**bottom**) Little Cottonwood stock titanites and zircon (solid white and light grey) and Ferguson stock titanite dates (dashed white and light grey). The heights of the KDE curves were normalized per grouping and the maximum abundances are labeled on the y-axis for each curve.

3.2.2. Trace Elements and Thermometry

The concentrations of all rare earth elements (REEs) in the Alta–Little Cottonwood titanites decreased with time (Figure 6A, Figure S1 and Table S1). Ytterbium, for example, decreased from

>10^3 times chondrite prior to 30 Ma to <10^2 times chondrite after 30 Ma. The oldest titanites defined a narrow range of REE contents with a slightly higher light (L)REE and lower heavy (H)REE contents. The europium anomaly (Eu/Eu*) in titanites ranged from 0.3 to 3, excluding outliers, with a mode of ~0.9. Pre-35-Ma titanites generally fell below 1.0, while 35–30-Ma titanites defined the full range of Eu/Eu* from 0.1–3. Analyses younger than 30 Ma also generally had an Eu/Eu* of less than 1.0. Many analyses of zircon had REE concentrations near or below the detection limit during LASS analysis and the spider plot does not include concentrations below 1× chondrite (Figure 6B). All zircon analyses showed a typical pattern of low LREE and middle (M)REE concentrations and increasing HREE content with a pronounced cerium anomaly (Ce/Ce*). Zircons increased in all REE with time, most notably Pr through Dy. The samarium content (Figure 6B) increased from <10^2 times chondrite prior to 30 Ma to generally >10^2 after 30 Ma. The Eu/Eu* (~0.1–7) in zircon became more scattered through time. Cerium anomalies (Ce/Ce*) in the Alta–Little Cottonwood zircons ranged from 10^1–10^4 and were positively correlated with Eu/Eu* ($p < 0.01$; $n = 22$), meaning zircons with a higher Ce/Ce* tended to have a less negative Eu/Eu* (closer to 1.0). The Ce^{4+} and Eu^{3+} ions were both favored at higher oxygen fugacity [90,91], which is suggestive of trends in the Alta–Little Cottonwood magma oxidation state, but there was no statistically significant relationship between either anomaly and the ^{207}Pb-corrected date.

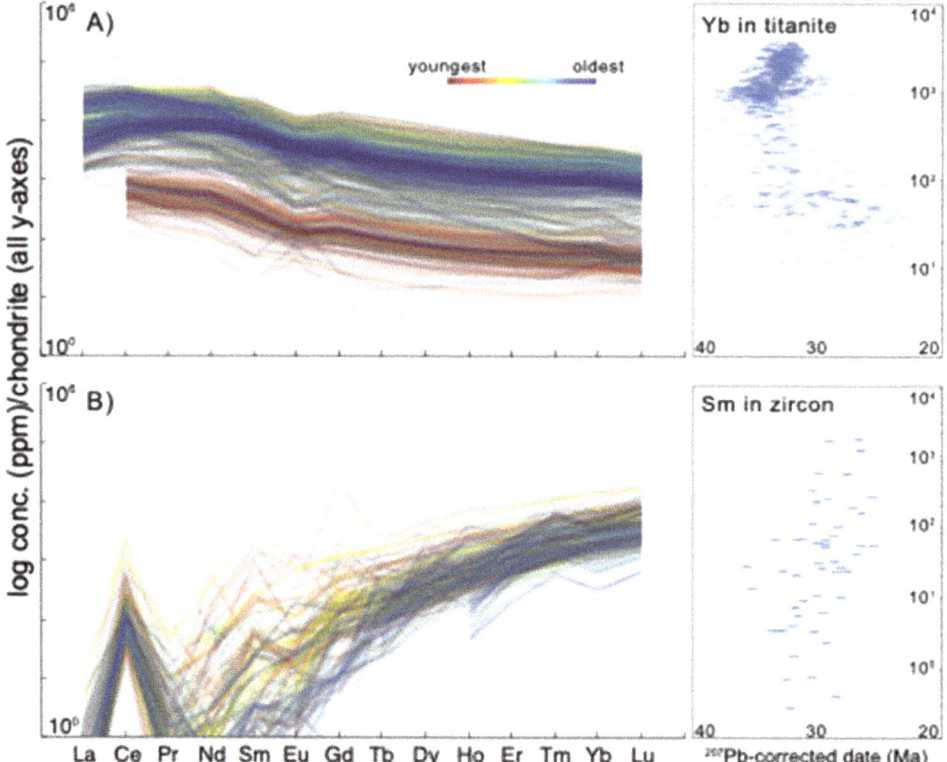

Figure 6. Semi-logarithmic chondrite normalized spider diagrams for all (**A**) titanite and (**B**) zircon analyses, colored using the ^{207}Pb-corrected dates. Older dates are blue and younger dates are red. Titanite analyses became depleted in lanthanides through time, while zircons became enriched through time. See the Data Repository File for the concentration and normalized data.

The titanium contents of Little Cottonwood stock zircons ranged from 3.8–10 ppm (Figure 7) with a dominant mode at 4.8 ppm. Zircons from the Alta border phase ranged from 2.7–8.4 ppm titanium with a mode at 4.0 ppm. The Ti contents of Alta central phase zircons ranged from 0–8.3 ppm with modes at ~3.0 and ~4.5 ppm Ti. Abnormally high Ti contents (>>10 ppm) were likely caused by the ablation of a Ti-rich phase like ilmenite, rutile, or titanite included in zircon. Inclusions were avoided during spot placement and none were observed during analysis, but mineral inclusions were observed during optical and SEM microscopy.

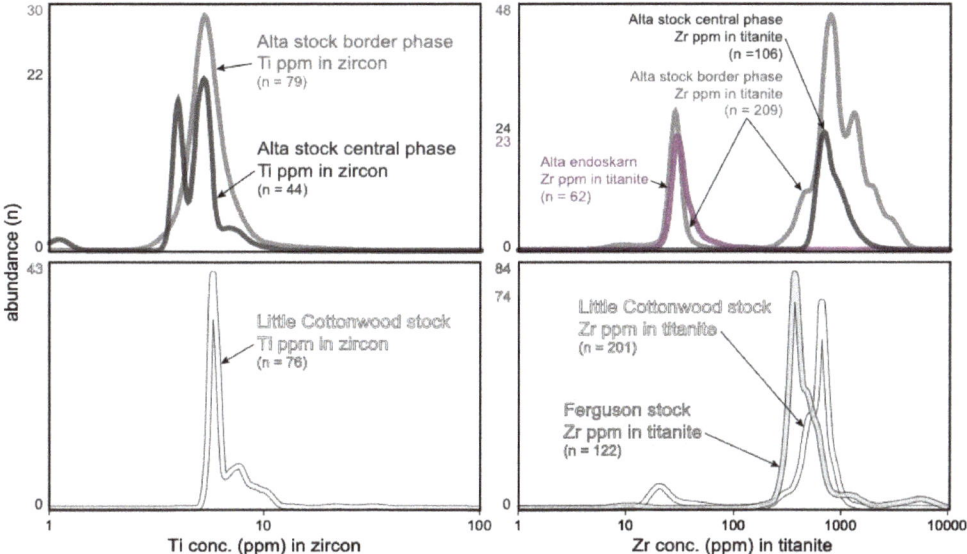

Figure 7. Kernel density estimates (KDEs) of the Ti content of zircon (left column) and Zr content in titanite (right column) colored by lithologic unit and calculated and plotted using IsoplotR [81] for the Alta endoskarn (purple), Alta stock border and central phases (dark greys), Little Cottonwood stock (white), and Ferguson stock (light grey). The heights of the KDE curves are normalized per grouping and the maximum abundances are labeled on the y-axis for each curve.

Zirconium contents of Little Cottonwood stock titanites ranged from 10 to >1000 ppm. The two units had different distributions. Ferguson stock titanites contained less Zr with a mode at ~400 ppm, while the Little Cottonwood stock titanites had a small mode at ~15 ppm and a large mode at ~700 ppm. Zirconium contents of the Alta stock border phase titanites ranged from ~10 to ~2000 ppm with a complex distribution. The two modes at ~15 ppm and ~750 ppm roughly corresponded to the Little Cottonwood stock titanite modes. The lower Zr mode in the Alta stock border phase titanites was found only in sample 88-I-9 from the southern margin of the pluton. The Alta stock central phase titanites had a more restricted range of Zr content from ~500–1500 ppm with a mode at ~700 ppm. The titanites from the Alta endoskarn sample 12-12-A2 contained 20–110 ppm Zr and largely overlapped with the mode defined by sample 88-I-9. Titanites from the wollastonite skarn contained anomalously high Zr contents (>>1000 ppm) that corresponded with sector zoning in the crystals. These titanites likely incorporated a non-equilibrium amount of Zr and other trace elements [92,93] and were not considered in the thermometry discussion below.

4. Discussion

4.1. Thermal History

Transforming the concentrations of tetravalent cations (Figure 7) into apparent temperatures and plotting against ^{207}Pb-corrected dates yielded a temperature–time path for the Alta–Little Cottonwood system (Figure 8). Ti-in-zircon thermometry produced temperatures of ~650–850 ± 40 °C (using α_{TiO2} = 0.5 ± 0.1 and α_{SiO2} = 0.8 ± 0.1) with a mode of ~718 °C for both intrusions (MSWD = 1.8). This mode was consistent with the zircon saturation temperature (~725 ± 25 °C), the likely solidus (675–725 °C [94,95]) for these rocks, and the petrographically determined crystallization sequence. Titanium-in-zircon apparent temperatures from the Alta–Little Cottonwood system are hotter than results from other felsic and intermediate rocks [62]. Other studies [62,96,97] have reported Ti-in-zircon apparent temperatures lower than that predicted by Ti-in-quartz, Zr saturation, and δ^{18}O quartz-magnetite pair thermometers, as well as temperatures of zircon paragenesis from phase relationships. The inconsistency reported in other studies and general sources of uncertainty could reflect: (1) misestimation (likely overestimation) of the TiO$_2$ and SiO$_2$ activities during zircon crystallization, (2) the pressure dependence of the thermometer, (3) the subsolidus open system behavior of zircon, and/or (4) prolonged Zr saturation in the magma and real low-temperature crystallization of zircons. Multiple indicators of the TiO$_2$ activity (see the Materials and Methods Section), including the absence of rutile from all samples, suggest that the TiO$_2$ activity of 0.5 ± 0.1 was reasonable, reflects the phase assemblage, and encompasses the mode (α_{TiO2} ~ 0.4) and 50% of the α_{TiO2} of experimental granitic melts [97]. TiO$_2$ activity may be lower (i.e., further from rutile saturation) during earlier crystallization of zircon and plagioclase in the magma [96], and it is unlikely that it is much higher than 0.6–0.7 at temperatures closer to the solidus. As previously noted in the Materials and Methods Section, lowering or raising the TiO$_2$ activity to a minimum of 0.3 or maximum of 0.9 changed the temperature by ±50 °C. This range of TiO$_2$ activity is possible over the entire span of zircon crystallization in magmas, but a TiO$_2$ activity near 1.0 is unlikely. Thus, a propagating uncertainty of ±0.1 (~±40 °C) encompassed nearly the entire range of temperature variation caused by any inaccuracy resulting from the α_{TiO2}. Schiller and Finger [97] suggest that the Ferry and Watson [83] Ti-in-zircon thermometer should not be applied to TiO$_2$ undersaturated rocks, such as the I-type Alta and Little Cottonwood stocks, since it underestimates temperatures for these magmas despite the α_{TiO2} term in Ferry and Watson's [83] calibration. Schiller and Finger [97] further suggest a temperature-dependent correction for the α_{TiO2} based on rhyolite-MELTS [86] or a general +70 °C upward correction of Ti-in-zircon temperatures (Figure 8) to account for the temperature underestimate. An ad hoc +70 °C correction raised the average Ti-in-zircon temperatures from the Alta and Little Cottonwood stocks above the range of T$_{Zr}$ temperatures (≥25 °C) and the mean of the Zr-in-titanite population interpreted as magmatic (see below). Although a stronger pressure dependence of 100 °C/GPa [87] has been suggested, the pressure variations possible in this shallow system would only serve to raise apparent temperatures an additional ~5–10 °C above the T$_{Zr}$ and mean Zr-in-titanite apparent temperatures (see the Data Repository File). Several observations further corroborated the Ti-in-zircon temperatures and suggest that zircons were not altered or were otherwise open systems following initial crystallization, including the low common-Pb content of most zircons, the near concordant and concordant nature of most of the U-Pb analyses, the dates that were consistent with the inclusion of zircon in titanite, the consistent intragrain isotopic dates and crystal growth zoning, and the oscillatory zoning as opposed to patchy or sector zoning. Persistent zirconium saturation is possible based on the bulk rock chemistry, agreement between the Ti-in-zircon and the Zr saturation thermometers, and the inherited zircons in both the Alta and Little Cottonwood stocks. Conversely, the resorbed cores and mantles in some zircons suggest that Zr saturation did not persist the entire time during the crystallization of the stocks. We conclude that accounting for a lower, non-unity α_{TiO2} in the original calibration [83] and propagating the uncertainty on the activity terms sufficiently reconciles the thermometry with likely zircon crystallization temperatures [96,97] and that the Ti-in-zircon apparent temperatures are likely accurate within the reported uncertainty.

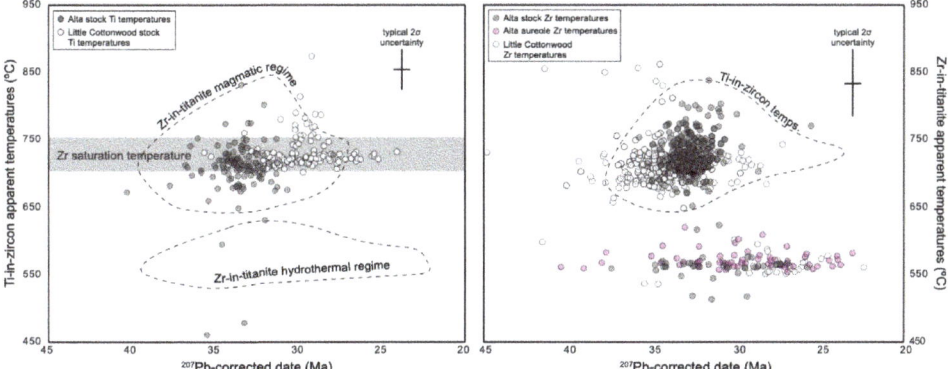

Figure 8. Plots of ^{207}Pb-corrected dates versus Ti-in-zircon (**left**) and Zr-in-titanite (**right**) apparent temperatures calculated separately for each sample (see the Data Repository File for thermometry inputs) and colored by lithologic unit (Alta = dark grey, Little Cottonwood = white, Alta endoskarn = purple). The crosses (+) are the data transformed +70 °C, as suggested by Schiller and Finger [97], to account for a low α_{TiO2}. The typical uncertainty bars are shown for each method but have been omitted from individual analyses for clarity. The Ti-in-zircon apparent temperatures calculated by Ferry and Watson [83] defined a unimodal population below the predicted zircon saturation temperature of ~725 °C (T_{Zr}; grey bar), while the Zr-in-titanite apparent temperatures defined a bimodal population that largely did not overlap the Ti thermometry data.

The Zr-in-titanite apparent temperatures formed two groups: (1) grains from both the Alta and Little Cottonwood stocks that recorded apparent temperatures ranging from ~650–800 °C with a mode at ~725 °C and were interpreted to have grown from a silicate melt, and (2) grains that recorded ~575 ± 50 °C conditions and were interpreted to have (re)crystallized in the presence of hydrothermal fluids in the solid state. These populations of Zr temperatures were consistent with the multiple textural populations of titanite observed in the Little Cottonwood and Alta stocks. The Zr-in-titanite thermometer is susceptible to similar sources of uncertainty as for the Ti-in-zircon thermometer discussed above, with a much stronger dependence on pressure. For the reasons previously outlined in the Materials and Methods and Discussion Sections, we interpreted the Zr-in-titanite temperatures as being accurate within the reported uncertainty. Samples included in the second group were LCS-02, 88-I-9, and 12-12-A2. The ranges of titanite dates from the two groups overlapped but the colder population was skewed slightly younger than the hotter population. This relationship was interpreted to record simultaneous titanite crystallization from silicate melt and fluid-mediated (re)crystallization of titanite in different parts of the system, with titanite (re)crystallization continuing after crystallization of the magmatic titanite had ceased.

4.2. Hydrothermal Permeability Structure through Time

The permeability structure surrounding intrusions fundamentally controls the process of fluid-infiltration-driven contact metamorphism [3,33,98]. The lack of pervasive hydrothermal titanite (re)crystallization in either the majority of the border phase or any of the central phase Alta stock samples suggests there was scarce infiltration of the hydrothermal fluids flowing through the abundant fractures and veins into unfractured volumes of the Alta stock [99] into the bulk of the non-fractured AS. Samples 88-I-9 and 12-12-A2 (Figure 4) were located at or proximal to the Alta stock wall-rock contact (0–40 m) and a recorded ≥11 Myr of titanite (re)crystallization, which is much longer than calcsilicate skarn sample 11-1 from the inner aureole (~8 Myr; ~37–29 Ma). These data suggest that the locus of hydrothermal infiltration migrated through time (Figure 9), and further suggests that infiltration in the Alta metamorphic aureole (for example, sample 11-1) and the calcsilicate and endoskarns adjacent to the Alta stock (sample 12-12-A2) reflected different and/or multiple stages of fluid infiltration during emplacement of Alta–Little Cottonwood magmas. A diachronous and perhaps stochastic process of infiltration may reflect a feedback between prograde mineral reactions and skarn mineralization, which both exploit pore space, and the permeability structure of the host rocks [33,100].

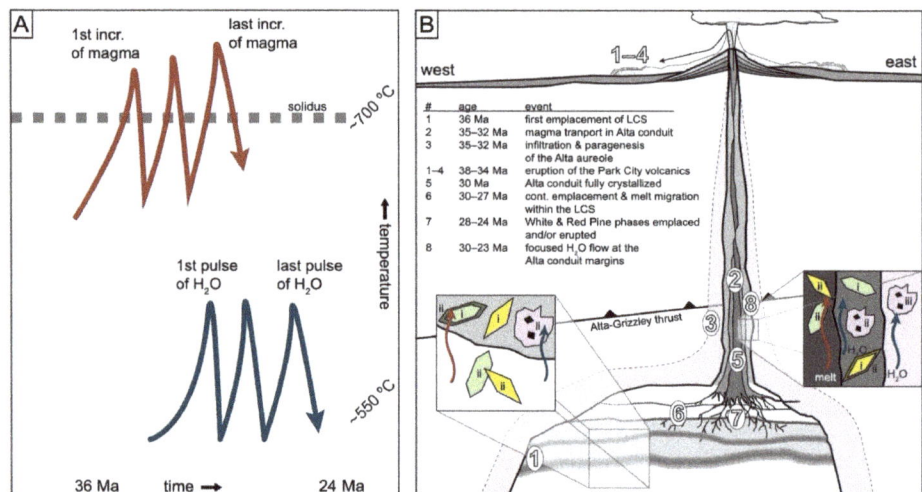

Figure 9. (**A**) Interpretation of the process behind the simultaneous (re)crystallization of titanite in both magmatic and hydrothermal processes suggested by the bimodal Zr-in-titanite thermometry. (**B**) A summary schematic illustrating the sequence of geologic events in the Alta stock–Little Cottonwood stock (AS–LCS) system that illustrates how incremental magmatism and pulsed hydrothermal activity could produce multiple populations (oldest = i, youngest = iii) of (re)crystallized titanite and zircon. The unit colors in (B) (dark grey = Alta, light grey = Ferguson stock, white = Little Cottonwood stock, and purple = Alta aureole) match Figures 1, 2, 4, 5, 7 and 8. The lithologic contacts and spatial extents of different units are diagrammatic and are meant to represent processes such as incremental magma emplacement, magma transport in a conduit, episodic hydrothermal infiltration, and protracted volcanic activity at the paleosurface.

4.3. Magma Accumulation vs. Eruptive Discharge

The total eruptive volumes of volcanic rocks related to the Alta and Little Cottonwood stocks are likely $\geq 10^3$ km^3 based on a minimum thickness of ~500 m [24] and a minimum outcrop area of ~1600 km^2 [101]. The areal exposure of the genetically related intrusive rocks is ~225 km^2, but the thicknesses of the intrusions are unknown. If the intrusions are assumed to be ~5–10 km thick [102], then the volume of the intrusive rocks (~1–2 × 10^3 km^3) is the same order of magnitude as the volume of erupted material. If this is the case, then the ~11 Myr intrusive duration implies a pluton growth rate that is 2–5× slower than the eruptive discharge from the system, which lasted only ~4.5 Myr (~36.5 to 32 Ma [23,24]). These rates are based on assumptions of the physical dimensions just presented and these estimates have large uncertainties and are therefore speculative. However, these very preliminary discharge estimates are consistent with other better-characterized systems [103,104] that suggest that magma discharge is a first-order control on the eruptibility of magma batches.

5. Conclusions

Petrochronology data indicate that pluton growth began in the structurally deepest parts of the Little Cottonwood stock at ~36 Ma. This roughly corresponds to the onset of volcanism [23,24,105] (Figures 9 and 10). From 36.5 to 31 Ma, most, if not all, of the Keetley/East Traverse volcanic sequence accumulated and the currently exposed level of the Alta stock grew, which we thus interpret to represent a conduit for magma to the earth's surface. The end of volcanism and the growth of the AS at ~31 Ma also correspond to the bulk of metamorphic titanite dates from the inner Alta aureole. However, the Little Cottonwood stock continued to grow until ~25 Ma, i.e., for several million years after the cessation of volcanism that exploited the Alta stock as a magma conduit.

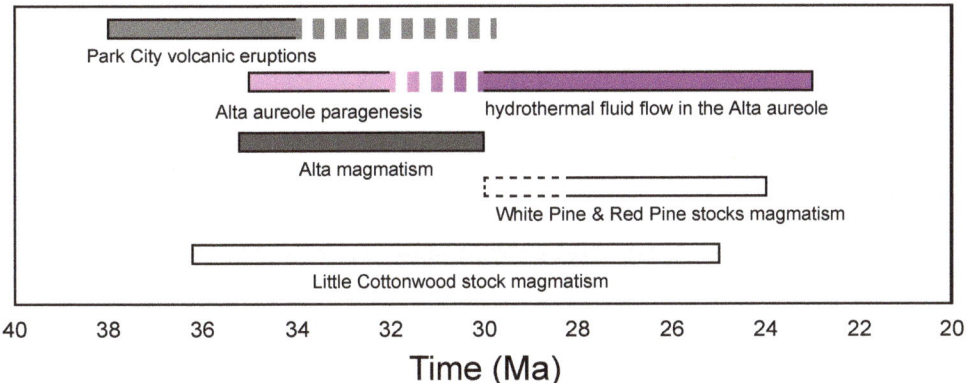

Figure 10. Timeline for the Alta–Little Cottonwood system based on preliminary U-Pb zircon and titanite petrochonology data. Sparse data (e.g., Park City volcanic units) and a lack of both zircon and titanite dates (e.g., Little Cottonwood stock) limits the interpretation and contributes significant uncertainty to the plotted durations. Defining the timing of these events is an ongoing research focus. The existing data also suggest these systems are episodic rather than continuous in their activity.

The isotopic dates and apparent temperatures defined a single population of zircon and two populations of titanite. The single population of apparent temperatures calculated by Ferry and Watson [83] Ti-in-zircon calibration were both consistent with the estimated T_{Zr} of the magma and the magmatic Zr-in-titanite apparent temperature population. The higher temperature population of titanites was interpreted to reflect crystallization from a silicate melt and the lower temperature population of titanites was interpreted to reflect growth, likely in the solid state, in the presence of hydrothermal fluid, possibly of magmatic origin; this is consistent with the range of titanite morphologies present in the rocks. The lower-temperature titanite group had a somewhat younger maximum age but the two populations overlapped in age by ~4 Myr. We thus interpreted the titanite dates to record temporally overlapping magmatic and hydrothermal regimes. However, the low-temperature population persisted to a much younger minimum age, indicating that hydrothermal activity outlasted crystallization of the Alta stock by several million years. Titanites from the southern margin of the Alta stock recorded hydrothermal (re)crystallization through this duration, suggesting the presence of a narrow hydrothermal conduit in the AS adjacent to its southern contact.

In summary, these relations suggest that the Alta stock served as a conduit for magma to the surface from 36–31 Ma, then as a conduit for hydrothermal fluids for another 6–8 Myr. After the magma conduit shut off at ~31 Ma, the Little Cottonwood stock continued to grow until at least 25 Ma and perhaps as late as 23 Ma.

The spatial coverage of present dates from the Little Cottonwood stock is insufficient to infer an emplacement pattern. We speculate that the youngest dates from the Little Cottonwood stock, which come from its structurally highest portions, may reflect upward melt migration and/or remobilization from younger, related intrusions.

Supplementary Materials: The following are available online at http://www.mdpi.com/2076-3263/10/4/129/s1, Figure S1: U-Pb Plots, Table S1: Petrochronology Data.

Author Contributions: Conceptualization: M.A.S., J.M.B., and J.R.B.; Data curation: M.A.S., C.W.F., and N.D.U.; Funding acquisition: M.A.S., J.M.B., and J.R.B.; Investigation: M.A.S., J.M.B., J.R.B., C.W.F., C.J.B., D.D.R., S.J.C., and N.D.U.; Methodology: M.A.S.; Visualization: M.A.S.; Writing—original draft: M.A.S., J.M.B., and J.R.B.; Writing—review and editing: M.A.S., J.M.B., J.R.B., C.W.F., C.J.B., D.D.R., S.J.C., and N.D.U. All authors have read and agreed to the published version of the manuscript.

Funding: This research was funded by the National Science Foundation, grant number 1853496.

Acknowledgments: We would like to acknowledge Allen Glazner, Drew Coleman, and Lukas Baumgartner for important discussions during the preparation of this manuscript. We would like to acknowledge Andrew Kylander-Clark and Diego Fernandez for assistance with the isotopic analysis, and thank the two reviewers for their constructive and helpful comments.

Conflicts of Interest: The authors declare no conflict of interest. The funders had no role in the design of the study; in the collection, analyses, or interpretation of data; in the writing of the manuscript, or in the decision to publish the results.

References

1. Annen, C. From Plutons to Magma Chambers: Thermal Constraints on the Accumulation of Eruptible Silicic Magma in the Upper Crust. *Earth Planet. Sci. Lett.* **2009**, *284*, 409–416. [CrossRef]
2. Watanabe, T.; Koyaguchi, T.; Seno, T. Tectonic Stress Controls on Ascent and Emplacement of Magmas. *J. Volcanol. Geotherm. Res.* **1999**, *91*, 65–78. [CrossRef]
3. Cui, X.; Nabelek, P.I.; Liu, M. Heat and Fluid Flow in Contact Metamorphic Aureoles with Layered and Transient Permeability, with Application to the Notch Peak Aureole, Utah. *J. Geophys. Res. Solid Earth* **2001**, *106*, 6477–6491. [CrossRef]
4. Balashov, V.N.; Yardley, B.W.D. Modeling Metamorphic Fluid Flow with Reaction-Compaction-Permeability Feedbacks. *Am. J. Sci.* **1998**, *298*, 441–470. [CrossRef]
5. Annen, C. Factors Affecting the Thickness of Thermal Aureoles. *Front. Earth Sci.* **2017**, *5*, 1–13. [CrossRef]
6. Reverdatto, V.V.; Sharapov, V.N.; Melamed, V.G. The Controls and Selected Peculiarities of the Origin of Contact Metamorphic Zonation. *Contrib. Mineral. Petrol.* **1970**, *29*, 310–337. [CrossRef]
7. Paterson, S.R.; Ardill, K.; Vernon, R.; Žák, J. A Review of Mesoscopic Magmatic Structures and Their Potential for Evaluating the Hypersolidus Evolution of Intrusive Complexes. *J. Struct. Geol.* **2019**, *125*, 134–147. [CrossRef]
8. Annen, C. Implications of Incremental Emplacement of Magma Bodies for Magma Differentiation, Thermal Aureole Dimensions and Plutonism-Volcanism Relationships. *Tectonophysics* **2011**, *500*, 3–10. [CrossRef]
9. Floess, D.; Baumgartner, L.P. Constraining Magmatic Fluxes through Thermal Modelling of Contact Metamorphism. *Geol. Soc. Spec. Publ.* **2015**, *422*, 41–56. [CrossRef]
10. Parry, W.T.; Bruhn, R.L. Fluid Inclusion Evidence for Minimum 11 Km Vertical Offset on the Wasatch Fault, Utah. *Geology* **1987**, *15*, 67–70. [CrossRef]
11. Constenius, K.N. Late Paleogene Extensional Collapse of the Cordilleran Foreland Fold and Thrust Belt. *Bull. Geol. Soc. Am.* **1996**, *108*, 20–39. [CrossRef]
12. Vogel, T.A.; Cambray, F.W.; Constenius, K.N. Origin and Emplacement of Igneous Rocks in the Central Wasatch Mountains, Utah. *Rocky Mt. Geol.* **2001**, *36*, 119–162. [CrossRef]
13. DeCelles, P.G.; Coogan, J.C. Regional Structure and Kinematic History of the Sevier Fold-and-Thrust Belt, Central Utah. *GSA Bull.* **2006**, *118*, 841–864. [CrossRef]
14. Copeland, P.; Currie, C.A.; Lawton, T.F.; Murphy, M.A. Location, Location, Location: The Variable Lifespan of the Laramide Orogeny. *Geology* **2017**, *45*, 223–226. [CrossRef]
15. Vogel, T.A.; Cambray, F.W.; Feher, L.; Constenius, K.N.; Group WIB Research. Petrochemistry and Emplacment History of the Wasatch Igneous Belt. In *Society of Economics Geologists Guidebook 29*, 2nd ed.; Society of Economic Geologists: Littleton, CO, USA, 1998; pp. 35–46.

16. Paulsen, T.; Marshak, S. Origin of the Uinta Recess, Sevier Fold-Thrust Belt, Utah: Influence of Basin Architecture on Fold-Thrust Belt Geometry. *Tectonophysics* **1999**, *312*, 203–216. [CrossRef]
17. Armstrong, P.A.; Taylor, A.R.; Ehlers, T.A. Is the Wasatch Fault Footwall (Utah, United States) Segmented over Million-Year Time Scales? *Geology* **2004**, *32*, 385–388. [CrossRef]
18. Parry, W.; Wilson, P.N.; Bruhn, R. Pore-Fluid Chemistry and Chemical Reactions on the Wasatch Normal Fault, Utah. *Geochim. Cosmochim. Acta* **1988**, *52*, 2053–2063. [CrossRef]
19. Armstrong, P.A.; Ehlers, T.A.; Chapman, D.S.; Farley, K.A.; Kamp, P.J.J. Exhumation of the Central Wasatch Mountains, Utah: 1. Patterns and Timing of Exhumation Deduced from Low-Temperature Thermochronology Data. *J. Geophys. Res. Solid Earth* **2003**, *108*. [CrossRef]
20. Ehlers, T.A.; Willett, S.D.; Armstrong, P.A.; Chapman, D.S. Exhumation of the Central Wasatch Mountains, Utah: 2. Thermokinematic Model of Exhumation, Erosion, and Thermochronometer Interpretation. *J. Geophys. Res. Solid Earth* **2003**, *108*. [CrossRef]
21. Friedrich, A.M.; Wernicke, B.P.; Niemi, N.A.; Bennett, R.A.; Davis, J.L. Comparison of Geodetic and Geologic Data from the Wasatch Region, Utah, and Implications for the Spectral Character of Earth Deformation at Periods of 10 to 10 Million Years. *J. Geophys. Res. Solid Earth* **2003**, *108*. [CrossRef]
22. John, D.A. Geologic Setting, Depths of Emplacement, and Regional Distribution of Fluid Inclusions in Intrusions of the Central Wasatch Mountains, Utah. *Econ. Geol.* **1989**, *84*, 386–409. [CrossRef]
23. Crittenden, M.D.; Stuckless, J.S.; Kistler, R.W.; Stern, T.W. Radiometric Dating of Intrusive Rocks in the Cottonwood Area, Utah. *US Geol. Surv. J. Res.* **1973**, *1*, 173–178.
24. Bromfield, C.S.; Erickson, A.J.; Haddadin, M.A.; Mehnert, H.H. Potassium-Argon Ages of Intrusion, Extrusion, and Associated Ore Deposits, Park City Mining District, Utah. *Econ. Geol.* **1977**, *72*, 837–848. [CrossRef]
25. Kowallis, B.J.; Ferguson, J.; Jorgensen, G.J. Uplift along the Salt Lake Segment of the Wasatch Fault from Apatite and Zircon Fission Track Dating in the Little Cottonwood Stock. *Int. J. Radiat. Appl. Instrum. Part D Nucl. Tracks Radiat. Meas.* **1990**, *17*, 325–329. [CrossRef]
26. Wison, J.C. Geology of the Alta Stock, Utah. Ph.D. Thesis, California Institute of Technology, Pasadena, CA, USA, 1961. [CrossRef]
27. Kemp, W.M. A Stable Isotope and Fluid Inclusion Study of the Contact Al(Fe)-Ca-Mg-Si Skarns in the Alta Stock Aureole, Alta, Utah. Ph.D. Thesis, University of Utah, Salt Lake City, UT, USA, 1985.
28. Baker, A.A.; Calkins, F.C.; Crittenden, M.D.; Bromfield, C.S. Geologic Map of the Brighton Quadrangle, Utah. US Geol. Surv. Geol. Quad. Map GQ-534; 1966. Available online: https://www.sciencebase.gov/catalog/item/4f4e4b0be4b07f02db69d912 (accessed on 2 April 2020).
29. Moore, J.N.; Kerrick, D.M. Equilibria in Siliceous Dolomites of the Alta Aureole, Utah. *Am. J. Sci.* **1976**, *276*, 502–524. [CrossRef]
30. Cook, S.J.; Bowman, J.R. Contact Metamorphism Surrounding the Alta Stock: Thermal Constraints and Evidence of Advective Heat Transport from Calcite + Dolomite Geothermometry. *Am. Mineral.* **1994**, *79*, 513–525.
31. Cook, S.J.; Bowman, J.R. Mineralogical Evidence for Fluid-Rock Interaction Accompanying Prograde Contact Metamorphism of Siliceous Dolomites: Alta Stock Aureole, Utah, USA. *J. Petrol.* **2000**, *41*, 739–757. [CrossRef]
32. Bowman, J.R.; Willett, S.D.; Cook, S.J. Oxygen Isotopic Transport and Exchange during Fluid Flow; One-Dimensional Models and Applications. *Am. J. Sci.* **1994**, *294*, 1–55. [CrossRef]
33. Cook, S.J.; Bowman, J.R.; Forster, C.B. Cook and Bowman 1997.Pdf. *Am. J. Sci.* **1997**, *297*, 1–55. [CrossRef]
34. Marchildon, N.; Dipple, G.M. Irregular Isograds, Reaction Instabilities, and the Evolution of Permeability during Metamorphism. *Geology* **1998**, *26*, 15–18. [CrossRef]
35. Wohlers, A.; Baumgartner, L.P. Melt Infiltration into Quartzite during Partial Melting in the Little Cottonwood Contact Aureole (UT, USA): Implication for Xenocryst Formation. *J. Metamorph. Geol.* **2013**, *31*, 301–312. [CrossRef]
36. Bankey, V.; Cueves, A.; Daniels, D.; Finn, C.A.; Hernandez, I. Digital Data Grids for the Magnetic Anomaly Map of North America. US Geol. Surv. Open File Rep. 414; 2002. Available online: https://pubs.usgs.gov/of/2002/ofr-02-414/ (accessed on 2 April 2020).
37. Biek, R.F. Interim Geologic Map of the Park City East Quadrangle, Summit and Wasatch Counties, Utah. Utah Geol. Surv. Open File Rep. 677; 2017. Available online: https://pubs.er.usgs.gov/publication/ofr6924 (accessed on 2 April 2020).
38. Bartley, J.M.; Coleman, D.S.; Glazner, A.F. Incremental Pluton Emplacement by Magmatic Crack-Seal. *Earth Environ. Sci. Trans. R. Soc. Edinb.* **2008**, *97*, 383–396. [CrossRef]

39. Hanson, S.L. *Mineralogy, Petrology, Geochemistry and Crystal Size Distribution of Tertiary Plutons of the Central Wasatch Mountains, Utah*; University of Utah: Salt Lake City, UT, USA, 1995.
40. Marsh, A.J.; Smith, R.K. The Oligocene Little Cottonwood Stock, Central Wasatch Mountains, Utah: An Example of Compositional Zoning by Side-Wall Fractional Crystallization of an Arc-Related Intrusion. *Mt. Geol.* **2005**, *34*, 83–95.
41. Crittenden, M.D. Geology of the Draper Quadrangle, Utah. US Geol. Surv. Geol. Quadrang. Map GQ-377 Scale 124,000; 1965. Available online: https://pubs.er.usgs.gov/publication/gq377 (accessed on 2 April 2020).
42. Johnson, B.R.; Glazner, A.F. Formation of K-Feldspar Megacrysts in Granodioritic Plutons by Thermal Cycling and Late-Stage Textural Coarsening. *Contrib. Mineral. Petrol.* **2010**, *159*, 599–619. [CrossRef]
43. Lundstrom, C.C. The Role of Thermal Migration and Low-Temperature Melt in Granitoid Formation: Can Granite Form without Rhyolitic Melt? *Int. Geol. Rev.* **2016**, *58*, 371–388. [CrossRef]
44. Whitney, J.A. The Origin of Granite: The Role and Source of Water in the Evolution of Granitic Magmas. *Spec. Pap. Geol. Soc. Am.* **1990**, *253*, 387–398. [CrossRef]
45. Higgins, M.D. Quantitative Petrological Evidence for the Origin of K-Feldspar Megacrysts in Dacites from Taapaca Volcano, Chile. *Contrib. Mineral. Petrol.* **2011**, *162*, 709–723. [CrossRef]
46. Glazner, A.F.; Johnson, B.R. Late Crystallization of K-Feldspar and the Paradox of Megacrystic Granites. *Contrib. Mineral. Petrol.* **2013**, *166*, 777–799. [CrossRef]
47. Johnson, B.W. *Oxygen Isotope, Cathodoluminescence, and Titanium in Quartz Geothermometry in the Alta Stock, UT: Geothermical Insights into Pluton Assembly and Early Cooling History*; University of Utah: Salt Lake City, UT, USA, 2009.
48. Ackerson, M.R.; Mysen, B.O.; Tailby, N.D.; Watson, E.B. Low-Temperature Crystallization of Granites and the Implications for Crustal Magmatism. *Nature* **2018**, *559*, 94–97. [CrossRef]
49. Stearns, M.A.; Hacker, B.R.; Ratschbacher, L.; Rutte, D.; Kylander-Clark, A.R.C. Titanite Petrochronology of the Pamir Gneiss Domes: Implications for Middle to Deep Crust Exhumation and Titanite Closure to Pb and Zr Diffusion. *Tectonics* **2015**, *34*, 784–802. [CrossRef]
50. Garber, J.M.; Hacker, B.R.; Kylander-Clark, A.R.C.; Stearns, M.; Seward, G. Controls on Trace Element Uptake in Metamorphic Titanite: Implications for Petrochronology. *J. Petrol.* **2017**, *58*, 1031–1057. [CrossRef]
51. Holder, R.M.; Hacker, B.R. Fluid-Driven Resetting of Titanite Following Ultrahigh-Temperature Metamorphism in Southern Madagascar. *Chem. Geol.* **2019**, *504*, 38–52. [CrossRef]
52. Bartley, J.M.; Glazner, A.F.; Coleman, D.S. Dike Intrusion and Deformation during Growth of the Half Dome Pluton, Yosemite National Park, California. *Geosphere* **2018**, *14*, 1283–1297. [CrossRef]
53. Smirnov, S.Z. ScienceDirect the Fluid Regime of Crystallization of Water-Saturated Granitic and Pegmatitic Magmas: A Physicochemical Analysis. *RGG* **2016**, *56*, 1292–1307. [CrossRef]
54. Kylander-Clark, A.R.C.; Hacker, B.R.; Cottle, J.M. Laser-Ablation Split-Stream ICP Petrochronology. *Chem. Geol.* **2013**, *345*, 99–112. [CrossRef]
55. Kylander-Clark, A.R.C. Petrochronology by Laser-Ablation Inductively Coupled Plasma Mass Spectrometry. *Rev. Mineral. Geochem.* **2017**, *83*, 183–198. [CrossRef]
56. Kohn, M.J. Titanite Petrochronology. *Rev. Mineral. Geochem.* **2017**, *83*, 419–441. [CrossRef]
57. Pupin, J.P. Zircon and Granite Petrology. *Contrib. Mineral. Petrol.* **1980**, *73*, 207–220. [CrossRef]
58. Frost, B.R.; Chamberlain, K.R.; Schumacher, J.C. Sphene (Titanite): Phase Relations and Role as a Geochronometer. *Chem. Geol.* **2001**, *172*, 131–148. [CrossRef]
59. Watson, E.B.; Harrison, T.M. Zircon Saturation Revisited: Temperature and Composition Effects in a Variety of Crustal Magma Types. *Earth Planet. Sci. Lett.* **1983**, *64*, 295–304. [CrossRef]
60. Boehnke, P.; Watson, E.B.; Trail, D.; Harrison, T.M.; Schmitt, A.K. Zircon Saturation Re-Revisited. *Chem. Geol.* **2013**, *351*, 324–334. [CrossRef]
61. Miller, C.F.; McDowell, S.M.; Mapes, R.W. Hot and Cold Granites: Implications of Zircon Saturation Temperatures and Preservation of Inheritance. *Geology* **2003**, *31*, 529–532. [CrossRef]
62. Schaltegger, U.; Brack, P.; Ovtcharova, M.; Peytcheva, I.; Schoene, B.; Stracke, A.; Marocchi, M.; Bargossi, G.M. Zircon and Titanite Recording 1.5 Million Years of Magma Accretion, Crystallization and Initial Cooling in a Composite Pluton (Southern Adamello Batholith, Northern Italy). *Earth Planet. Sci. Lett.* **2009**, *286*, 208–218. [CrossRef]
63. Troitzsch, U.; Ellis, D.J. Thermodynamic Properties and Stability of AlF-Bearing Titanite $CaTiOSiO_4$–$CaAlFSiO_4$. *Contrib. Mineral. Petrol.* **2002**, *142*, 543–563. [CrossRef]

64. Hemley, J.J.; Meyer, C.; Hodgson, C.J.; Thatcher, A.B. Sulfide Solubilities in Alteration-Controlled Systems. *Science* **1967**, *158*, 1580–1582. [CrossRef]
65. Markl, G.; Piazolo, S. Stability of High-Al Titanite from Low-Pressure Calcsilicates in Light of Fluid and Host-Rock Composition. *Am. Mineral.* **1999**, *84*, 37–47. [CrossRef]
66. Aleksandrov, S.M.; Troneva, M.A. Composition, Mineral Assemblages, and Genesis of Titanite and Malayaite in Skarns. *Geochem. Int.* **2007**. [CrossRef]
67. Watters, W.A. Petrography and Genesis of a Wollastonite Body and Its Associated Rocks at Holyoake Valley, Nelson, New Zealand. *N. Z. J. Geol. Geophys.* **1995**, *38*, 315–323. [CrossRef]
68. Ferry, J.M.; Wing, B.A.; Rumble, D. Formation of Wollastonite by Chemically Reactive Fluid Flow during Contact Metamorphism, Mt. Morrison Pendant, Sierra Nevada, California, USA. *J. Petrol.* **2001**, *42*, 1705–1728. [CrossRef]
69. Woodford, D.T.; Sisson, V.B.; Leeman, W.P. Boron Metasomatism of the Alta Stock Contact Aureole, Utah: Evidence from Borates, Mineral Chemistry, and Geochemistry. *Am. Mineral.* **2001**, *86*, 513–533. [CrossRef]
70. Stearns, M.A.; Cottle, J.M.; Hacker, B.R.; Kylander-Clark, A.R.C. Extracting Thermal Histories from the Near-Rim Zoning in Titanite Using Coupled U-Pb and Trace-Element Depth Profiles by Single-Shot Laser-Ablation Split Stream (SS-LASS) ICP-MS. *Chem. Geol.* **2016**, *422*, 13–24. [CrossRef]
71. Eggins, S.M.; Kinsley, L.P.J.; Shelley, J.M.G. Deposition and Element Fractionation Processes during Atmospheric Pressure Laser Sampling for Analysis by ICP-MS. *Appl. Surf. Sci.* **1998**, *127*, 278–286. [CrossRef]
72. Aleinikoff, J.N.; Wintsch, R.P.; Fanning, C.M.; Dorais, M.J. U–Pb Geochronology of Zircon and Polygenetic Titanite from the Glastonbury Complex, Connecticut, USA: An Integrated SEM, EMPA, TIMS, and SHRIMP Study. *Chem. Geol.* **2002**, *188*, 125–147. [CrossRef]
73. Wiedenbeck, M.; Allé, P.; Corfu, F.; Griffin, W.L.; Meier, M.; Oberli, F.; von Quadt, A.; Roddick, J.C.; Spiegel, W. Three Natural Zircon Standards for U-Th-Pb, Lu-Hf, Trace Element and REE Analyses. *Geostand. Geoanal. Res.* **1995**, *19*, 1–23. [CrossRef]
74. Wiedenbeck, M.; Hanchar, J.M.; Peck, W.H.; Sylvester, P.; Valley, J.; Whitehouse, M.; Kronz, A.; Morishita, Y.; Nasdala, L.; Fiebig, J.; et al. Further Characterisation of the 91,500 Zircon Crystal. *Geostand. Geoanal. Res.* **2004**, *28*, 9–39. [CrossRef]
75. Mazdab, F.K. Characterization of Flux-Grown Trace-Element-Doped Titanite Using the High-Mass-Resolution Ion Microprobe (SHRIMP-RG). *Can. Mineral.* **2009**, *47*, 813–831. [CrossRef]
76. Spandler, C.; Hammerli, J.; Sha, P.; Hilbert-Wolf, H.; Hu, Y.; Roberts, E.; Schmitz, M. MKED1: A New Titanite Standard for in Situ Analysis of Sm-Nd Isotopes and U-Pb Geochronology. *Chem. Geol.* **2016**, *425*, 110–126. [CrossRef]
77. Sláma, J.; Košler, J.; Condon, D.J.; Crowley, J.L.; Gerdes, A.; Hanchar, J.M.; Horstwood, M.S.A.; Morris, G.A.; Nasdala, L.; Norberg, N.; et al. Plešovice Zircon–A New Natural Reference Material for U-Pb and Hf Isotopic Microanalysis. *Chem. Geol.* **2008**, *249*, 1–35. [CrossRef]
78. Petrus, J.A.; Kamber, B.S. VizualAge: A Novel Approach to Laser Ablation ICP-MS U-Pb Geochronology Data Reduction. *Geostand. Geoanal. Res.* **2012**, *36*, 247–270. [CrossRef]
79. Paton, C.; Hellstrom, J.; Paul, B.; Woodhead, J.; Hergt, J. Iolite: Freeware for the Visualisation and Processing of Mass Spectrometric Data. *J. Anal. At. Spectrom.* **2011**, *26*, 2508–2518. [CrossRef]
80. Ludwig, K.R. Isoplot/Ex: A Geochronological Toolkit for Microsoft Excel. *Berkeley Geochronol. Cent. Spec. Publ.* **2001**, *1a*, 55.
81. Vermeesch, P. Geoscience Frontiers IsoplotR: A Free and Open Toolbox for Geochronology. *Geosci. Front.* **2018**, *9*, 1479–1493. [CrossRef]
82. Ludwig, K.R. On the Treatment of Concordant Uranium-Lead Ages. *Geochim. Cosmochim. Acta* **1998**, *62*, 665–676. [CrossRef]
83. Ferry, J.M.; Watson, E.B. New Thermodynamic Models and Revised Calibrations for the Ti-in-Zircon and Zr-in-Rutile Thermometers. *Contrib. Mineral. Petrol.* **2007**, *154*, 429–437. [CrossRef]
84. Hayden, L.A.; Watson, E.B.; Wark, D.A. A Thermobarometer for Sphene (Titanite). *Contrib. Mineral. Petrol.* **2008**, *155*, 529–540. [CrossRef]
85. Ghiorso, M.S.; Sack, O. Fe-Ti Oxide Geothermometry: Thermodynamic Formulation and the Estimation of Intensive Variables in Silicic Magmas. *Contrib. Mineral. Petrol.* **1991**, *108*, 485–510. [CrossRef]

86. Gualda, G.A.R.; Ghiorso, M.S.; Lemons, R.V.; Carley, T.L. Rhyolite-MELTS: A Modified Calibration of MELTS Optimized for Silica-Rich, Fluid-Bearing Magmatic Systems. *J. Petrol.* **2012**, *53*, 875–890. [CrossRef]
87. Ferriss, E.D.A.; Essene, E.J.; Becker, U. Computational Study of the Effect of Pressure on the Ti-in-Zircon Geothermometer. *Eur. J. Mineral.* **2008**, *20*, 745–755. [CrossRef]
88. Wendt, I.; Carl, C. The Statistical Distribution of the Mean Squared Weighted Deviation. *Chem. Geol. Isot. Geosci. Sect.* **1991**, *86*, 275–285. [CrossRef]
89. Storey, C.D.; Jeffries, T.E.; Smith, M. Common Lead-Corrected Laser Ablation ICP–MS U–Pb Systematics and Geochronology of Titanite. *Chem. Geol.* **2006**, *227*, 37–52. [CrossRef]
90. Trail, D.; Bruce Watson, E.; Tailby, N.D. Ce and Eu Anomalies in Zircon as Proxies for the Oxidation State of Magmas. *Geochim. Cosmochim. Acta* **2012**, *97*, 70–87. [CrossRef]
91. Zhong, S.; Seltmann, R.; Qu, H.; Song, Y. Characterization of the Zircon Ce Anomaly for Estimation of Oxidation State of Magmas: A Revised Ce/Ce* Method. *Mineral. Petrol.* **2019**, *113*, 755–763. [CrossRef]
92. Piccoli, P.; Candela, P.; Rivers, M. Interpreting Magmatic Processes from Accessory Phases: Titanite—A Small-Scale Recorder of Large-Scale Processes. *Earth Environ. Sci. Trans. R. Soc. Edinb.* **2000**, *91*, 257–267. [CrossRef]
93. Shtukenberg, A.G.; Punin, Y.O.; Artamonova, O.I. Effect of Crystal Composition and Growth Rate on Sector Zoning in Solid Solutions Grown from Aqueous Solutions. *Mineral. Mag.* **2009**, *73*, 385–398. [CrossRef]
94. Piwinskii, A.J. Experimental Studies of Igneous Rock Series Central Sierra Nevada Batholith, California. *J. Geol.* **1968**, *76*, 548–570. [CrossRef]
95. Naney, M.T. Phase Equilibria of Rock-Forming Ferromagnesian Silicates in Granitic Systems. *Am. J. Sci.* **1983**, *283*, 993–1033. [CrossRef]
96. Fu, B.; Page, F.Z.; Cavosie, A.J.; Fournelle, J.; Kita, N.T.; Lackey, J.S.; Wilde, S.A.; Valley, J.W. Ti-in-Zircon Thermometry: Applications and Limitations. *Contrib. Mineral. Petrol.* **2008**, *156*, 197–215. [CrossRef]
97. Schiller, D.; Finger, F. Application of Ti-in-zircon thermometry to granite studies: Problems and possible solutions. *Contrib. Mineral. Petrol.* **2019**, *174*, 51. [CrossRef]
98. Gerdes, M.L.; Baumgartner, L.P.; Person, M.; Rumble, D. One-and Two-Dimensional Models of Fluid Flow and Stable Isotope Exchange at an Outcrop in the Adamello Contact Aureole, Southern Alps, Italy. *Am. Mineral.* **1995**, *80*, 1004–1019. [CrossRef]
99. John, D.A. *Evolution of Hydrothermal Fluids in the Alta Stock, Central Wasatch Mountains, Utah*; US Geological Survey: Reston, VA, USA, 1991; Volume 1977.
100. Brooks Hanson, R. The Hydrodynamics of Contact Metamorphism. *Geol. Soc. Am. Bull.* **1995**, *107*, 595–611. [CrossRef]
101. Hintze, L.F. Geologic Map of Utah. Utah Geol. Mineral. Surv. Map A-1 Scale 1,500,000. 1980. Available online: https://mrdata.usgs.gov/geology/state/state.php?state=UT (accessed on 2 April 2020).
102. Davis, J.W.J.W.; Coleman, D.S.D.S.; Gracely, J.T.J.T.; Gaschnig, R.; Stearns, M.A. Magma Accumulation Rates and Thermal Histories of Plutons of the Sierra Nevada Batholith, CA. *Contrib. Mineral. Petrol.* **2012**, *163*, 449–465. [CrossRef]
103. White, S.M.; Crisp, J.A.; Spera, F.J. Long-Term Volumetric Eruption Rates and Magma Budgets. *Geochem. Geophys. Geosyst.* **2006**, *7*. [CrossRef]
104. Schöpa, A.; Annen, C. The Effects of Magma Flux Variations on the Formation and Lifetime of Large Silicic Magma Chambers. *J. Geophys. Res. Solid Earth* **2013**, *118*, 926–942. [CrossRef]
105. Nelson, M.E. Age and Stratigraphic Relations of the Fowkes Formation, Eocene, of Southwestern Wyoming and Northeastern Utah. *Rocky Mt. Geol.* **1973**, *12*, 27–31.

© 2020 by the authors. Licensee MDPI, Basel, Switzerland. This article is an open access article distributed under the terms and conditions of the Creative Commons Attribution (CC BY) license (http://creativecommons.org/licenses/by/4.0/).

Article

The Granite Aqueduct and Autometamorphism of Plutons

John M. Bartley [1],*, Allen F. Glazner [2], Michael A. Stearns [3] and Drew S. Coleman [2]

[1] Department of Geology and Geophysics, University of Utah, Salt Lake City, UT 84112-0101, USA
[2] Department of Geological Sciences, University of North Carolina, Chapel Hill, North Carolina, NC 27599, USA; afg@unc.edu (A.F.G.); dcoleman@email.unc.edu (D.S.C.)
[3] Department of Earth Science, Utah Valley University, Orem, UT 84058, USA; mstearns@uvu.edu
* Correspondence: john.bartley@utah.edu

Received: 29 February 2020; Accepted: 8 April 2020; Published: 10 April 2020

Abstract: Ian Carmichael wrote of an "andesite aqueduct" that conveys vast amounts of water from the magma source region of a subduction zone to the Earth's surface. Diverse observations indicate that subduction zone magmas contain 5 wt % or more H_2O. Most of the water is released from crystallizing intrusions to play a central role in contact metamorphism and the genesis of ore deposits, but it also has important effects on the plutonic rocks themselves. Many plutons were constructed incrementally from the top down over million-year time scales. Early-formed increments are wall rocks to later increments; heat and water released as each increment crystallizes pass through older increments before exiting the pluton. The water ascends via multiple pathways. Hydrothermal veins record ascent via fracture conduits. Pipe-like conduits in Yosemite National Park, California, are located in or near aplite–pegmatite dikes, which themselves are products of hydrous late-stage magmatic liquids. Pervasive grain-boundary infiltration is recorded by fluid-mediated subsolidus modification of mineral compositions and textures. The flood of magmatic water carries a large fraction of the total thermal energy of the magma and transmits that energy much more rapidly than conduction, thus enhancing the fluctuating postemplacement thermal histories that result from incremental pluton growth. The effects of water released by subduction zone magmas are central not only to metamorphism and mineralization of surrounding rocks, but also to the petrology and the thermal history of the plutons themselves.

Keywords: incremental intrusion; hydrothermal fluid; microstructure; dissolution; precipitation; textural coarsening

1. Introduction

It has long been appreciated that water released by arc plutons plays a central role in the genesis of ore deposits as well as in contact metamorphism. Carmichael [1] used phase equilibria and thermodynamic arguments to estimate the water contents of volcanic arc magmas, and concluded that subduction-related intermediate magmas typically contain several percent water. He also estimated the plutonic/volcanic ratio in arcs to be approximately 5 and thus suggested that more than 80% of the water carried upward by arc magmas is released beneath the Earth's surface by crystallization of intrusive bodies. Although a small proportion of the water released by subsurface crystallization is fixed in hydrous minerals, most must be released to the surface at volcanic vents and via diffuse flow.

Effects of the prodigious amount of water carried by arc magmas on the plutonic rocks themselves have generally been underappreciated. Magma bodies solidify predominantly from the top down and from the outside in, and geochronology indicates that granitic plutons commonly grow incrementally by downward stacking of sheets (Figure 1; [2,3]). The majority of the water released by crystallizing magma therefore must pass through already solidified plutonic rock by some combination of fracture flow, grain-boundary flow, and diffusion before it encounters wall rocks (Figure 2). In this paper, we argue that passage of this water promotes textural modification and re-equilibration of minerals to lower temperatures via dissolution–precipitation and net-transfer reactions, and that such postmagmatic modification is ubiquitous and a key element in understanding the textural development of granitic rocks and the assembly of granitic plutons.

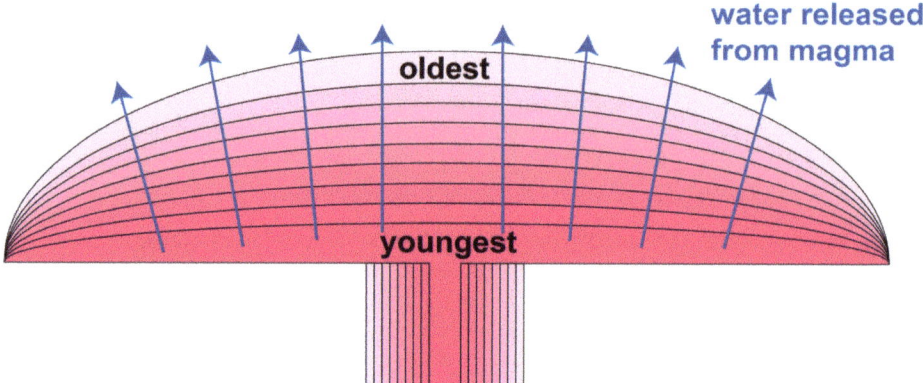

Figure 1. Schematic diagram illustrating incremental pluton growth by downward stacking of intrusive increments. Space for each increment is made by opening of a subhorizontal crack and accommodated by some combination of roof uplift and floor subsidence. Arrows represent migrating water released by crystallization. Buoyancy will cause much of the water to ascend through overlying older increments but migration paths will be strongly affected by the spatial pattern of permeability.

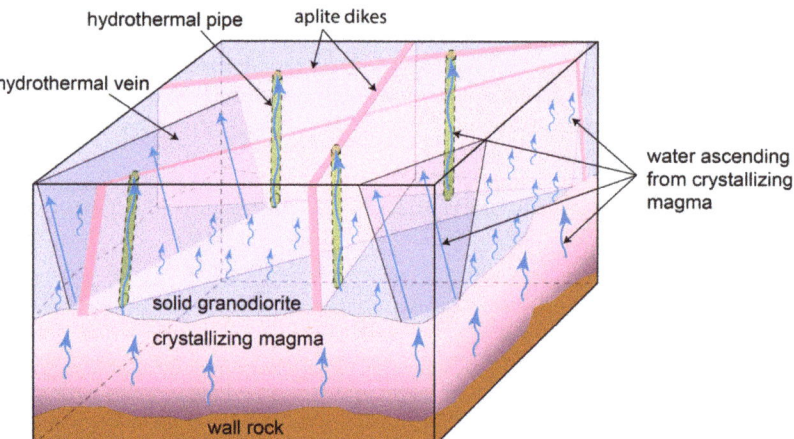

Figure 2. Schematic diagram illustrating fluid ascent paths in a growing pluton. The pluton is assumed to grow and to crystallize downward from its roof (not shown). Aplite dikes are fed by felsic melt that accumulates at the top of the active magma body [4,5].

2. Amount of Water in Arc Magmas and Arc Rocks

Several lines of evidence indicate that arc magmas typically contain several wt % H_2O. Sisson and Layne [6] reported water contents up to approximately 6 wt % in glass inclusions from mafic arc magmas. Measured water contents of 134 melt inclusions in arc rocks ranging from basalt to dacite [7] average 3.7 wt %; compilation for basalts and basaltic andesites averages somewhat less, around 2.5 wt % [8]. Plank et al. [9] reported an average of about 4 wt % H_2O for mafic arc magmas. Water contents of inclusions do not directly measure the content in the magma owing to degassing before entrapment, crystallization after entrapment, and other phenomena, so these measurements are not definitive. Estimates based on phase equilibria (e.g., water contents needed to reproduce a given phase assemblage) lead to significantly higher estimated water contents of 6 wt % to greater than 8 wt % (e.g., [1,10,11]).

Water contents of rocks crystallized from arc magmas are significantly lower than the estimated magmatic compositions. Granites and granodiorites from the EarthChem database have median H_2O^+ contents of 0.6 wt % (Figure 3). Given the magma water contents above and these data, for the purposes of argument we assume that typical intermediate arc magmas release 5 wt % H_2O upon full crystallization.

Five wt % H_2O is a surprisingly large volume. If 1 kg of magma releases 5 wt % H_2O while crystallizing to a rock with density 2700 kg/m^3, the water, if condensed, has roughly 15% of the volume of the crystallized rock. Consider an end-member scenario of vertical ascent of 100% of the magmatic water released from crystallization of a 3 km thick stack of sheets emplaced in top-down fashion. Roughly 375,000 kg of magmatic H_2O per square meter must ascend through the roof of the pluton—enough to form a layer of condensed water 450 m thick. Cumulative water/rock ratio will decrease passing downward, but the substantial majority of the pluton nonetheless must have interacted with a large volume of water released from deeper magma.

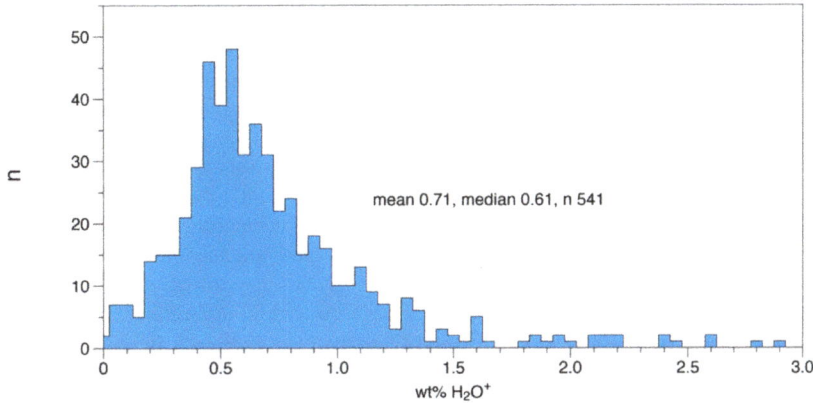

Figure 3. Weight percent H_2O^+ in rocks labeled as granite or granodiorite, with determinations by USGS rapid rock, wet chemistry, or gravimetry, in the EarthChem database.

3. Incremental Pluton Growth

It is now widely accepted that large plutons commonly grow incrementally over million-year time scales (e.g., [2,12,13]). The most compelling support for incremental growth comes from geochronological results that record growth durations far longer than durations indicated by thermal modeling of plutons emplaced as single magma bodies (e.g., [14,15]). Field evidence from some plutons corroborates incremental growth. Plutons that grew by magmatic crack-seal [16] can contain arrays of wall-rock inclusions that are concordant with wall-rock structure adjacent to the pluton (e.g., [14,17,18]). The Half Dome Granodiorite of the Tuolumne Intrusive Suite in Yosemite National Park, California, contains km-scale cyclical variations in composition that are interpreted to record episodic freezing owing to waning of the magma flux during incremental growth [4]. Incremental growth of the felsic parts of Half Dome compositional cycles is indicated by multiple generations of crosscutting dikes that range in composition from granodiorite to leucogranite [5].

4. Downward Growth of Plutons

Spatial patterns of U-Pb zircon dates in the Tuolumne Intrusive Suite [2] and field and geochronologic evidence from other examples (e.g., [19–23]) indicate that large plutons commonly grow by downward stacking of intrusive sheets. In the Tuolumne Intrusive Suite, this pattern applies both to the relations between major map units (Kuna Crest, Half Dome, Cathedral Peak) and to the internal structure of the Half Dome, where crosscutting relations in the field confirm that each deeper compositional cycle is younger than the overlying one [4]. Downward stacking of intrusive increments requires that water released by crystallization of younger increments ascends through older increments that already have crystallized before encountering wall rocks. Some of the water will likely move outward as well as upward, depending on permeability structure. A horizontal fluid conduit overlain by a flow barrier can force a significant amount of water to flow outward and form a wide hydrothermal aureole, as is seen around the Alta Stock, Utah [24,25]. However, even in such an example, it is likely that fluid is mainly transported upward toward the surface [13].

5. Fluid Transport Paths

Field and petrographic observations indicate that aqueous fluid released from crystallizing magma ascends via multiple paths (Figure 2). Perhaps the most obvious paths are fracture conduits which in a pluton are typically cooling joints where the infiltrating fluid leaves behind mineral precipitates to form hydrothermal veins (Figure 4a,b). Fractures undoubtedly are efficient fluid conduits and thus represent an important transport mode, but sealing of fractures by mineral precipitates gives them a limited lifetime unless deformation regenerates fracture permeability.

Focused fluid ascent also occurs via pipe-like conduits. In the Tuolumne Intrusive Suite, hydrothermal alteration is concentrated along pipes that range in diameter from a few decimeters to a meter or more (Figure 4c,d). The pipes are spatially associated with aplite and/or pegmatite dikes, in some instances confined to a dike and in others extending into adjacent granodiorite. The pipes are easily recognized in the field because they weather preferentially to form pits and even tunnels. Lines of pits along felsic dikes are common and thus, in locations where such dikes are abundant, pipe-like conduits are undoubtedly important fluid transport paths (Figure 4c,d). We do not yet know the nature of the porosity that governs formation of such hydrothermal pipes. We speculate that pipe formation involves open-system fluid–rock chemical reactions that reduce the volume of solid phases, increasing permeability where reactions take place and producing a positive feedback loop that further focuses fluid infiltration.

Not all fluid ascent through a pluton is focused along fractures and pipes, however. Hydrothermal veins are surrounded by haloes of altered rock that are visible in the field (Figure 4a,b). The haloes indicate that fluid penetrated into mesoscopically unfractured rock, apparently along microcracks and grain boundaries. Diffuse infiltration is locally reflected in the formation of endoskarn, that is, granitic rock in which primary texture persists but primary minerals have been extensively replaced by minerals such as albite, epidote, chlorite, and titanite (e.g., [13]). A key question, however, is whether grain-scale fluid infiltration is more pervasive but leaves behind a more subtle record than the bleached zones in Figure 3. Detailed petrography indicates that this is the case.

Thin rims or selvages of albite of near-end-member composition are commonly present along feldspar grain boundaries in granitic rocks (Figure 5). Rogers [26] attributed the albite to precipitation from an extremely fractionated late-stage intergranular melt. However, this origin is excluded by ternary feldspar phase equilibria which indicate that Na-rich feldspar precipitated from minimum-melt granite is anorthoclase rather than albite [27]. Phillips [28] attributed the albite to exsolution from primary igneous feldspars. However, the albite is found exclusively at contacts between plagioclase and K-feldspar or between K-feldspar grains. If exsolution were the process, albite should also be present at contacts between feldspar and other minerals such as quartz, but this is not observed. Moreover, perthite in which Na-rich and K-rich phases are intimately intermixed is the common product of K-feldspar exsolution; migration of exsolved albite to the external surface of a K-feldspar grain would be highly anomalous.

Figure 4. Field evidence of focused fluid transport in granitic plutons. (**a**) Fracture filled with hydrothermal epidote; diorite near Mack Lake, Sierra Nevada, California, USA. Note bleached zone around the vein. (**b**) Epidote-filled fracture network and surrounding bleached zone, Piute Creek, Sierra Nevada, California. (**c**) Pits weathered into large hydrothermal pipes, in and adjacent to felsic dike, Yosemite National Park, USA. (**d**) Line of pits weathered into hydrothermal pipes along contact of felsic dike, Yosemite National Park, USA (compass with 5 cm ring indicates scale).

The spatial distribution of albite thus is more consistent with O'Neil and Taylor's [29] inference that dissolution–precipitation is the likely mechanism for down-temperature re-equilibration of plutonic feldspars. Indeed, mineralogical, kinetic, isotopic, and microstructural evidence indicates that dissolution–precipitation is a widespread mechanism by which mineralogical changes take place in diverse geologic settings [30]. Localization of albite exclusively along feldspar–feldspar grain boundaries suggests interface-coupled precipitation of albite from a grain-boundary fluid (i.e., precipitation of albite occurred where a feldspar grain boundary provided a favorable nucleation surface). More speculatively, plagioclase-K-feldspar contacts may have been particularly favored for albite precipitation because albite-plagioclase and albite-K-feldspar contacts have lower surface energies than plagioclase-K-feldspar contacts [31].

Other microstructural observations from granitic rocks also support the importance of modification by interaction with a grain-boundary fluid. Cathodoluminescence (CL) imaging of feldspar in leucogranite from the Tuolumne Intrusive Suite in Yosemite (Figure 5b) reveals patchy and irregular internal zoning that is invisible in backscattered electron (BSE) imaging as well as in optical petrography. Unlike in quartz in which CL intensity variations largely reflect Ti content (e.g., [32]), the multiple trace-element activators that govern CL in feldspar are difficult to identify. However, the CL variations seen in Figure 5b,d indicate that different parts of a single crystal differ in their trace-element content and grew under different conditions. Such CL images are interpreted to record fracturing and healing of primary igneous grains that, if observed only optically or in BSE images, would likely be interpreted as unmodified after magmatic crystallization. Black spots on the BSE image in Figure 5a are micropores that probably reflect incomplete refilling of fractures [33]. Valley and Graham [32] presented similar CL images of quartz from granite on the Isle of Skye. Coupled with in situ oxygen isotopic analyses, these textures demonstrate that " ... the flow of hydrothermal fluids [that mediated dissolution–precipitation of quartz] was heterogeneous, anisotropic and crack controlled." Johnson et al. [34] imaged similar quartz textures in the Alta Granodiorite, Utah (Figure 5d), and also found that Ti contents of the quartz range down to values indicative of subsolidus crystallization.

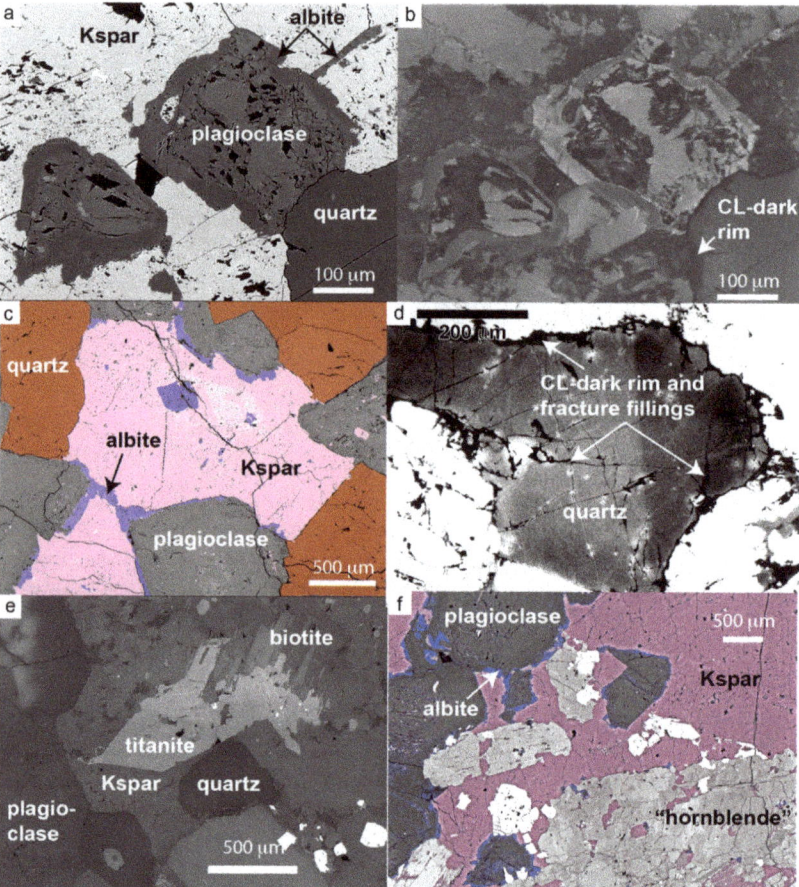

Figure 5. Backscattered electron (BSE) and cathodoluminescence (CL) images showing evidence of grain-scale modification by fluid infiltration. (**a**) BSE image of leucogranite, Tuolumne Intrusive Suite. BSE-dark albite selvages everywhere separate plagioclase from K-feldspar. (**b**) CL image of the same area as (**a**). Irregular patchy zoning of both feldspars reflects fracture, infiltration, and healing. The CL-dark (Ti-poor, low-temperature) rim on quartz reflects precipitation from grain-boundary fluid. (**c**) False-colored BSE image of Half Dome Granodiorite, Tuolumne Intrusive Suite, again showing albite along plagioclase-K-feldspar grain boundaries and locally within K-feldspar. (**d**) CL image of Alta Granodiorite, Utah [34]. CL-dark quartz rim and fracture fillings record precipitation from infiltrating fluid. (**e**) BSE image of Alta Granodiorite. Cores of euhedral titanite grains yield U-Pb dates of 36-31 Ma in agreement with U-Pb zircon dates from the same samples, whereas rims of euhedral grains and anhedral titanite intergrown with biotite give U-Pb dates as young as 23 Ma [13]. See text and reference [13] for further details. (**f**) False-colored BSE image of Half Dome Granodiorite. In addition to albite selvages, this image shows patchy zoning of "hornblende", which actually ranges in composition from hornblende to actinolite and records greenschist facies replacement (see text and ref. [35] for petrological details).

Textures seen in Figure 5a–d record interaction between minerals and fluids both at grain boundaries and in transgranular cracks. The micromechanics of grain-scale fractures are beyond the scope of this paper, but grain boundaries are mechanically favorable sites for cracks to form, both because they are weak and because differences in elastic properties across them result in stress concentrations. In rocks as inherently impermeable as granite, it is very likely that infiltration along grain boundaries was in cracks rather than in primary grain-boundary pore spaces of the sort seen in sandstone, for example. However, the grain-boundary stress concentrations also can produce transgranular fractures, particularly in minerals that contain cleavage planes. We therefore interpret the microstructures in Figure 5a–d to reflect crack-controlled fluid infiltration.

Textures, U-Pb dates, and Zr contents of titanite in the Alta and Little Cottonwood stocks reinforce the inference that the Alta Granodiorite was pervasively modified by an infiltrating aqueous fluid (Figure 5e; see [13] for full details). The cores of euhedral titanite grains yield in situ U-Pb dates that range from ca. 36 to 31 Ma, in agreement with U-Pb zircon dates from the same samples. Zirconium concentrations, determined simultaneously with U-Pb dates by split-stream analysis, indicate crystallization at approximately 650 °C (i.e., near the nominal granite solidus). In contrast, rims of these grains, as well as granular titanite that forms rims on Fe oxides and is intergrown with biotite, yield dates as young as 23 Ma and subsolidus Zr temperatures around 550 °C. We suggest that postmagmatic titanite in the Alta stock reflects reaction with infiltrating fluid released by crystallization of the underlying Little Cottonwood stock, which is a deeper part of the same magmatic system and grew incrementally through the time span recorded by titanite in the Alta stock [13].

Hornblende phenocrysts in the Half Dome Granodiorite commonly have been extensively modified by an infiltrating fluid under greenschist facies conditions [35]. Although euhedral and apparently intact when viewed in hand specimen, the patchy zoning seen in Figure 4f reflects a compositional range from actinolite to magmatic hornblende. The phenocrysts thus are pseudomorphs of the primary igneous amphibole. The phenocrysts host up to 50% inclusions of every mineral found in the rock including hydrous phases such as chlorite and epidote. Analyses of entire pseudomorphs by X-ray fluorescence yield a typical magmatic hornblende. Challener and Glazner [34] thus inferred replacement of primary magmatic hornblende that was isochemical except for the addition of water.

In summary, evidence from phase compositions and microstructures indicates pervasive subsolidus interaction with an aqueous fluid. In most cases, presently available evidence does not demand that that fluid was derived from crystallization of a deeper, younger part of the same pluton, although this is strongly suggested by data from the Alta and Little Cottonwood plutons [13].

6. Thermal Energy Carried by Escaping H_2O

In addition to facilitating dissolution–precipitation reactions and other forms of textural modification, upward-flowing H_2O carries a significant amount of thermal energy. As flow of vapor in fractures is orders of magnitude faster than heat conduction, this upward transfer of energy can have a significant effect on the thermal history of the crystallizing system.

The effect can be approximated using an enthalpy-composition (*H-X*) diagram constructed with thermodynamic data for water [36] and relevant minerals [37]. Construction of *H-X* diagrams was outlined by Ussler and Glazner [38]. Unlike familiar *T-X* diagrams, which involve the intensive parameter *T*, *H-X* diagrams use the extensive parameter *H* (specific enthalpy) and obey mass balance.

Figure 6 is a partial and partly schematic *H-X* diagram for the pseudobinary system haplogranite-H_2O at 200 MPa, constructed from these datasets relative to a standard state of 25 °C and 100 kPa (1 bar). Isotherms show *T* variation, and the eutectic, a point in a *T-X* diagram, is a triangle in the *H-X* rendering [38]. The diagram is based on data at and below the eutectic. Consequently, phase boundaries of the granite + melt and melt fields are purely schematic. The position of the melt apex of the eutectic is given by the water content of eutectic melt [39] and enthalpy calculations for water-saturated albite melt [40].

We assume a water-saturated granitic melt at 200 MPa with a liquidus at 950 °C and a solidus at 650 °C ([39], p. 52). Below the eutectic (which is the solidus), isotherms tie the solid quartz–feldspar assemblage at 0 wt % H_2O with the vapor phase, which we idealize as pure H_2O. In our simplified model, the magma carries 5.5 wt % H_2O, releasing 5 wt % upon crystallization, and it reaches vapor saturation at the liquidus and releases H_2O linearly with temperature during crystallization; thus, half of the H_2O has been released as vapor at 800 °C.

If such a magma crystallizes to the solidus at 650 °C and retains its water as trapped vapor, the vapor phase has a specific enthalpy H of 2830 J/g; the solid assemblage has H = 750 J/g. This large difference results from the higher heat capacity of H_2O and its large enthalpy of vaporization. If this vapor escapes via fractures and invades cooler rock above without cooling, it carries a significant amount of thermal energy into the cooler rock volume (for simplicity we assume that depth, and thus pressure, differences are small so that we ignore them in the calculations). For example, in the case of rapid transport, idealized as an adiabatic case, if this vapor thermally equilibrates with an equal mass of 300 °C rock around the fractures, the temperature of the equilibrated volume will be about 570 °C (Figure 6, point A). If vapor from crystallizing magma at 800 °C invades and equilibrates with rock at 500 °C, it will induce melting if the mass fraction of H_2O is approximately 0.25 or greater (point B).

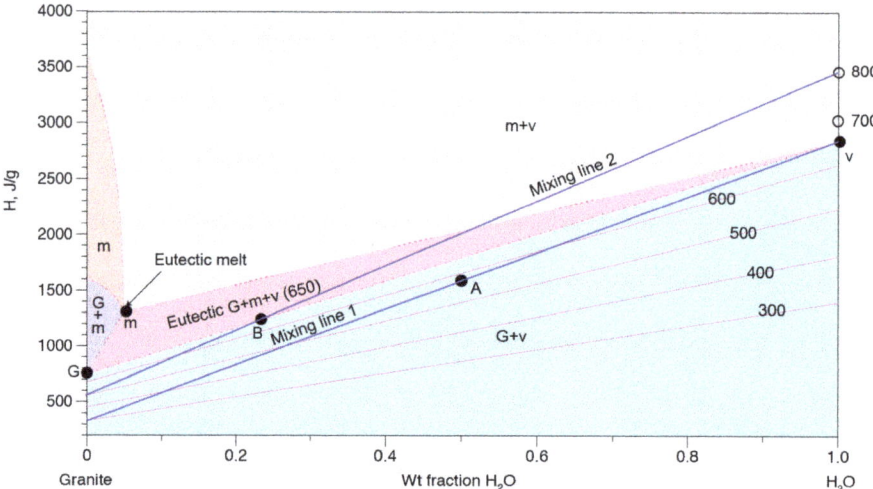

Figure 6. Schematic enthalpy-composition (H-X) diagram for the system haplogranite–H_2O. H_2O has a much higher specific enthalpy than minerals, and so it carries a significant amount of heat (see text). At and below the eutectic triangle, the diagram is based on thermodynamic data, but phase boundaries on the left side are just schematic. Diagonal magenta lines are isotherms. For clarity, isotherms are not shown above the eutectic. The open circles on the right-hand vertical axis locate the enthalpy of pure H_2O at 700° and 800 °C; the corresponding isotherms extend from the open circles in orientations subparallel to the lower-temperature isotherms. Blue lines are mixing lines between H_2O and the solid quartz–feldspar assemblage. Note that the 800 °C mixing line crosses the solidus (i.e., a water-rock ratio > ca. 0.33 at this temperature) will cause remelting.

7. Other Evidence of Late-Stage Textural Modification

A number of studies report evidence of late- to postmagmatic modification of plutonic minerals and textures that is not demonstrably tied to fluid infiltration, but it is probable that pore fluid was involved. Diffusion modeling of Ti zoning in quartz from the Tuolumne Intrusive Suite led Ackerson et al. [41] to conclude that up to 80% of the quartz in these granitic rocks crystallized at temperatures 100–200 °C below the nominal solidus. Because the modal proportion of quartz does not differ from its magmatic abundance, this demands recrystallization of the primary quartz. Such recrystallization might not have happened in the presence of an aqueous fluid, but dissolution–reprecipitation in a grain-boundary fluid probably is the most feasible mechanism.

Higgins [42] and Johnson and Glazner [43,44] presented textural and mineralogical evidence that K-feldspar megacrysts in granitic rocks reflect late-stage textural coarsening in which smaller crystals are consumed and their constituents are transported to feed the growth of megacrysts. The precise mechanism is uncertain, as is the presence or absence of a melt phase when it happened. Higgins [42] proposed Ostwald ripening, that is, coarsening driven by minimization of surface free energy, but this is unlikely if recrystallization was driven only by surface free energy [45,46]. However, the assertion that crystal coarsening does not happen in coarsely crystalline materials because Ostwald ripening is ineffective (e.g., [47,48]) is clearly contradicted by diverse studies in materials science including petrology (e.g., [49–52]).

Textural coarsening in plutonic rocks likely results from dissolution–precipitation caused by fluctuating temperature, fluid infiltration, and concomitant mineral reactions. The consistently K-rich composition of the megacrysts (typically Or_{85-90} from core to rim [43,44]) indicates equilibration with plagioclase at ca. 400 °C, far below the nominal solidus. The megacrysts also are extremely Ca-poor (ca. $An_{0.1}$), which not only corroborates a low equilibration temperature but also excludes diffusive exchange as an equilibration mechanism owing to the low diffusivity of Ca in feldspar [44]. The change of magmatic alkali feldspar to its present low-temperature composition need not be directly linked to textural coarsening, but this mineral reaction has invariably affected K-feldspar megacrysts and it would contribute energy to drive the textural change. The common presence of overgrown corrosion surfaces inside the megacrysts indicates that megacryst growth was not monotonic and suggests a specific process. Primary magmatic K-feldspar grains were episodically corroded by an increase in temperature, infiltration of a fluid, or both. The larger surviving grains regrow preferentially because they provide nucleation surfaces onto which components in the fluid can crystallize. Repeated fluctuations ultimately lead to the demise of small grains, a process that is clearly observed in real time in thermal cycling experiments [49–52]. Pulses of both heat and fluids released by crystallization of later increments into already-crystallized granite thus are strong candidates for facilitating both textural coarsening and modification of primary igneous feldspar compositions.

8. Summary

Subduction-related magmas transport and release large quantities of water into the crust. Water released by crystallizing magma probably escapes along multiple paths through previously crystallized plutonic rock. It is difficult to quantify how much of the water follows each path, but it is likely that a large fraction is channeled along fractures and pipes. However, pervasive infiltration of aqueous fluid along grain boundaries and microfractures commonly has significantly modified primary mineral compositions and textures. We suggest that the infiltrating fluid was derived from crystallization of younger, deeper magmatic increments added to the growing pluton. This suggestion is clearly speculative; more evidence is needed to test whether the grain-boundary fluid was derived from cogenetic magma.

Fluid release from crystallizing magma does not affect the temperature of the magma but the fluid advects a large fraction of the magma's energy into adjacent already-solidified granite. This implies that the sawtooth postsolidus thermal histories predicted for early intrusive increments by conductive incremental-growth thermal models (e.g., [15]) substantially underestimate the actual amplitude of the thermal fluctuations. In some instances, infiltration of magmatic fluid may produce not only mineral reactions and textural modification but also remelting of previously solidified portions of the pluton. Remelting of already solidified granite by this mechanism would be difficult to conclusively demonstrate because the result will be a granitic pore melt in a matrix of granite; therefore, cooling will merely redeposit quartz and feldspars on crystals that survived remelting. The result is likely to be textural modifications that would be difficult to distinguish from subsolidus textural modifications.

The sparseness of mappable internal contacts in plutons is a major reason that the incremental growth of plutons over million-year time scales was not recognized until the development of high-precision geochronology. We suggest that an important reason that increment contacts are rarely observable in the field is extensive postemplacement textural modification. Modification is a practically inevitable consequence of incremental growth because much, if not all, of the modification is caused by pulses of water and heat released by magmatic crystallization entering already-crystallized portions of the pluton.

Author Contributions: Conceptualization, J.M.B., A.F.G., M.A.S., D.S.C.; methodology, J.M.B., A.F.G., M.A.S., D.S.C.; formal analysis, A.F.G., M.A.S.; investigation, J.M.B., A.F.G., M.A.S.; resources, J.M.B., A.F.G., M.A.S.; data curation, A.F.G., M.A.S.; writing—original draft preparation, J.M.B.; writing—review and editing, A.F.G., M.A.S., D.S.C.; visualization, J.M.B., A.F.G., M.A.S.; project administration, J.M.B., A.F.G., M.A.S., D.S.C.; funding acquisition, J.M.B., A.F.G., M.A.S., D.S.C. All authors have read and agreed to the published version of the manuscript.

Funding: This research was funded by the U.S. National Science Foundation through grant numbers EAR-0538094 and EAR-1853496 to J.B.; grant number EAR-1250505 to A.G.; grant number EAR-0538129 to A.G. and D.C.; and grant number EAR-1853496 to M.S. A.G. also gratefully acknowledges support from the Mary Lily Kenan Flagler Bingham Professorship.

Acknowledgments: Discussions with Ryan D. Mills and John R. Bowman contributed to development of the ideas presented in this paper. Input from two anonymous reviewers helped us to improve the clarity of the paper. U.S. National Park Service personnel, especially Greg Stock, Jan van Wagtendonk, and Peggy Moore, have been supportive and helpful to our field studies in Yosemite. John Bowman provided the CL image from the Alta Granodiorite in Figure 4d.

Conflicts of Interest: The authors declare no conflict of interest.

References

1. Carmichael, I.S.E. The andesite aqueduct: Perspectives on the evolution of intermediate magmatism in west-central (105–99°W) Mexico. *Contr. Miner. Petrol.* **2002**, *143*, 641–663. [CrossRef]
2. Coleman, D.S.; Gray, W.; Glazner, A.F. Rethinking the emplacement and evolution of zoned plutons: Geochronologic evidence for the incremental assembly of the Tuolumne Intrusive Suite, California. *Geology* **2004**, *32*, 433–436. [CrossRef]
3. Michel, J.; Baumgartner, L.; Putlitz, B.; Schaltegger, U.; Ovtcharova, M. Incremental growth of the Patagonian Torres del Paine laccolith over 90 k.y. *Geology* **2008**, *36*, 459–462. [CrossRef]
4. Coleman, D.S.; Bartley, J.M.; Glazner, A.F.; Pardue, M.J. Is chemical zonation in plutonic rocks driven by changes in source magma composition or shallow-crustal differentiation? *Geosphere* **2012**, *8*, 1568–1587. [CrossRef]
5. Bartley, J.M.; Glazner, A.F.; Coleman, D.S. Dike intrusion and deformation during growth of the Half Dome pluton, Yosemite National Park, California. *Geosphere* **2018**, *14*, 1–15. [CrossRef]
6. Sisson, T.W.; Layne, G.D. H_2O in basalt and basaltic andesite glass inclusions from four subduction-related volcanoes. *Earth Plan. Sci. Lett.* **1993**, *117*, 619–635. [CrossRef]
7. Wallace, P.J. Volatiles in subduction zones magmas: Concentrations and fluxes based on melt inclusion and volcanic gas data. *J. Volc. Geotherm. Res.* **2005**, *140*, 217–240. [CrossRef]
8. Metrich, N.; Wallace, P.J.; Putirka, K.D.; Frank, J., III. Volatile abundances in basaltic magmas and their degassing paths tracked by melt inclusions. *Rev. Miner. Geochem.* **2008**, *69*, 363–402. [CrossRef]
9. Plank, T.; Kelley, K.A.; Zimmer, M.M.; Hauri, E.H.; Wallace, P.J. Why do mafic arc magmas contain ~4 wt % water on average? *Earth Plan. Sci. Lett.* **2013**, *364*, 168–179. [CrossRef]
10. Grove, T.; Parman, S.; Bowring, S.; Price, R.; Baker, M. The role of an H_2O-rich fluid component in the generation of primitive basaltic andesites and andesites from the Mt. Shasta region, N California. *Contr. Miner. Petrol.* **2002**, *143*, 641–663. [CrossRef]
11. Laumonier, M.; Gaillard, F.; Muir, D.; Blundy, J.; Unsworth, M. Giant magmatic water reservoirs at mid-crustal depth inferred from electrical conductivity and growth of the continental crust. *Earth Plan. Sci. Lett.* **2016**, *457*, 173–180. [CrossRef]
12. Matzel, J.E.P.; Bowring, S.A.; Miller, R.B. Time scales of pluton construction at differing crustal levels; examples from the Mount Stuart and Tenpeak Intrusions, north Cascades, Washington. *Geol. Soc. Am. Bull.* **2006**, *118*, 1412–1430. [CrossRef]

13. Stearns, M.A.; Bartley, J.M.; Bowman, J.R.; Forster, C.W.; Beno, C.J.; Riddle, D.D.; Callis, S.J.; Udy, N.D. Simultaneous magmatic and hydrothermal regimes recorded by multiphase U-Pb petrochronology, Alta-Little Cottonwood stocks, Utah, USA. *Geosciences* **2020**, *10*, 129. [CrossRef]
14. Glazner, A.F.; Bartley, J.M.; Coleman, D.S.; Gray, W.; Taylor, R.Z. Are plutons assembled over millions of years by amalgamation of small magma chambers? *GSA Today* **2004**, *14*, 4–11. [CrossRef]
15. Annen, C.; Scaillet, B.; Sparks, R.S.J. Thermal constraints on the emplacement rate of a large intrusive complex: The Manaslu Leucogranite, Nepal Himalaya. *J. Petrol.* **2006**, *47*, 71–95. [CrossRef]
16. Bartley, J.M.; Coleman, D.S.; Glazner, A.F. Incremental emplacement of plutons by magmatic crack-seal. *Trans. R. Soc. Edinb.* **2006**, *97*, 383–396. [CrossRef]
17. Pitcher, W.S.; Berger, A.R. *The Geology of Donegal: A Study of Granite Emplacement and Unroofing*; Wiley-Interscience: New York, NY, USA, 1972.
18. Mahan, K.H.; Bartley, J.M.; Coleman, D.S.; Glazner, A.F.; Carl, B.S. Sheeted intrusion of the synkinematic McDoogle pluton, Sierra Nevada, California. *Geol. Soc. Am. Bull.* **2003**, *115*, 1570–1582. [CrossRef]
19. Leuthold, J.; Muentener, O.; Baumgartner, L.P.; Putlitz, B.; Ovtcharova, M.; Schaltegger, U. Time resolved construction of a bimodal laccolith (Torres del Paine, Patagonia). *Earth Plan. Sci. Lett.* **2012**, *325*, 85–92. [CrossRef]
20. Davis, J.W.; Coleman, D.S.; Gracely, J.T.; Gaschnig, R.; Stearns, M. Magma accumulation rates and thermal histories of plutons of the Sierra Nevada batholith, CA. *Contr. Miner. Petrol.* **2012**, *163*, 449–465. [CrossRef]
21. Horsman, E.; Tikoff, B.; Morgan, S. Emplacement-related fabric and multiple sheets in the Maiden Creek sill, Henry Mountains, Utah, USA. *J. Struct. Geol.* **2005**, *27*, 1426–1444. [CrossRef]
22. Farina, F.; Dini, A.; Innocenti, F.; Rocchi, S.; Westerman, D.S. Rapid incremental assembly of the Monte Capanne pluton (Elba Island, Tuscany) by downward stacking of magma sheets. *Geol. Soc. Am. Bull.* **2010**, *122*, 1463–1479. [CrossRef]
23. Gaynor, S.P.; Coleman, D.S.; Rosera, J.M.; Tappa, M.J. Geochronology of a Bouguer gravity low. *J. Geophys. Res.* **2018**, *124*, 1–12. [CrossRef]
24. Cook, S.J.; Bowman, J.R. Contact metamorphism surrounding the Alta stock; thermal constraints and evidence of advective heat transport from calcite + dolomite geothermometry. *Am. Min.* **1994**, *79*, 513–525.
25. Cook, S.J.; Bowman, J.R.; Forster, C.B. Contact metamorphism surrounding the Alta Stock; finite element model simulation of heat- and $^{18}O/^{16}O$ mass-transport during prograde metamorphism. *Am. J. Sci.* **1997**, *297*, 1–55. [CrossRef]
26. Rogers, J.J.W. Origin of albite in granitic rocks. *Am. J. Sci.* **1961**, *259*, 186–193. [CrossRef]
27. Carmichael, I.S.E. The crystallization of feldspar in volcanic acid liquids. *J. Geol. Soc. Lond.* **1963**, *119*, 95–130. [CrossRef]
28. Phillips, E.R. Myrmekite and albite in some granites of the New England Batholith, New South Wales. *J. Geol. Soc. Austr.* **1964**, *11*, 49–60. [CrossRef]
29. O'Neil, J.R.; Taylor, H.P.J. The oxygen isotope and cation exchange. *Am. Min.* **1967**, *52*, 1414–1437.
30. Putnis, A.; John, T. Replacement processes in the Earth's Crust. *Elements* **2010**, *6*, 159–164. [CrossRef]
31. Glazner, A.F. K-feldspar and cooties; or why K-feldspar likes to grow into big crystals. In Proceedings of the GSA Annual Meeting, Phoenix, AZ, USA, 22–25 September 2019.
32. Valley, J.W.; Graham, C.M. Ion microprobe analysis of oxygen isotope ratios in quartz from Skye granite; healed micro-cracks, fluid flow, and hydrothermal exchange. *Contr. Miner. Petrol.* **1996**, *124*, 225–234. [CrossRef]
33. Glazner, A.F.; Bartley, J.M.; Coleman, D.S.; Lindgren, K. Aplite diking and infiltration: A differentiation mechanism restricted to plutonic rocks. *Cont. Min. Petrol.* **2020**, *175*, 37. [CrossRef]
34. Johnson, B.W.; Bowman, J.R.; Nash, B.P.; Valley, J.W.; Bartley, J.M. Oxygen isotope, TitaniQ, and cathodoluminescence analyses of the Alta stock, UT: Preliminary insights into pluton assembly. *Geol. Soc. Am. Abstr.* **2009**, *41*, 43.
35. Challener, S.C.; Glazner, A.F. Igneous or metamorphic? Hornblende phenocrysts as greenschist facies reaction cells in the Half Dome Granodiorite, California. *Am. Min.* **2017**, *102*, 436–444. [CrossRef]
36. Burnham, C.W.; Holloway, J.R.; Davis, N.F. *Thermodynamic Properties of Water to 1000 °C and 10,000 Bars*; Geological Society of America: Boulder, CO, USA, 1969.

37. Robie, R.A.; Hemingway, B.S.; Fisher, J. Thermodynamic properties of Minerals and Related Substances at 298.15 K and 1 Bar (105 Pascals) Pressure and at Higher Temperatures. 1978. Available online: https://pubs.er.usgs.gov/publication/b2131 (accessed on 10 April 2020).
38. Ussler, W., III; Glazner, A.F. Graphical analysis of enthalpy-composition relationships in mixed magmas. *J. Volc. Geotherm. Res.* **1992**, *51*, 23–40. [CrossRef]
39. Johannes, W.; Holtz, F. *Petrogenesis and Experimental Petrology of Granitic Rocks*; Springer: New York, NY, USA, 1996.
40. Glazner, A.F. The ascent of water-rich magma and decompression heating: A thermodynamic analysis. *Am. Min.* **2019**, *104*, 890–896. [CrossRef]
41. Ackerson, M.R.; Mysen, B.O.; Tailby, N.D.; Watson, E.B. Low-temperature crystallization of granites and the implications for crystal magmatism. *Nature* **2018**, *559*, 94–97. [CrossRef] [PubMed]
42. Higgins, M.D. Origin of megacrysts in granitoids by textural coarsening; a crystal size distribution (CSD) study of microcline in the Cathedral Peak Granodiorite, Sierra Nevada, California. *Geol. Soc. Spec. Pubs.* **1999**, *168*, 207–219. [CrossRef]
43. Johnson, B.R.; Glazner, A.F. Formation of K-feldspar megacrysts in granodioritic plutons by thermal cycling and late-stage textural coarsening. *Contr. Miner. Petrol.* **2010**, *159*, 599–619. [CrossRef]
44. Glazner, A.F.; Johnson, B.R. Late crystallization of K-feldspar and the paradox of megacrystic granites. *Contr. Min. Petrol.* **2013**, *166*, 777–799. [CrossRef]
45. Cabane, H.; Laporte, D.; Provost, A. Experimental investigation of the kinetics of Ostwald ripening of quartz in silicic melts. *Contrib. Miner. Petrol.* **2001**, *142*, 361–373. [CrossRef]
46. Cabane, H.; Laporte, D.; Provost, A. An experimental study of Ostwald ripening of olivine and plagioclase in silicate melts; implications for the growth and size of crystals in magmas. *Contrib. Miner. Petrol.* **2005**, *150*, 37–53. [CrossRef]
47. Holness, M.B.; Vernon, R.H. The influence of interfacial energies on igneous microstructures. In *Layered Intrusions*; Charlier, B., Namur, O., Latypov, R., Tegner, C., Eds.; Springer: Dordrecht, The Netherlands, 2015; pp. 183–227.
48. Gualda, G.A.R. On the origin of alkali feldspar megacrysts in granitoids: The case against textural coarsening. *Contr. Miner. Petrol.* **2019**, *174*, 88. [CrossRef]
49. Mills, R.D.; Glazner, A.F. Experimental study on the effects of temperature cycling on coarsening of plagioclase and olivine in an alkali basalt. *Contrib. Miner. Petrol.* **2013**, *166*, 97–111. [CrossRef]
50. Erdmann, M.; Koepke, J. Experimental temperature cycling as a powerful tool to enlarge melt pools and crystals at magma storage conditions. *Am. Miner.* **2016**, *101*, 960–969. [CrossRef]
51. Donhowe, D.P.; Hartel, R.W. Recrystallization of ice cream during controlled accelerated storage. *Int. Dairy J.* **1996**, *6*, 1191–1208. [CrossRef]
52. Mills, R.D.; Ratner, J.J.; Glazner, A.F.; Mills, R.D. Experimental evidence for crystal coarsening and fabric development during temperature cycling. *Geology* **2011**, *39*, 1139–1142. [CrossRef]

© 2020 by the authors. Licensee MDPI, Basel, Switzerland. This article is an open access article distributed under the terms and conditions of the Creative Commons Attribution (CC BY) license (http://creativecommons.org/licenses/by/4.0/).

Article

Climate and the Development of Magma Chambers

Allen F. Glazner

Department of Geological Sciences, University of North Carolina, Chapel Hill, NC 27599-3315, USA; afg@unc.edu

Received: 2 February 2020; Accepted: 26 February 2020; Published: 1 March 2020

Abstract: Whether magma accumulating in the crust develops into a persistent, eruptible magma body or an incrementally emplaced pluton depends on the energy balance between heat delivered to the bottom in the form of magma and heat lost out the top. The rate of heat loss to the surface depends critically on whether heat transfer is by conduction or convection. Convection is far more efficient at carrying heat than conduction, but requires both abundant water and sufficient permeability. Thus, all else being equal, both long-term aridity and self-sealing of fractures should promote development of persistent magma bodies and explosive silicic volcanism. This physical link between climate and magmatism may explain why many of the world's great silicic ignimbrite provinces developed in arid environments, and why extension seems to suppress silicic caldera systems.

Keywords: igneous petrology; tectonics; heat flow; glaciation; climate

1. Introduction

> ... we speculate that hot springs were relatively active during wet periods in contrast to the present situation, and that the sinter and travertine along the fault zone at the east edge of the horst on which the rhyolite lies were deposited during one or more pluvial periods associated with glacial stages. The present climate sustains no thermal springs other than an ephemeral flow at Coso Hot Springs following local precipitation.

In the quotation above Duffield et al. [1] noted that climate and long-term precipitation trends may affect the activity of thermal springs, in this case at the Plio-Pleistocene Coso volcanic field in eastern California. Thermal springs are a major conduit by which heat is removed from geothermal systems [2], and so increased thermal spring activity means more rapid cooling of the underlying magmatic system. The purpose of this paper is to show that such interactions between magma and climate may play a role in whether magma intruded into the crust accumulates fast enough to form a large magma body rather than an incrementally emplaced pluton, and thus whether climate can affect the development of caldera-forming eruptions.

A parcel of magma in the crust moving upward through a fracture has several possible fates. It may (1) reach the surface, contributing to a volcanic eruption; (2) freeze en route, showing up in the geologic record as a dike; (3) freeze where earlier parcels did, forming an incrementally emplaced pluton; or (4) reach a site of persistent magma accumulation, contributing to a magma body. The geologic record contains evidence for all of these scenarios.

It was long thought fates 3 and 4 were essentially the same in that plutons, with volumes on the order of 10^3–10^4 km^3, are just the frozen remains of large magma chambers, but a growing body of evidence shows that many plutons were assembled incrementally over timescales of 10^5–10^6 years and never existed as large bodies of magma [3,4]. However, ignimbrite eruptions with volumes on the order of 1000 km^3 are proof that large magma bodies can exist in the crust, if only ephemerally.

In this paper I examine the conditions that differentiate fates 3 and 4; specifically, factors that tip the balance in favor of incremental pluton assembly versus development of a large magma chamber.

The critical balance is whether magma supply to the bottom of the system is sufficiently rapid to outpace loss of heat out the top. The former is presumably governed by magma supply rate or tectonic control on magma supply from the mantle or lower crust, and the latter by heat-transfer processes in the crust above the site of magma accumulation. I use simple physical arguments to show that long-term climate—specifically, precipitation—can play a role in the ultimate fate of upper-crustal magma accumulations.

2. Energy Balance

To form a large magma body, the rate of heat input via magma injection must exceed the rate of heat loss via conduction, convection, and advection. For a one-dimensional system (e.g., a horizontal sheet of magma sufficiently broad that it can be considered one-dimensional; Figure 1), this is simply thermal energy in the form of magma put into the bottom of the system balanced against loss of heat out the top. The latter quantity must average out to surface heat flow, because heat loss is ultimately to the surface.

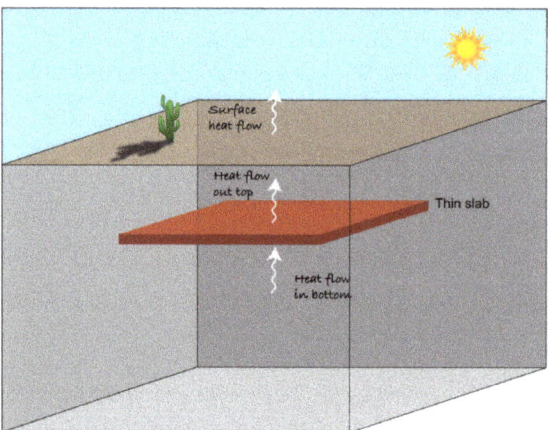

Figure 1. Schematic view of one-dimensional thermal model of a magma sheet with thickness much less than horizontal dimensions. Whether or not the magma body grows depends on the balance between heat in the bottom (as magma) and heat out the top. As heat eventually escapes to the surface, over the long term heat out the top of the sheet averages out to surface heat flow.

2.1. Conductive Heat Flow

Heat flow by conduction obeys Fourier's Law:

$$q = k\frac{dT}{dz} \tag{1}$$

where q is heat flux, k is thermal conductivity, T is temperature, and z is depth (positive downward). Surface heat flow in non-magmatic areas of the world is typically ~35–70 mW/m^2 [5], and the heat flux into the base of the crust is typically about half of this [6]. In this paper I use 40 mW/m^2 as a typical surface heat-flow value in non-magmatic continental areas. Given a typical k of 2 W/m K [7], a corresponding geothermal gradient, assuming conduction alone, would be 20 °C/km.

2.2. Convection and the Nusselt Number

Convection of fluids around a magma body can extract heat much more efficiently than conduction alone. In geothermal areas convection is accomplished predominantly by circulation of meteoric water through fractures in rocks above and within a body of cooling magma. The ratio of total heat transfer by convection, advection and conduction to that of conduction alone is the Nusselt number, Nu:

$$Nu = \frac{total\ heat\ flux}{conductive\ heat\ flux} \qquad (2)$$

In geothermal areas, Nu can vary from unity for purely conductive systems to 100 or higher for systems with vigorous convection [8–10].

In magmatic areas regional heat flow can exceed 150 mW/m^2. Individual thermal springs and fumaroles in the U.S. portion of the Cascade Range release up to several tens of MW over small areas [2] and collectively release ~1 GW, mostly in the Oregon and northern California segments. The Yellowstone hydrothermal system produces ~5 GW, corresponding to heat flows of ~2 W/m^2 averaged over the caldera system [11]. The far smaller (~100 km^2) Grímsvötn geothermal area under Vatnajökull in Iceland produces comparable power, leading to the astounding heat flow estimate of ~50 W/m^2 [12].

Such high heat flows and power outputs indicate that the majority of heat transport is by convection of geothermal fluids rather than by conduction. Fournier [13] noted that the conductive geothermal gradient needed to sustain heat flow of 2 W/m^2 at Yellowstone would be ~1000 °C/km, requiring molten rock <1 km below the surface. This is contradicted by drill holes showing temperatures ~300 °C at 1 km [14] and by lack of seismic evidence for significant molten rock at such shallow depths [15]. Thus, high heat flow in hydrothermal systems such as Yellowstone are obvious evidence of the role that convection plays in moving heat.

2.3. Magma Accretion

For a sheet of partially molten rock such as that in Figure 1, the energy balance between heat input into the bottom and released out the top can be approximated by a dimensionless magma accretion number

$$M = \frac{\rho L V}{(q - q_{bkg})} \qquad (3)$$

where q is the heat flux out the top of the body, q_{bkg} is the background heat flux, ρ is magma density, L is latent heat of crystallization, and V is the rate at which new magma is accreted to the body. For $M < 1$ new magma freezes as it arrives; for $M > 1$ the body of partially molten rock grows.

Using $\rho = 2700$ kg/m^3, $L = 3 \times 10^5$ J/kg, and $q_{bkg} = 40$ mW/m^2, we can estimate magma accretion rates that would be necessary for $M > 1$, and thus high enough to enlarge a growing magma body, as a function of surface heat flow, assuming steady-state (Figure 2). Although the relationship is linear, it is plotted on logarithmic scales to better show the relationship at lower q. Required vertical magma accretion rates are on the order of 1–10 mm/a (= 1–10 km/Ma) for some of the geothermal areas, but on the order of 100–1000 mm/a for the larger ones. Millimeter per year displacements of the Earth's surface are within the range of observed rock uplift and erosion rates (e.g., [16,17]), whereas the higher rates, involving injection of a pile of magma of thickness comparable to that of the continental crust in 1 Ma, are not. Geothermal areas typically have lifetimes on the order of 10^5–10^6 years [18], and sustained intrusion rates on the order of 100–1000 km/Ma are clearly incompatible with the geologic record.

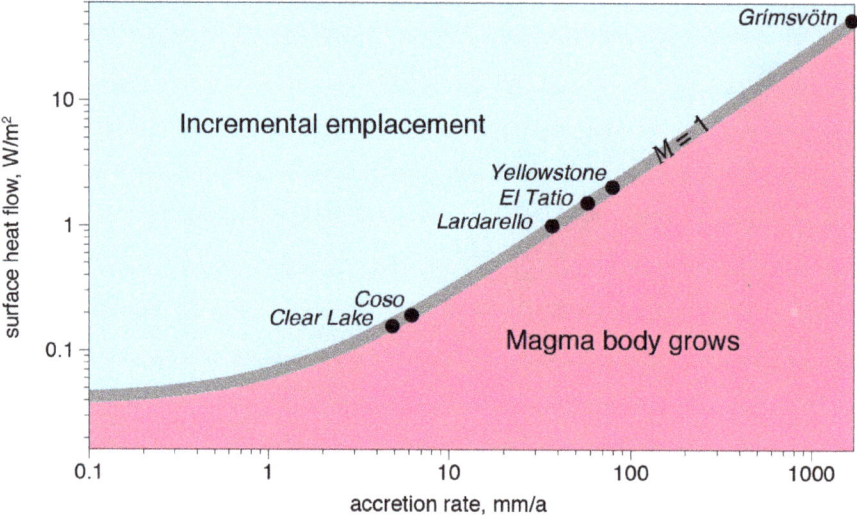

Figure 2. Plot of values of accretion rate and surface heat flow for which the magma accretion number $M = 1$ (gray boundary). For accretion rates below this, cooling is fast enough that an incrementally emplaced pluton grows; for higher accretion rates, a body of partially molten rock grows. Black dots give estimated surface heat flow at selected geothermal areas, showing the magma accretion rates that would be needed to sustain those heat flows by conduction alone. The fields with high surface heat flow require magma accretion rates far in excess of geodetically observed rates.

3. Cooling Effects of Circulating Fluids: Nusselt Number

The Nusselt number Nu has a strong effect on M because it controls surface heat flow q. We can rewrite Equation (3) as

$$M = \frac{\rho L V}{\left(q_s \cdot Nu - q_{bkg}\right)} \quad (4)$$

where q_s is surface heat flow under purely conductive conditions. M is greatly decreased for $Nu > 1$; for $Nu = 10$, M is roughly 9 times smaller than for $Nu = 1$. Thus, a Nusselt number on the order of 10–100 has a drastic effect on cooling of an accumulating magma body.

4. Thermal Models

Thermal modeling provides insight into the effects of conduction versus convection on development of bodies of partially molten rock. Models appropriate to intermediate-composition (e.g., dacitic) magma were run using the Matlab partial differential equation solver; details of thermal modeling are in Appendix A. Enhanced heat flow owing to convection was simulated by increasing thermal conductivity k by a factor of Nu in the rock above the magma injections (e.g., [19]).

Results of a series of thermal models, with magma injected in sheets 50 m thick at time-averaged vertical accretion rates V of 5–20 mm/a and with Nu = 1 and 10, are presented in Figure 3. Models ran until 10 km of magma had accumulated. This is an unreasonable amount of magma to inject over a time on the order of 1 Ma, but the models were run this long to see if the temperatures stabilized. I monitored the maximum temperature just prior to each new injection as a proxy for thermal maturation of the system. For injections at 5 km depth (Figure 3a) and Nu = 1, at an injection rate of 5 mm/a, the maximum temperature leveled out at <600 °C at 1 Ma; 10 mm/a brought the temperature to ~750 °C after 1 Ma; and 20 mm/a brought the temperature to 900 °C, forming a significant body of eruptible magma, in 500 ka. However, with convection turned on (Nu = 10), none of the models approached magmatic temperatures; at 20 mm/a, the temperature leveled out at <400 °C. Models run with injections at 10 km depth (Figure 3b) reached temperatures 50–100 °C higher, but again none reached magmatic temperatures when convection was turned on.

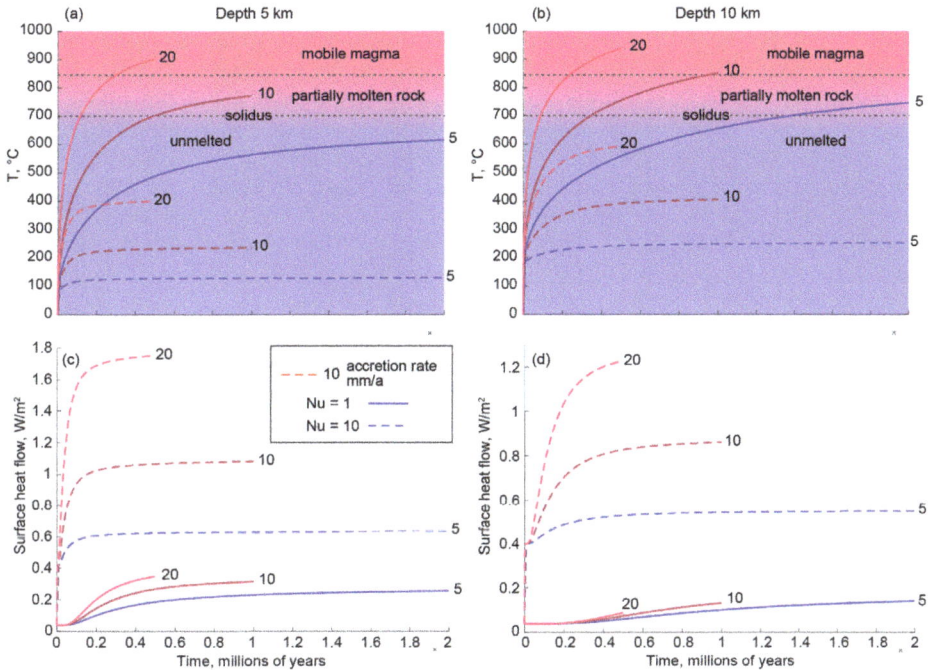

Figure 3. Thermal models of emplacement of 50 m sills at depths of 5 km (**a**,**c**) and 10 km (**b**,**d**), at rates of 5, 10, and 20 mm/a and Nusselt numbers of 1 and 10. (**a**,**b**) Maximum temperatures attained just before next injection of magma as a function of depth. Dotted lines at 700 and 850 °C are model solidus and 50 wt% melt, respectively; the latter is the approximate threshold for a mobile magma body. Models were run until 10 vertical km of magma was intruded. At 5 km, only the highest accretion rate produces a mobile magma body; at 10 km, both 10 and 20 mm/a models reach the 850 °C threshold. Heat flow (**c**,**d**) is far higher for the convective models with $Nu = 10$ than for conductive models with $Nu = 1$, reaching 1 W/m² or more for the faster accretion rates.

Surface heat flow is dramatically enhanced when $Nu = 10$ (Figure 3c,d). For example, at an accretion rate of 10 mm/a, surface heat flow by conduction levels out at ~300 mW/m², whereas at $Nu = 10$ it reaches >1000 mW/m².

5. Climatic Effects on Volcanism and Hydrothermal Systems

The magma accretion number arguments and modeling above predict that growth of large magma bodies is enhanced when heat transfer to the surface via convection is inhibited—that is, when Nu is small. The main factors that can inhibit convection include (1) lack of circulating water and (2) decreased permeability owing to precipitation of minerals in fractures or to particularly tight rock.

Although it has long been known that volcanism affects climate, an ever-growing body of work shows that the reverse is also true: climate can affect volcanism and hydrothermal systems over both short (seasonal) and long (glacial-interglacial) time scales. For example, a number of studies have shown a weak but persistent link between intensity of volcanism and waxing and waning of glacial cycles (e.g., [20–24]). These correlations have been tied to crustal stresses induced by changes in the mass of ice on land and water in the oceans. Climate and the state of the hydrologic system affect the intensity of hydrothermal activity; in the Kenya rift valley, ages of deposits from inactive geothermal systems correlate with periods of high lake level in the rift [25], and deposition of geothermal travertine in Italy, Turkey, and New Mexico correlates with global and regional paleoclimate [26–28]. At Yellowstone, the pressure increase caused by thick glacial ice apparently increased subsurface temperatures by shifting the boiling-point curve [29]. At shorter time scales, geyser periodicity in Yellowstone is weakly tied to both seasonal and decadal precipitation changes [30], and the power output of the El Tatio hydrothermal system in Chile, estimated from chloride flux, is twice as high in wet periods as in dry ones [31].

In many regions, fossil deposits indicate the presence of more vigorous and widespread geothermal activity in the geologically recent past. At the Coso field in California, extensive siliceous sinter deposits [32] underlie a 234 ka basalt flow that is unaffected by thermal activity, indicating more widespread activity in the Pleistocene [1]. Such extensive deposits of siliceous sinter are common in geothermal areas [33–36], and are typically interpreted as evidence of more widespread activity in the past, although migration of thermal vents over time can spread sinter over a wider area than the currently active one. Extensive thermal travertine plateaus and ridges [26,28] may also be relics of more widespread thermal activity in the past, presumably during wetter periods.

6. Climate and the Growth of Persistent Magma Bodies

Sufficiently rapid magma influx can always produce a large body of partially molten rock if the magma is trapped, but for systems near the critical balance between magma input and heat loss (Fig. 2), climate can play a role in whether the body attains large size because the rate of cooling depends on whether heat transport occurs by conduction or convection. In particular, in areas where precipitation is not sufficient to replenish water vapor emitted and lost during hydrothermal activity, convection can be starved, changing the mechanism of heat release from dominantly convective to dominantly conductive and dropping Nu from values of 10 or more to near unity. Thermal modeling (Figure 3) demonstrates that such a change can have a profound effect on the thermal state of an accumulating magma body, and that long-term dry climatic conditions should favor the development of caldera-forming magma bodies.

Coso exemplifies a geothermal field in an area that has undergone radical climate swings in the Quaternary. The field is currently quite arid, receiving ~200 mm/a annual precipitation, and is bordered on the south by China Lake playa, a desiccated remnant of a pluvial lake, on the north by Owens Lake playa, a former saline lake, and on the west by a dry valley once occupied by the pluvial Owens River. During the last glacial maximum, however, Owens Lake overflowed into the pluvial Owens River and fed a chain of lakes, including China Lake, that eventually spilled into Death Valley. Glaciers came to the valley floor in Owens Valley, which was occupied by tree species that indicate a far wetter climate [37]:

> *It is suggested that meltwater from the retreating glacial ice inundated the Owens River Lake chain causing pluvial Owens Lake to reach its highstand. This caused an increase in effective moisture, due to high groundwater, allowing the mesophytic Rocky Mountain juniper to exist at the site.*

7. Implications of Precipitation Control on Magmatism

7.1. Paleogeographic Settings of Silicic Caldera Complexes

The hypothesis that climate can affect magmatic systems by modifying the rate at which they cool may explain the paleogeographic settings of a number of large explosive silicic centers, as several of the world's largest silicic caldera provinces (Figure 4) formed in arid to hyperarid environments. The most obvious contemporary example of this is the Altiplano-Puna area [38] of the high Andes. Along the Andes major calderas are largely concentrated in the arid Central Volcanic Zone, with the greatest concentration in the driest part near the Atacama Desert (Figure 5). Extensive silicic ignimbrites of the

Early Cretaceous Paraná-Etendeka province of South America and Africa are intercalated with thick aeolian sandstones that indicate arid conditions throughout the eruptive sequence [39,40], consistent with the hot, arid, equatorial setting of central Gondwana during the Early Cretaceous [41]. Extensive Oligocene ignimbrites in Colorado, Utah, New Mexico, and Arizona are similarly intercalated with thick eolianites that are interpreted as remnants of a widespread erg [42] and corresponding arid conditions.

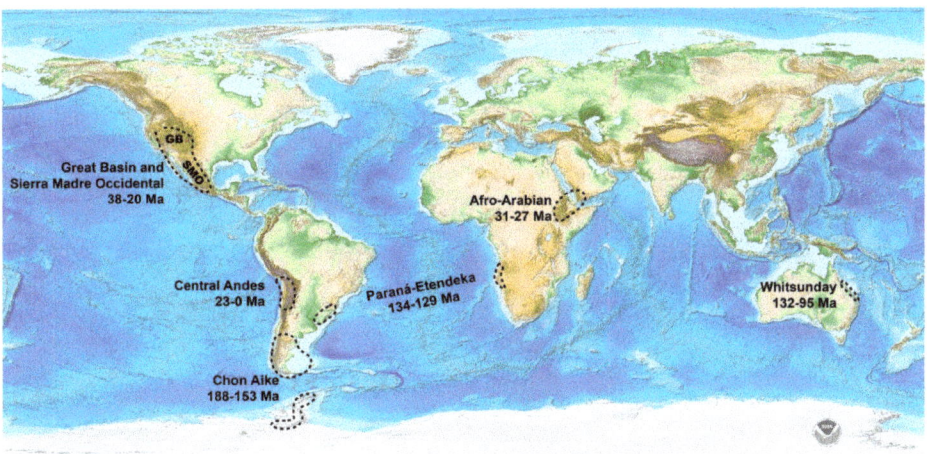

Figure 4. Locations of several large silicic ignimbrite provinces that formed in arid or plausibly arid environments, along with the three silicic large igneous provinces identified by Bryan [43]. All but the central Andean province formed during continental rifting, but most of these also formed in arid or hyperarid environments. Although there is a strong global correlation between magmatism and lithospheric extension, in detail extension may actually work to suppress growth of large magma bodies and caldera formation by opening fractures and promoting hydrothermal circulation.

Several other major silicic ignimbrite provinces formed in regions that were transitional from temperate to arid belts on paleoclimatic maps of Boucot et al. [44]. These include the Oligocene Sierra Madre Occidental province of Mexico [45–47] and explosive silicic volcanism of the contemporaneous Afro-Arabian province in Ethiopia and Yemen [48]. The latter accompanied eruption of the Ethiopian Traps during rifting that formed the Red Sea and Gulf of Aden. The Chon Aike/West Antarctica silicic province formed in the Jurassic during rifting of Gondwana [49,50]. The paleogeographic settings of these regions during explosive volcanism are not fully understood, but they were plausibly arid.

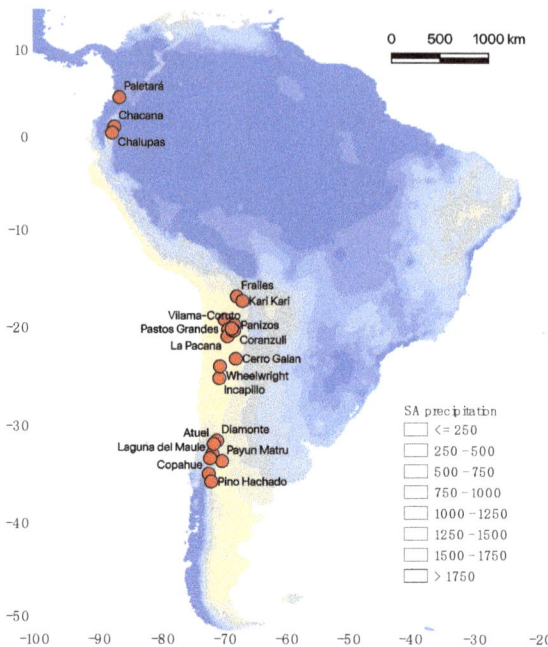

Figure 5. Locations of major late Cenozoic calderas (generally >10 km in diameter) in the Andes, from [51,52]. Map shows mean annual rainfall in mm from 190–2000, from worldclim.org. Calderas are dominantly found in the arid to hyperarid parts of the range.

7.2. Role of Extension and An Explanation for Really Obvious Counterexamples

The list of significant silicic caldera systems in non-arid environments is large and includes Toba [53], Krakatau [54], and numerous calderas in Japan [55] and New Zealand [56]. These clearly formed in regions with abundant water in the form of precipitation or seawater, but this observation does not negate the basic physics presented here. There are two obvious ways to promote development of an eruptible magma body in an area with abundant hydrothermal fluid. The first is to supply magma to the bottom of the system fast enough that even a vigorous hydrothermal system with high Nu cannot remove it fast enough to prevent solidification. However, modeling presented above indicates that extremely high magma accretion rates, on the order of 150 mm/a (= 150 km/Ma) are required to build up an eruptible magma body. This may be possible, but evidence for sustained crustal inflation at such extreme rates has not been recognized. This is a frontier area that deserves further investigation [57,58].

A second, more realistic way to increase M in a hydrothermal system is to seal up permeability so that convection cannot take place even when abundant working fluid is present. Self-sealing of hydrothermal systems is widespread and leaves behind clear physical evidence in the form of veins choked with precipitated minerals [59,60]. Wisian and Blackwell [61] noted that

> At least one major reason to associate geothermal systems with young [extensional] faulting is self-sealing, the process whereby cooling, ascending fluids precipitate minerals in pores and fractures (thereby reducing permeability). This process will eventually limit or eliminate flow in high-temperature cases. There is abundant evidence of self-sealing in many fossil geothermal systems . . .

Worldwide there is an undeniable, strong link between magmatism and lithospheric extension, but in detail there are anomalies in this relationship that suggest a role for extension in suppressing development of large silicic magma bodies. By reasoning above, opening of sealed fractures is favored by extensional faulting; thus, a system in which self-sealed fractures are episodically broken open by extensional faulting should have lower M than one in which fractures are simply sealed up, and systems without extensional faulting should be more prone to sealing, high M, and development of crustal magma bodies and large silicic eruptions. Analysis of the stress state of calderas around the world is beyond the scope of this paper, but, based on surface geology and earthquake focal mechanisms, both Sumatra and Honshu are undergoing shortening [62–64]. The North Island of New Zealand, which is rifting [65], is a major counterexample to the hypothesis proposed here.

The process of seal-breaking extensional faulting may explain the curious anticorrelation between extension and caldera formation in the Great Basin of the western United States, where eruption of major ignimbrite sheets tended to occur in between periods of extension and stratal tilting [66–68]. Gans and Bohrson [69] explained this by proposing that rapid extension can suppress volcanism by a number of mechanisms, including allowing access of meteoric water to midcrustal depths. Extension may suppress caldera formation by continually reopening permeability, promoting hydrothermal convection and magma body cooling. This process may also explain the paucity of calderas in arcs with backarc spreading [52].

7.3. Possible Orographic Control of Caldera Formation

Voluminous explosive silicic eruptions blanketed Nevada and western Utah in the Oligocene and Miocene during the southward-sweeping ignimbrite flareup [70,71], leaving behind 23 recognized calderas in Nevada and a nearly equal number of ignimbrite sheets that were probably erupted from unrecognized calderas (Figure 6). Magmatism swept southward across Nevada from ~45 Ma to ~15 Ma, but the caldera belt is bounded by a line that runs NNW, parallel to the Sierra Nevada Range, 100–200 km east of its crest. West of this line only the only recognized calderas are the 760 ka Long Valley caldera [72], 9.5 Ma Little Walker caldera [73,74], and 16 Ma Woods Mountains caldera [75].

It is possible that this sharp southwestern boundary to the caldera belt was caused by the westward transition from the arid interior West to the wetter continental margin in California. North America has been at roughly the same paleolatitude for the past 100 Ma [76], with the Great Basin in the belt of westerlies. Today, orographic precipitation wrings east-moving storms of their moisture over the coastal mountains and Sierra Nevada, producing the deserts of southern California, Nevada, and Arizona, including hyperarid Death Valley. In the mid-Cenozoic Nevada was likely a high plateau; regardless of whether the current Sierra Nevada rose above this plateau or were its western slope, the plateau was likely arid [77], promoting magma accumulation in the crust and caldera-forming eruptions.

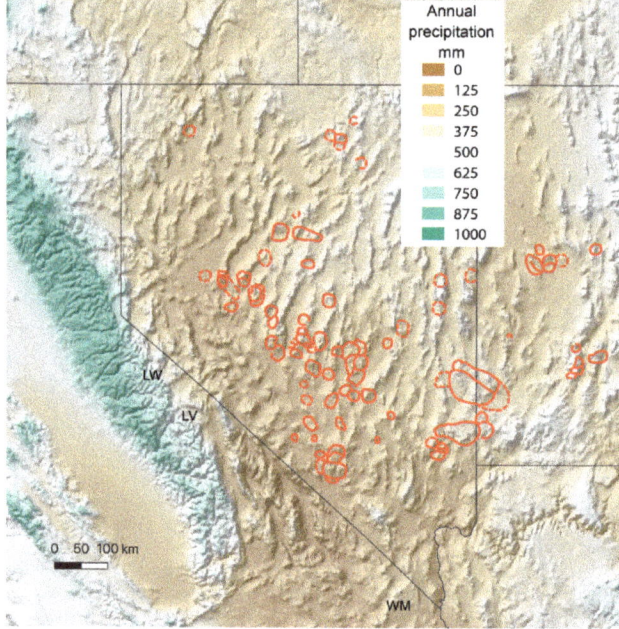

Figure 6. Map of Nevada and surroundings, showing calderas from Henry and John [70] and mean annual precipitation. The ignimbrite flareup started at ~45 Ma in Idaho and swept south and southwest to southern Nevada ~15 Ma, but the voluminous caldera-forming events stopped as they approached the California border, perhaps fearful of high taxes or that their fruit would be confiscated. One explanation for this western border is that caldera-forming eruptions were favored in the arid high plateau of Nevada but suppressed in wetter regions closer to the ocean. LW, LV, and WM refer to Little Walker, Long Valley, and Woods Mountains calderas.

7.4. A Final Speculation

Cather et al. [77] proposed that large silicic eruptions can force world climate into a colder phase by fertilizing the oceans with iron, leading to carbon dioxide drawdown. Such a mechanism could interact with the processes discussed here in interesting ways. For example, if a warm climate leads to drier conditions in magmatic areas, volcanism could shift from effusive to explosive, triggering cooling that could dampen the climate and suppress such effusive volcanism.

8. Conclusions

The rate at which heat is removed from crustal magma accumulations determines whether they grow into persistent magma bodies or instead form an incrementally emplaced pluton. Hydrothermal convection is far more efficient than conduction at moving heat to the surface, and regional heat flow in many geothermal areas is so high that convection must be the dominant mode of heat transfer. However, convection can only occur if there is sufficient water in the shallow crust to sustain it, and if self-sealing of fractures is overcome by extensional faulting.

The necessity of water recharge for convection suggests that long-term aridity could force systems into conduction, dramatically slowing removal of heat and promoting growth of zones of partially molten rock. This process may explain why many of Earth's greatest silicic ignimbrite provinces, such as the central Andes, Oligocene Great Basin, and Cretaceous Paraná-Etendeka, developed in arid and hyperarid areas. Many geothermal areas that are currently in arid regions sit among sinter and travertine deposits that indicate significantly greater activity in the past, presumably during wetter and cooler glacial cycles when groundwater was more abundant.

Arid conditions are clearly not a requirement for formation of silicic calderas as counterexamples are abundant, but lack of permeability caused by self-sealing may shut down convection in areas not undergoing extension. This could explain why caldera formation appears to be suppressed in areas undergoing extension or in arcs with backarc extension.

Funding: This research received no external funding

Acknowledgments: Ken Wohletz and his Heat3D software were important to early development of the ideas in this paper. I thank all the scientists who worked on the Coso geothermal system under the aegis of the Geothermal Program Office and Frank Monastero, and their presentations at numerous Coso meetings, for opening my eyes to the links between petrology, structural geology, hydrology, and geophysics. Chuck Stern enlightened me about the Andes. Craig Magee, John Eichelberger, and two anonymous reviewers provided constructive and challenging reviews that significantly improved the presentation, and John Bartley provided an early review and contributed to this study in many ways.

Conflicts of Interest: The authors declare no conflict of interest.

Appendix A

The heat conduction equation was solved in Matlab using the Partial Differential Equation Toolbox. The parabolic heat equation solver uses the Method of Lines, wherein the spatial domain is discretized on a mesh and the partial differential equation is converted into a set of ordinary differential equations. Boundary conditions were a surface temperature of 0 °C, a basal heat flux of 0.04 W/m^2, a model depth of 25 km and width of 1000 m, insulated right and left boundary conditions, and a sheet thickness of 50 m. Sheets were added at time intervals specified by the accretion rate, at 5 or 10 km depth, and latent heat was accounted for by adding 200 °C to the nominal magma temperature (1000 °C, appropriate for magma of intermediate composition). Downward displacement of rocks under the injection site was not accounted for. Thermal conductivity in the layer above the injection site was multiplied by Nu.

References

1. Duffield, W.A.; Bacon, C.R.; Dalrymple, G.B. Late Cenozoic volcanism, geochronology, and structure of the Coso Range, Inyo County, California. *J. Geophys. Res.* **1980**, *85*, 2381–2404. [CrossRef]
2. Ingebritsen, S.E.; Mariner, R.H. Hydrothermal heat discharge in the Cascade Range, northwestern United States. *J. Volcanol. Geotherm. Res.* **2010**, *196*, 208–218. [CrossRef]
3. Coleman, D.S.; Gray, W.; Glazner, A.F. Rethinking the emplacement and evolution of zoned plutons: Geochronologic evidence for incremental assembly of the Tuolumne Intrusive Suite, California. *Geology* **2004**, *32*, 433–436. [CrossRef]
4. Glazner, A.F.; Bartley, J.M.; Coleman, D.S.; Gray, W.; Taylor, R.Z. Are plutons assembled over millions of years by amalgamation from small magma chambers? *GSA Today* **2004**, *14*. [CrossRef]
5. Pollack, H.N.; Hurter, S.J.; Johnson, J.R. Heat flow from the Earth's interior: Analysis of the global data set. *Rev. Geophys.* **1993**, *31*, 267–280. [CrossRef]
6. Pollack, H.N.; Chapman, D.S. On the regional variation of heat flow, geotherms, and lithospheric thickness. *Tectonophysics* **1977**, *38*, 279–296. [CrossRef]
7. Ingebritsen, S.E.; Geiger, S.; Hurwitz, S. Numerical simulation of magmatic hydrothermal systems. *Rev. Geophys.* **2010**, *48*, 1–33. [CrossRef]
8. Wooding, R.A. Large-Scale geothermal field parameters and convection theory. In *Proceedings of the Second Workshop Geothermal Reservoir Engineering*; Kruger, P., Ramey, H.J., Eds.; Stanford Geothermal Program Workshop Report SGP-TR-20; Applied Mathematics Division (DSIR): Wellington, New Zealand, 1976; pp. 339–344.
9. Kilty, K.; Chapman, D.S.; Mase, C.W. Forced convective heat transfer in the Monroe Hot Springs geothermal system. *J. Volcanol. Geotherm. Res.* **1979**, *6*, 257–277. [CrossRef]

10. Stimac, J.; Goff, F.; Wohletz, K. *Thermal Modeling of the Clear Lake Magmatic System, California: Implications for Conventional and Hot Dry Rock Geothermal Development*; Los Alamos National Laboratory: Los Alamos, NM, USA, 1997; p. 38.
11. Lowenstern, J.B.; Hurwitz, S. Monitoring a supervolcano in repose: Heat and volatile flux at the yellostone caldera. *Elements* **2008**, *4*, 35–40. [CrossRef]
12. Björnsson, H.; Björnsson, S.; Sigurgeirsson, T. Penetration of water into hot rock boundaries of magma at Grímsvötn. *Nature* **1982**, *295*, 580–581. [CrossRef]
13. Fournier, R.O. Geochemistry and dynamics of the Yellowstone National Park hydrothermal system. *Annu. Rev. Earth Planet. Sci.* **1989**, *17*, 13–53. [CrossRef]
14. White, D.E.; Muffler, L.J.P.; Truesdell, A.H. Vapor-Dominated hydrothermal systems compared with hot-water systems. *Econ. Geol.* **1971**, *66*, 75–97. [CrossRef]
15. Huang, H.-H.; Lin, F.-C.; Schmandt, B.; Farrell, J.; Smith, R.B.; Tsai, V.C. The Yellowstone magmatic system from the mantle plume to the upper crust. *Science* **2015**, *348*, 773–776. [CrossRef] [PubMed]
16. Riebe, C.S.; Kirchner, J.W.; Granger, D.E.; Finkel, R.C. Minimal climatic control on erosion rates in the Sierra Nevada, California. *Geology* **2001**, *29*, 447–450. [CrossRef]
17. Nocquet, J.M.; Sue, C.; Walpersdorf, A.; Tran, T.; Lenôtre, N.; Vernant, P.; Cushing, M.; Jouanne, F.; Masson, F.; Baize, S.; et al. Present-Day uplift of the western Alps. *Sci. Rep.* **2016**, *6*, 1–6. [CrossRef]
18. Browne, P.R.L. Hydrothermal alteration in active geothermal fields. *Annu. Rev. Earth Planet. Sci.* **1978**, *6*, 229–250. [CrossRef]
19. Muñoz, M.; Hamza, V. Heat flow and temperature gradients in Chile. *Stud. Geophys. Geod.* **1993**, *37*, 315–348. [CrossRef]
20. Glazner, A.F.; Manley, C.R.; Marron, J.S.; Rojstaczer, S. Fire or ice: Anticorrelation of volcanism and glaciation in California over the past 800,000 years. *Geophys. Res. Lett.* **1999**, *26*. [CrossRef]
21. Watt, S.F.L.; Pyle, D.M.; Mather, T.A. The volcanic response to deglaciation: Evidence from glaciated arcs and a reassessment of global eruption records. *Earth Sci. Rev.* **2013**, *122*, 77–102. [CrossRef]
22. Jellinek, A.M.; Manga, M.; Saar, M.O. Did melting glaciers cause volcanic eruptions in eastern California? Probing the mechanics of dike formation. *J. Geophys. Res. Earth* **2004**, *109*. [CrossRef]
23. Huybers, P.; Langmuir, C. Feedback between deglaciation, volcanism, and atmospheric CO_2. *Earth Planet. Sci. Lett.* **2009**, *286*, 479–491. [CrossRef]
24. McGuire, W.J.; Howarth, R.J.; Firth, C.R.; Solow, A.R.; Pullen, A.D.; Saunders, S.J.; Stewart, I.S.; Vita-Finzi, C. Correlation between rate of sea-level change and frequency of explosive volcanism in the Mediterranean. *Nature* **1997**, *389*, 473–476. [CrossRef]
25. Sturchio, N.C.; Dunkley, P.N.; Smith, M. Climate-Driven variations in geothermal activity in the northern Kenya rift valley. *Nature* **1993**, *362*, 233–234. [CrossRef]
26. De Filippis, L.; Faccenna, C.; Billi, A.; Anzalone, E.; Brilli, M.; Soligo, M.; Tuccimei, P. Plateau versus fissure ridge travertines from Quaternary geothermal springs of Italy and Turkey: Interactions and feedbacks between fluid discharge, paleoclimate, and tectonics. *Earth Sci. Rev.* **2013**, *123*, 35–52. [CrossRef]
27. Piper, J.D.A.; Mesci, L.B.; Gürsoy, H.; Tatar, O.; Davies, C.J. Palaeomagnetic and rock magnetic properties of travertine: Its potential as a recorder of geomagnetic palaeosecular variation, environmental change and earthquake activity in the Si{dotless}cak Çermik geothermal field, Turkey. *Phys. Earth Planet. Inter.* **2007**, *161*, 50–73. [CrossRef]
28. Goff, F.; Shevenell, L. Travertine deposits of Soda Dam, New Mexico, and their implications for the age and evolution of the Valles caldera hydrothermal system (USA). *Geol. Soc. Am. Bull.* **1987**, *99*, 292–302. [CrossRef]
29. Bargar, K.E.; Fournier, R.O. Effects of glacial ice on subsurface temperatures of hydrothermal systems in Yellowstone National Park, Wyoming: Fluid-Inclusion evidence. *Geology* **1988**, *16*, 1077–1080. [CrossRef]
30. Hurwitz, S.; Kumar, A.; Taylor, R.; Heasler, H. Climate-Induced variations of geyser periodicity in Yellowstone National Park, USA. *Geology* **2008**, *36*, 451–454. [CrossRef]
31. Munoz-Saez, C.; Manga, M.; Hurwitz, S. Hydrothermal discharge from the El Tatio basin, Atacama, Chile. *J. Volcanol. Geotherm. Res.* **2018**, *361*, 25–35. [CrossRef]
32. Ross, C.P.; Yates, R.G. The Coso quicksilver district Inyo County, California. *U. S. Geol. Surv. Bull.* **1943**, *936-Q*, 395–416.
33. Fernandez-Turiel, J.L.; Garcia-Valles, M.; Gimeno-Torrente, D.; Saavedra-Alonso, J.; Martinez-Manent, S. The hot spring and geyser sinters of El Tatio, northern Chile. *Sediment. Geol.* **2005**, *180*, 125–147. [CrossRef]

34. Lynne, B.Y.; Campbell, K.A.; Moore, J.N.; Browne, P.R.L. Diagenesis of 1900-year-old siliceous sinter (opal-A to quartz) at Opal Mound, Roosevelt Hot Springs, Utah, U.S.A. *Sediment. Geol.* **2005**, *179*, 249–278. [CrossRef]
35. Howald, T.; Person, M.; Campbell, A.; Lueth, V.; Hofstra, A.; Sweetkind, D.; Gable, C.W.; Banerjee, A.; Luijendijk, E.; Crossey, L.; et al. Evidence for long timescale (>10^3 years) changes in hydrothermal activity induced by seismic events. *Geofluids* **2015**, *15*, 252–268. [CrossRef]
36. Sibbett, B.S. Geology of the Tuscarora geothermal prospect, Elko County, Nevada. *Geol. Soc. Am. Bull.* **1982**, *93*, 1264–1272. [CrossRef]
37. Koehler, P.A.; Anderson, R.S. Full-Glacial shoreline vegetation during the maximum highstand at Owens Lake, California. *Gt. Basin Nat.* **1994**, *54*, 142–149.
38. De Silva, S.L. Altiplano-Puna volcanic complex of the central Andes. *Geology* **1989**, *17*, 1102–1106. [CrossRef]
39. Peate, D.W. The Paraná-Etendeka Province. *Geophys. Monogr.* **1997**, *100*, 217–245.
40. Milner, S.C.; Duncan, A.R.; Whittingham, A.M.; Ewart, A. Trans-Atlantic correlation of eruptive sequences and individual silicic volcanic units within the Parana-Etandeka igneous province. *J. Volcanol. Geotherm. Res.* **1995**, *69*, 137–157. [CrossRef]
41. Hay, W.W.; Floegel, S. New thoughts about the Cretaceous climate and oceans. *Earth Sci. Rev.* **2012**, *115*, 262–272. [CrossRef]
42. Cather, S.M.; Connell, S.D.; Chamberlin, R.M.; McIntosh, W.C.; Jones, G.E.; Potochnik, A.R.; Lucas, S.G.; Johnson, P.S. The Chuska erg: Paleogeomorphic and paleoclimatic implications of an Oligocene sand sea on the Colorado Plateau. *Bull. Geol. Soc. Am.* **2008**, *120*, 13–33. [CrossRef]
43. Bryan, S. Silicic large igneous provinces. *Episodes* **2007**, *30*, 20–31. [CrossRef] [PubMed]
44. Boucot, A.J.; Xu, C.; Scotese, C.R.; Morley, R.J. Phanerozoic paleoclimate; an atlas of lithologic indicators of climate. *Concepts Sedimentol. Paleontol.* **2013**, *11*, 478.
45. Ferrari, L.; Valencia-Moreno, M.; Bryan, S. Magmatismo y tectonica en la Sierra Madre Occidental y su relacion con la evolucion de la margen occidental de Norteamerica TT—Magmatism and tectonics of the Sierra Madre Occidental and its relation with the evolution of western North America. *Bol. Soc. Geol. Mex.* **2005**, *57*, 343–378. [CrossRef]
46. Wark, D.A. Oligocene ash flow volcanism, northern Sierra Madre Occidental; role of mafic and intermediate-composition magmas in rhyolite genesis. *J. Geophys. Res.* **1991**, *96*, 411. [CrossRef]
47. McDowell, F.W.; Clabaugh, S.E. Ignimbrites of the Sierra Madre Occidental and their relation to the tectonic history of western Mexico. *Spec. Pap. Geol. Soc. Am.* **1979**, 113–124. [CrossRef]
48. Ukstins Peate, I.; Baker, J.A.; Kent, A.J.R.; Al-Kadai, M.; Al-Subbary, A.; Ayalew, D.; Menzies, M. Correlation of Indian Ocean tephra to individual Oligocene silicic eruptions from Afro-Arabian flood volcanism. *Earth Planet. Sci. Lett.* **2003**, *211*, 311–327. [CrossRef]
49. Kay, S.M.; Ramos, V.A.; Mpodozis, C.; Sruoga, P. Late Paleozoic to Jurassic silicic magmatism at the Gondwana margin: Analogy to the Middle Proterozoic in North America? *Geology* **1989**, *17*, 324–328. [CrossRef]
50. Pankhurst, R.J.; Leat, P.T.; Sruoga, P.; Rapela, C.W.; Marquez, M.; Storey, B.C.; Riley, T.R. The Chon Aike province of Patagonia and related rocks in West Antarctica; a silicic large igneous province. *J. Volcanol. Geotherm. Res.* **1998**, *81*, 113–136. [CrossRef]
51. Stern, C.R. Active Andean volcanism: Its geologic and tectonic setting. *Rev. Geol. Chile* **2004**, *31*, 161–206. [CrossRef]
52. Hughes, G.R.; Mahood, G.A.; Hughes, G.R. Silicic calderas in arc settings: Characteristics, distribution, and tectonic controls. *Bull. Geol. Soc. Am.* **2011**, *123*, 1577–1595. [CrossRef]
53. Chesner, C.A. The Toba Caldera complex. *Quat. Int.* **2012**, *258*, 5–18. [CrossRef]
54. Self, S.; Rampino, M.R. The 1883 eruption of Krakatau. *Nature* **1981**, *294*, 699–704. [CrossRef]
55. Matumoto, T. Caldera volcanoes and pyroclastic flows of Kyûsyû. *Bull. Volcanol.* **1963**, *26*, 401–413. [CrossRef]
56. Wilson, C.J.N.; Rogan, A.M.; Smith, I.E.M.; Northey, D.J.; Nairn, I.A.; Houghton, B.F. Caldera volcanoes of the Taupo volcanic zone, New Zealand. *J. Geophys. Res.* **1984**, *89*, 8463–8484. [CrossRef]
57. Magee, C.; Bastow, I.D.; de Vries, B.V.W.; Jackson, C.A.L.; Hetherington, R.; Hagos, M.; Hoggett, M. Structure and dynamics of surface uplift induced by incremental sill emplacement. *Geology* **2017**, *45*. [CrossRef]
58. Magee, C.; Hoggett, M.; Jackson, C.A.L.; Jones, S.M. Burial-Related compaction modifies intrusion-induced forced folds: Implications for reconciling roof uplift mechanisms using seismic reflection data. *Front. Earth Sci.* **2019**. [CrossRef]

59. Parry, W.T.; Hedderly-Smith, D.; Bruhn, R.L. Fluid inclusions and hydrothermal alteration on the Dixie Valley fault, Nevada. *J. Geophys. Res.* **1991**, *96*, 19733–19748. [CrossRef]
60. Minissale, A. The Larderello geothermal field: A review. *Earth Sci. Rev.* **1991**, *31*, 133–151. [CrossRef]
61. Wisian, K.W.; Blackwell, D.D. Numerical modeling of Basin and Range geothermal systems. *Geothermics* **2004**, *33*, 713–741. [CrossRef]
62. Mount, V.S.; Suppe, J. Present-Day stress orientations adjacent to active Strike-Slip faults—California and sumatra. *J. Geophys Res. Solid Earth* **1992**, *97*, 11995–12013. [CrossRef]
63. Tikoff, B.; Teyssier, C. Strain modeling of displacement-field partitioning in transpressional orogens. *J. Struct. Geol.* **1994**, *16*, 1575–1588. [CrossRef]
64. Heidbach, O.; Rajabi, M.; Cui, X.; Fuchs, K.; Müller, B.; Reinecker, J.; Reiter, K.; Tingay, M.; Wenzel, F.; Xie, F.; et al. The world stress map database release 2016: Crustal stress pattern across scales. *Tectonophysics* **2018**, *744*, 484–498. [CrossRef]
65. Wilson, C.J.N.; Houghton, B.F.; Briggs, R.M. Volcanic and structural evolution of Taupo Volcanic Zone, New Zealand: A review. *J. Volcanol. Geotherm. Res.* **1995**, *68*, 1. [CrossRef]
66. Taylor, W.J.; Bartley, J.M.; Lux, D.R.; Axen, G.J. Timing of tertiary extension in the Railroad Valley-Pioche transect, Nevada: Constraints from 40Ar/39Ar ages of volcanic rocks. *J. Geophys. Res.* **1989**, *94*, 7757–7774. [CrossRef]
67. Best, M.G.; Christiansen, E.H. Limited extension during peak tertiary volcanism, Great Basin of Nevada and Utah. *J. Geophys Res. Solid Earth* **1991**, *96*, 13509–13528. [CrossRef]
68. Gans, P.B.; Bohrson, W.A. Suppression of volcanism during rapid extension in the Basin and Range province, United States. *Science* **1998**, *279*, 66–68. [CrossRef]
69. Henry, C.D.; John, D.A. Magmatism, ash-flow tuffs, and calderas of the ignimbrite flareup in the western Nevada volcanic field, Great Basin, USA. *Geosphere* **2013**, *9*, 951–1008. [CrossRef]
70. Best, M.G.; Christiansen, E.H.; Gromme, S. Introduction: The 36–18 Ma southern Great Basin, USA, ignimbrite province and flareup: Swarms of subduction-related supervolcanoes. *Geosphere* **2013**, *9*, 260–274. [CrossRef]
71. Bailey, R.A.; Dalrymple, G.B.; Lanphere, M.A. Volcanism, structure, and geochronology of Long Valley caldera, Mono County, California. *J. Geophys. Res.* **1976**, *81*, 725–744. [CrossRef]
72. Noble, D.C.; Slemmons, D.B.; Korringa, M.K.; Dickinson, W.R.; Al-Rawi, Y.; McKee, E.H. Eureka Valley Tuff, east-central California and adjacent Nevada. *Geology* **1974**, *2*, 139–142. [CrossRef]
73. Pluhar, C.J.; Deino, A.L.; King, N.M.; Busby, C.; Hausback, B.P.; Wright, T.; Fischer, C. Lithostratigraphy, magnetostratigraphy, and radiometric dating of the Stanislaus Group, CA, and age of the Little Walker Caldera. *Int. Geol. Rev.* **2009**, *51*, 873–899. [CrossRef]
74. McCurry, M. Geology and petrology of the Woods Mountains volcanic center, southeastern California: Implications for the genesis of peralkaline rhyolite ash flow tuffs. *J. Geophys. Res.* **1988**, *93*, 14835–14855. [CrossRef]
75. Seton, M.; Müller, R.D.; Zahirovic, S.; Gaina, C.; Torsvik, T.; Shephard, G.; Talsma, A.; Gurnis, M.; Turner, M.; Maus, S.; et al. Global continental and ocean basin reconstructions since 200 Ma. *Earth Sci. Rev.* **2012**, *113*, 212–270. [CrossRef]
76. Best, M.G.; Barr, D.L.; Christiansen, E.H.; Gromme, S.; Deino, A.L.; Tingey, D.G. The Great Basin Altiplano during the middle Cenozoic ignimbrite flareup: Insights from volcanic rocks. *Int. Geol. Rev.* **2009**, *51*, 589–633. [CrossRef]
77. Cather, S.M.; Dunbar, N.W.; McDowell, F.W.; McIntosh, W.C.; Scholle, P.A. Climate forcing by iron fertilization from repeated ignimbrite eruptions: The icehouse -silicic large igneous province (SLIP) hypothesis. *Geosphere* **2009**, *5*, 315–324. [CrossRef]

© 2020 by the author. Licensee MDPI, Basel, Switzerland. This article is an open access article distributed under the terms and conditions of the Creative Commons Attribution (CC BY) license (http://creativecommons.org/licenses/by/4.0/).

Article

New Conceptual Model for the Magma-Hydrothermal-Tectonic System of Krafla, NE Iceland

Knútur Árnason

ÍSOR, Iceland GeoSurvey, 108 Reykjavík, Iceland; Knutur.Arnason@isor.is

Received: 11 December 2019; Accepted: 16 January 2020; Published: 19 January 2020

Abstract: The complexity of the Krafla volcano and its geothermal system(s) has puzzled geoscientists for decades. New and old geoscientific studies are reviewed in order to shed some light on this complexity. The geological structure and history of the volcano is more complex than hitherto believed. The visible 110 ka caldera hosts, now buried, an 80 ka inner caldera. Both calderas are bisected by an ESE-WNW transverse low-density structure. Resistivity surveys show that geothermal activity has mainly been within the inner caldera but cut through by the ESE-WNW structure. The complexity of the geothermal system in the main drill field can be understood by considering the tectonic history. Isotope composition of the thermal fluids strongly suggests at least three different geothermal systems. Silicic magma encountered in wells K-39 and IDDP-1 indicates a hitherto overlooked heat transport mechanism in evolved volcanos. Basaltic intrusions into subsided hydrothermally altered basalt melt the hydrated parts, producing a buoyant silicic melt which migrates upwards forming sills at shallow crustal levels which are heat sources for the geothermal system above. This can explain the bimodal behavior of evolved volcanos like Krafla and Askja, with occasional silicic, often phreatic, eruptions but purely basaltic in-between. When substantial amounts of silicic intrusions/magma have accumulated, major basalt intrusion(s) may "ignite" them causing a silicic eruption.

Keywords: Krafla volcano; geothermal systems; conceptual models; volcanology

---

1. Introduction

The Krafla volcano is the most studied volcano in Iceland. The onset of the Krafla Fires 1975–1984, often referred to as the Krafla Rifting Episode, initiated intensive volcanological research which greatly increased the understanding of volcanism in extensional rift settings. The Krafla volcano has been closely monitored since the Krafla Fires. Geothermal exploration and drilling have also built-up extensive knowledge on the volcano and its geothermal system(s).

The Krafla region has long been known for its geothermal activity. The first geothermal exploration was conducted in 1969 and was continued the following years [1,2]. The first two exploration wells were drilled in 1974. Based on the findings, it was decided to build the first major geothermal power plant in Iceland in Krafla. The construction of a 60 MWe power plant and production drilling started in 1975 and were continued concurrently with the Krafla Fires that started in December 1975. It soon became apparent that the Krafla Fires caused contamination by magmatic gases, CO_2 and H_2S [3–5]. In the deeper part of the wellfield at that time, to the west and south-west of Mt. Krafla, the volcanic gases increased dramatically in the deeper part of the reservoir and massive precipitation of pyrite and pyrrhotite clogged the wells. The shallower colder part of the reservoir was, however, not affected. Drilling activities were therefore shifted to the southern slopes of Mt. Krafla and Hvíthólar and in 1978, the power plant started the production of 7 MWe. Production drilling continued and, in 1984, the production was up to 30 MWe. In 1990, a new exploration and drilling phase started. The contamination of magmatic gases was diminishing and, by 1999, the power plant was fully operational and producing 60 MWe.

The complexity of the Krafla volcano and its geothermal activity has puzzled geoscientists for a long time and a convincing conceptual model of the volcano and geothermal activity has, in the opinion of the author, been lacking. In this paper, an attempt is made to shed some light on this complexity. A review of old and new information and data and numerical modelling are used to put forward a new, hopefully sensible, conceptual model for the Krafla volcano and its geothermal systems.

2. Tectonic Setting and Geology

The Krafla volcano is located within the Northern Volcanic Zone of Iceland (Figure 1). The Reykjanes Ridge (Mid-Atlantic Ridge) comes on shore at the tip of the Reykjanes peninsula, SW Iceland. Crustal spreading veers to the northeast because of an interaction with a mantle plume under central Iceland. The plate boundary coincides with the Reykjanes peninsula with a mixed strike–slip and extensional motion [6] to the Hengill volcano, which is a triple point with the Western Volcanic Zone (WVZ) and the South Iceland Seismic Zone (SISZ). The SISZ is a left lateral strike–slip zone that conveys most of the crustal spreading to the Eastern Volcanic Zone (EVZ) and connects to the Northern Volcanic Zone (NVZ). The spreading then veers north-westward to the Kolbeinsey Ridge, north of Iceland, through a right lateral transform zone, the Tjörnes Fracture Zone (TFZ).

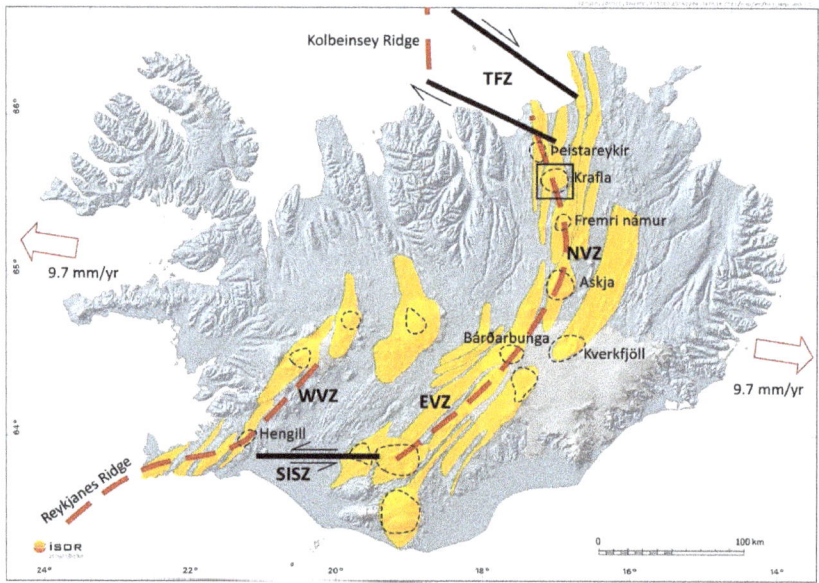

Figure 1. A simplified tectonic map of Iceland showing the location of Krafla (black square). Red broken lines represent spreading zones. WVZ is the Western Volcanic Zone, EVZ is the Eastern Volcanic Zone, NVZ is the Northern Volcanic Zone. SISZ is the South Iceland Seismic/Transform Zone and TFZ is the Tjörnes Fracture/Transform Zone. Thin black broken lines show central volcanoes and yellow coloured areas fissure swarms.

Figure 1 shows that the NVZ is arc-shaped towards the transform zones in the north. It hosts five volcanic centres, from south to north, Kverkfjöll, Askja, Fremri Námur, Krafla and Þeistareykir. Each volcanic centre has its own fissure swarm. The fissure swarms overlap and are arranged in a westward stepping en-echelon fashion.

The Krafla volcanic system is believed to have been active for about 200,000 years [7]. The volcanic system consists of a central volcano, approximately 20 km in diameter, bisected by an about 90 km long NNE-SSW trending fissure swarm (Figure 2). The fissure swarm takes up and accommodates most of the crustal spreading in the part of the northern volcanic zone around it. The central volcano is, generally speaking, characterised by gently sloping topographic high with a caldera in the middle. The caldera is about 8 to 10 km in diameter (W–E elongated) and is partly filled with volcanic products. It developed from an explosive eruption producing dacitic welded tuff about 110 ka ago [7]. The volcano remains active with recurring volcanic episodes. Krafla has a bimodal volcanic character: for long periods, it produces mainly basaltic fissure eruptions and dike injections into the fissure swarm, but intermittently, it erupts semi-silicic to silicic magma or tephra. Three such eruptions have been identified and dated [7,8]. The oldest is the phreatic eruption 110 ka ago forming the visible caldera, then subglacial lava eruptions about 80 ka old forming Hlíðarfjall SW of the caldera, Jörundur ESE of the caldera and Rani NW of the caldera and a 24 ka subglacial fissure eruption forming Hrafntinnuhryggur, an obsidian ridge south east of mount Mt. Krafla (for locations, see Figures 2 and 3). Since the occurrence of the 80 ka silicic formations outside the caldera, volcanism has mainly been centred in the eastern part of the fissure swarm. A simplified geological map of Krafla is shown in Figure 3.

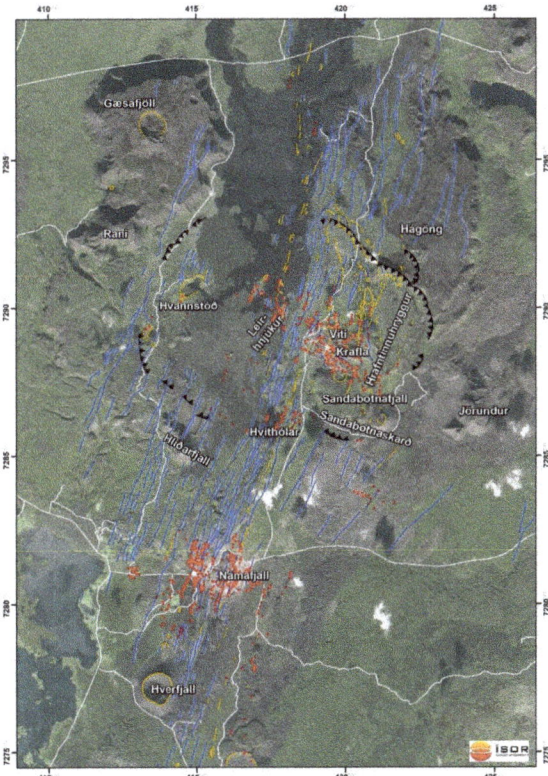

Figure 2. An overview of the Krafla area and the central part of the fissure swarm. The figure shows the visible 110 ka caldera (black), faults and fissures close to Krafla (blue), craters and eruptive fissures (orange) and geothermal manifestations (red). The figure also shows the subsidiary geothermal area in Námafjall, south of Krafla. Coordinates are UTM, WGS84, zone 28 in km.

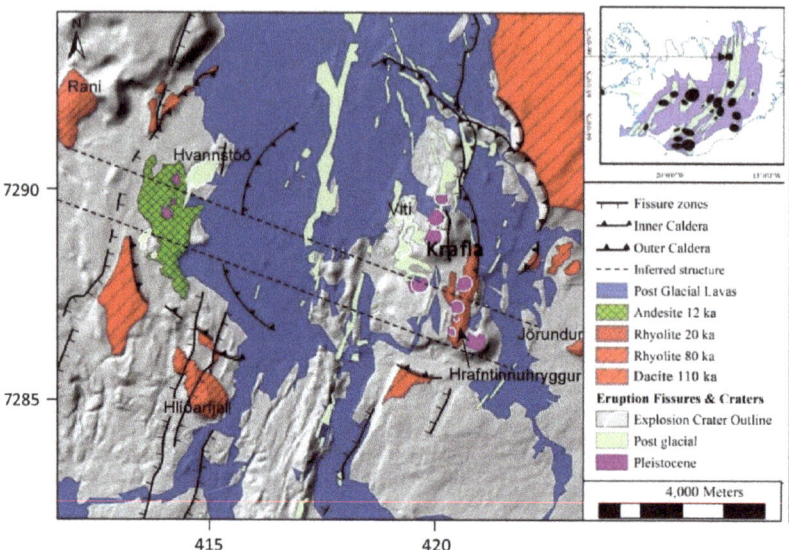

Figure 3. A simplified geological map of Krafla. The map also shows inferred structural features discussed in this paper. Note the abundant explosion craters in the western slopes and south of Mt. Krafla. In the upper right inlet, black dots are volcanos and green strips represent fissure swarms. Coordinates are UTM WGS84 in km. (Modified from [9]).

The Krafla volcanic system has also shown bimodal behaviour in tectonics and crustal spreading. About 8 ka ago, the spreading moved from the eastern part of the fissure swarm to the western part and back again about 3 ka ago [7] (Figure 4). The shifting of the spreading to the west did not result in increased volcanic activity in the western part of the caldera, only one eruption has been identified in Hvannstóð, which is about 5 ka old [7] (Figure 4). Resistivity survey shows that a mature high-temperature geothermal system never developed in the western part of the caldera (see Section 3.3 below) and only minor extinct geothermal manifestations south of Hvannstóð (Figure 4). Extrusive volcanism is almost exclusively found in the eastern part, before and after the western part of the fissure swarm was active [7], and a high-temperature geothermal system with extensive surface manifestations was developed in the eastern part (see Figure 2). The shifting back of the spreading from the western to the eastern part 3 ka ago had a profound influence on the geothermal system, as discussed below.

The fact that volcanic activity takes place dominantly in the eastern part of the 110 ka caldera could be explained by considering some details of the fissure swarm. Figure 5 shows that south of the Krafla caldera, the crust east of the fissure swarm is moving in an approximate direction of 22° south-east, while to the north it is moving approximately 4° south-east. These different spreading directions have been confirmed by GPS measurements [10]. The difference is about 18°, leading to a N–S opening component in the eastern part of the volcano of about one-third of the spreading motion. This opening component favours the ascent of mantle-derived magma and volcanism manifested by subglacial extrusives: Mt. Krafla, Dalfjall and Sandabotnafjall south of Mt. Krafla, abundant explosion craters in the western slopes of and south-east of Mt. Krafla (Figure 3) and post-glacial eruptions and dike injections centred at Leirhnjúkur.

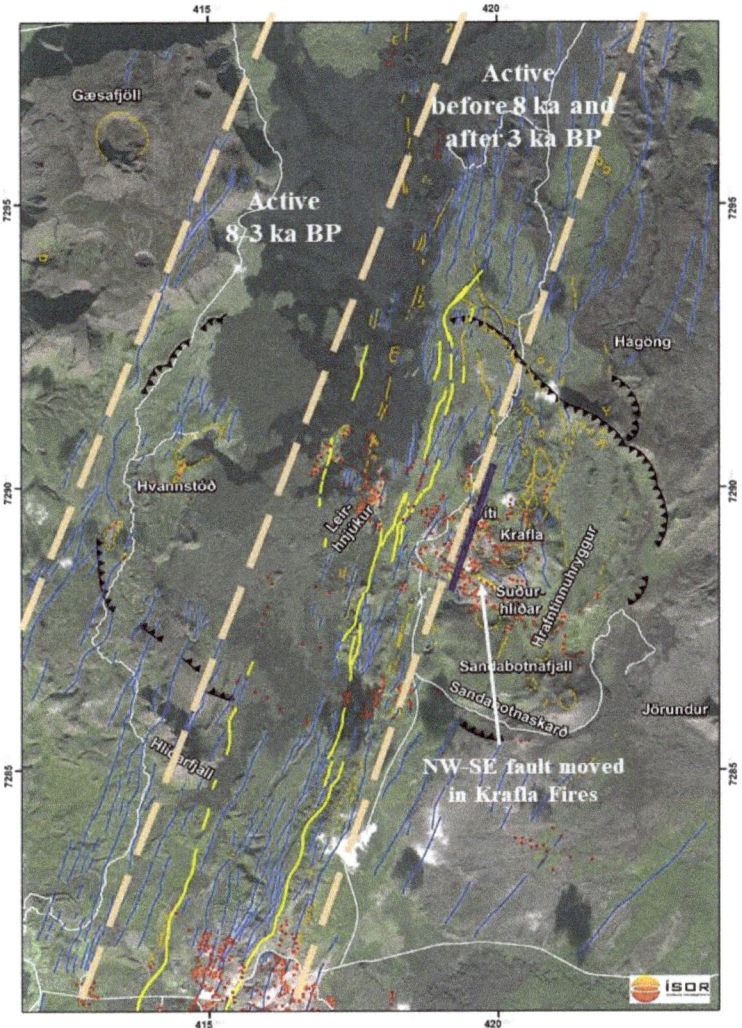

Figure 4. Different parts of the fissure swarm active at different times. Blue lines are faults and fissures, yellow lines show the eastern- and westernmost faults and fissures activated in the Krafla Fires and NW–SE fault south of Mnt. Krafla that moved in the Krafla Fires. The purple line marks the Hveragil gulley. Orange lines mark eruptive craters and fissures. Coordinates are UTM, WGS84, zone 28 in km.

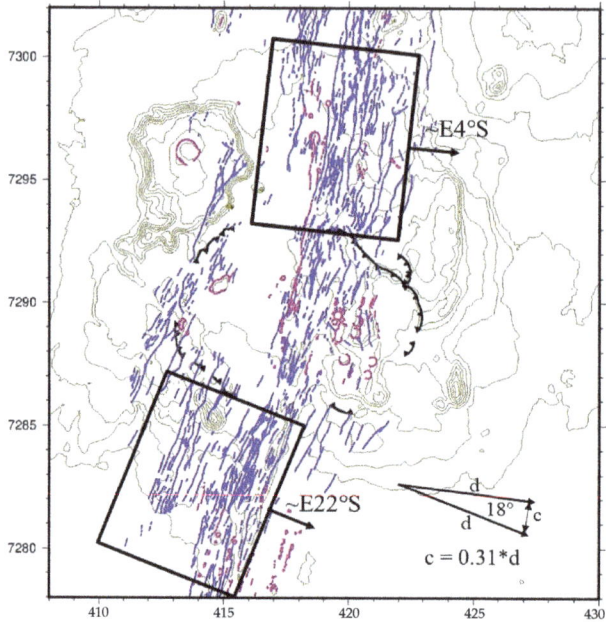

Figure 5. Different crustal spreading directions south and north of the Krafla caldera leading to an opening component in the eastern part. The inlet bottom right shows how spreading by, d, leads to opening component, c, in the east. Blue lines represent faults and fissures and purple lines eruptive fissures and craters. Coordinates are UTM WGS84 zone 28 in km.

During the past 3000 years, eruptions in Krafla have taken place every 300–1000 years [7]. Geothermal drilling in the central eastern and southern parts of the visible caldera has shown a pile of alternating extrusive hyaloclastites and lavas, underlain by intrusive rocks with similar bimodal compositional distribution of the volcanic and plutonic rocks in the substrata [11,12]. The depth to the intrusive rocks varies from about 800–1100 m in the central part of the caldera to about 1500–1600 m in the southern part (see discussion below).

The two latest eruptive phases of the Krafla volcano were the Mývatn Fires in 1724 to 1729 and the Krafla Fires in 1975 to 1984. The Mývatn Fires started with a phreatomagmatic eruption in the Víti crater (Figure 3) emitting glassy rhyolitic bombs and minor basaltic, felsite and gabbroic lithics, representing intrusive rocks at depth [13]. Repeated dike injections into the fissure swarm, centred at Leirhnjúkur about 2 km west of Víti, started after the initial explosion. Two main basaltic fissure eruptions took place in 1727 and 1729, mainly within the caldera but with a small eruption west of Námafjall, about 5 km south of the caldera [7].

The Krafla Fires started by a small basaltic fissure eruption within the caldera. Repeated dike injections, occasionally with small eruptions, took place until 1980. From 1981 to 1984, the four main fissure eruptions took place. All the eruptions and most of the dike injections were into the northern part of the caldera and fissure swarm, but a few were towards south, one all the way to Bjarnarflag, west of Námafjall [7,14]. During the Krafla Fires, the active fissure swarm widened by about 9 m close to the caldera and subsided by up to about 1–2 m, while the rift flanks were uplifted [15–17].

During the Krafla Fires, periodic uplift and subsidence took place in the caldera which were closely monitored by levelling and tilt metres [14,18–21]. In quiet periods, the ground was up-lifted by about 5 mm/day with a centre of up-lift shown approximately by the red star in Figure 6. During dike injections (and eruptions), a rapid subsidence took place and then up-lift again at the end of injection. Modelling of the deformation due to a single Mogi source indicated the centre of inflation/deflation at the depth interval of 3.9–7.5 km. Detailed study of ground deformation during and after a Krafla Fires eruption in September 1984 [18] and multiple magma reservoirs were suggested, where deeper reservoirs feed a shallow reservoir at about 2.6 km depth. This is an interesting idea and will be taken up below.

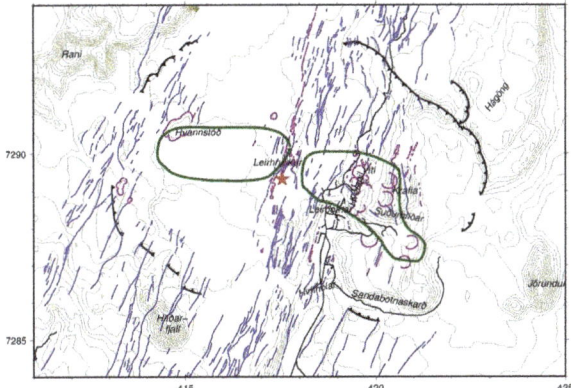

Figure 6. The location of S-wave "shadows" observed during the Krafla Fires (green). The red star shows approximate centre of uplift and subsidence [18]. Blue lines are faults and fissures and purple lines are craters and eruptive fissures. Coordinates are UTM, WGS84, zone 28 in km.

A study of seismic wave propagation from earthquakes during the Krafla Fires [22] revealed volumes within the caldera where seismic S-waves were highly attenuated or could not pass through. The estimated areal extent of these S-wave "shadows" is shown in Figure 6. The upper boundaries of the shadows were estimated to be at about 3 km depth and the lower boundaries, though poorly constrained, at about 7 km depth. These volumes have been interpreted to contain magma. This will be further discussed below.

3. Geophysics

3.1. Geodesy

During and after the Krafla Fires, crustal movements have been monitored by levelling and GPS measurements and later by Interferometric Synthetic Aperture Radar (InSAR). Inflation continued until 1989, when a deflation started. In the beginning, the subsidence was about 7 cm/year but decreased exponentially with time and by 2008 it was practically over [23]. After that, crustal deformation (subsidence) has been attributed mainly to pressure draw-down in the geothermal reservoir. From recent InSAR and GPS measurements, it appears that the deflation of the Krafla caldera since 1989 has reverted to a minor inflation at a rate of 10–15 mm/year in 2018–2019 [24]. Whether this is due to magma transport or re-injection of fluids into the geothermal reservoir needs to be worked out.

3.2. Gravity

An extensive gravity survey was carried out from 1967 to 1984, covering the Krafla area and its surroundings [25]. Figure 7 shows a de-trended residual Bouguer gravity map based on these data. The map shows a relative gravity high at and within the rims of the visible 110 ka caldera. This gravity high, more or less surrounds a gravity low inside the caldera. Superimposed on this gravity low is a relative gravity high at and east of Leirhnjúkur.

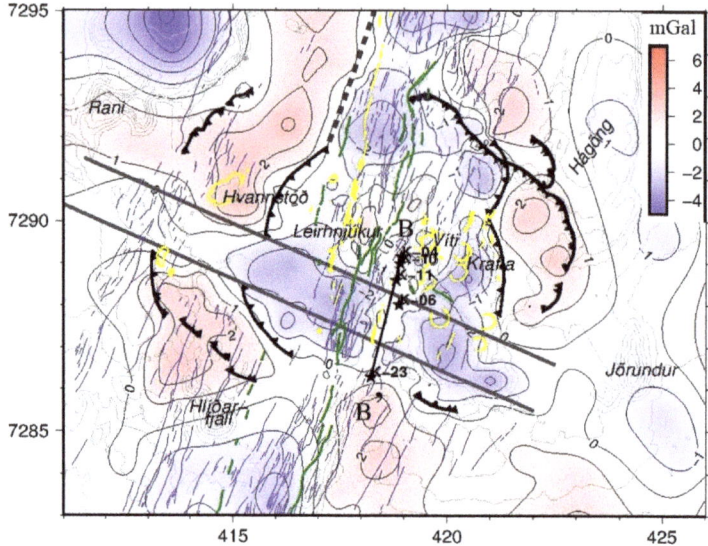

Figure 7. A de-trended residual Bouguer gravity map (mGals) of Krafla volcano. The visible 110 ka caldera rim is shown with heavy black hatched lines. An inferred buried inner caldera rim is shown with lighter black hatched lines, an inferred lower density transverse structure is shown with grey lines and a clear density boundary to the north is shown as grey broken line. The location of the lithological section in Figure 8 is shown by straight black line (black stars are wells on the section). Blue and green lines show faults and fissures, and craters and eruptive fissures and explosion craters are shown with yellow lines. Coordinates are UTM, WGS84, zone 28 in km.

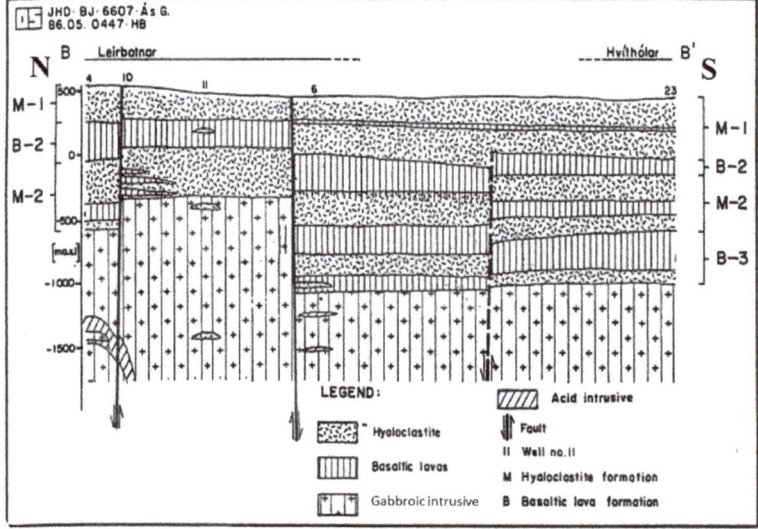

Figure 8. A lithological cross-section based on drilled wells on a N–S trending profile in Krafla [11]. The location of the profile is shown in Figure 7.

Figure 7 shows other interesting features. The caldera is bisected by two more or less linear gravity lows. One is along the part of the fissure swarm that was active in the Krafla Fires (bounded by green lines in Figure 7). The other is an ESE-WNW trending gravity low from west of Mt. Jörundur in the SE and to the valley Gæsadalur (south of Gæsafjöll) in the NW. Where these anomalies would cut through the caldera rim, the rim is not visible.

If the spreading rate is assumed to be about 1.8–2.0 cm/year [26], the total spreading distance since the formation of the 110 ka caldera is about 2 to 2.2 km. By assuming that 75%–100% of the spreading has taken place in the fissure swarm through the caldera, it should be torn apart by some 1.5–2.2 km. The gaps in the southern and northern caldera rims (as seen on surface) are about 3.5 to 4 km, therefore, parts of the rims are subsided and buried. The same might partly apply to the gaps in the visible eastern and western parts of the caldera rim, but the fact that they are cut through by a low-gravity anomaly suggests that the caldera might be torn apart by an ESE-WNW trending transverse structure with rocks of relatively low density.

The origin of the ESE-WNW gravity low is not clear. It is likely to be due to some transform tectonics where the spreading is gradually migrating westwards, towards the oceanic ridge north of Iceland. Similar structures or trenches are known in transform zones further to the north, where the crustal spreading is migrating westwards. The transverse structure could be of similar origin as Lake Botnsvatn in the Húsavík transform zone, i.e., a pull-apart-basin. The transverse structure in Krafla has almost exactly the same strike as the Husavík–Flatey transform at the southern margin of the TFZ.

There is a clue of the nature of the ESE-WNW low-gravity anomaly from drilling. Figure 8 shows a lithological section from the centre and towards the southern rim of the caldera [11]. The wells on which the section is based, and the location of the section are shown in Figure 7. North of the transverse structure the section shows an about 900–1100 m thick pile of hyaloclastite with interbedded lava flows and dominant intrusions below (wells K-11, K-10 and K-04). In well K-6, within the structure, the intrusions come at about 600–700 m greater depth and with correspondingly thicker extrusive less dense rocks. The higher gravity north of the structure, therefore, reflects intrusions at a shallower depth.

As stated earlier, the gravity is relatively high at and inside the visible caldera rims in the southwest, northwest and east (Figure 7). These high gravity anomalies are bounded by steep gravity gradients towards a gravity low in the centre of the caldera, reflecting less dense rocks. In the eastern part of the caldera, the gradient coincides with arc-shaped eruptive fissures from Hólseldar, about 2 ka old [11] (Figure 7). It might be tempting to argue that the high gravity at and inside the caldera rim is due to dense intrusions, but the steep gradients towards lower gravity show that the density contrasts are at shallow depth. This gravity low can be explained by that there is another caldera with low-density rocks buried inside the visible caldera. The estimated rims of this inferred inner caldera and the bisecting ESE-WNW transverse low-density structure are shown in Figure 7.

Even though the last glacial stage is normally considered to have started at about 110 ka BP, the results of geological studies show that the Krafla area was not glaciated until about 80 ka ago [27]. In the 30 ka between the formation of the outer caldera and until glaciation, it has been mostly filled with subaerial lava flows, up to the lowest parts of its rims in the rift graben. The inner caldera was probably formed shortly after the area was glaciated. It is suggested here that the inner caldera was formed 80 ka ago in sub-glacial eruption(s) forming the rhyolitic mountains Hlíðarfjall, Jörundur and Rani outside the 110 ka caldera (Figures 2 and 7). In [9], several examples of caldera formation are discussed where drained rhyolitic magma is erupted far outside the caldera subsidence. The caldera was later filled with subglacial hyaloclastite of considerably lower density than the subaerial lavas filling the outer caldera, resulting in the gravity low. Any visible signs of the inner caldera are now completely masked by Holocene lavas. The presence of this buried inner caldera and the ESE-WNW transverse structure get support from a resistivity survey discussed below. Some bounds can be put on the age of the transverse structure. It cuts through the inner caldera, so it is younger than 80 ka. It is, however, not seen cutting through Sandabotnafjall, just south of Mt. Krafla, which is estimated to be 35–40 ka old [7], so the age of the transverse structure is somewhere between 80 and 35–40 ka.

The Bouguer gravity map in Figure 7 shows yet another feature worth mentioning. There is a sharp gravity change at a line in, and parallel to the fissure swarm to the north (grey broken line in Figure 7). The fissure swarm hosts less dense rocks east of this line than to the west. This indicates that after the glaciation, the spreading and subsidence have mainly taken place in the eastern part of the fissure swarm.

3.3. Resistivity

Extensive resistivity surveys for geothermal exploration (dating back to the early 1970's) have been carried out in Krafla. High-temperature geothermal systems have a characteristic resistivity structure due to geothermal alteration [28,29]. Below resistive near-surface unaltered rocks, they have a shallow low-resistivity "cap" (clay cap) with conductive smectite and zeolite alteration minerals formed in the temperature range of 50–230 °C. The low-resistivity cap is underlain by a resistive "core" with resistive chlorite and epidote alteration minerals formed at temperatures above 230 °C.

Figure 9 shows a resistivity map at 200 m a.sl. (at a depth of about 300 m) based on 1D inversion of central-loop TEM soundings [30]. Areas where high resistivity (resistive core) has appeared below the conductive clay cap are gridded red. The figure also shows the 110 ka caldera, the inferred inner caldera and the transverse low-density structure.

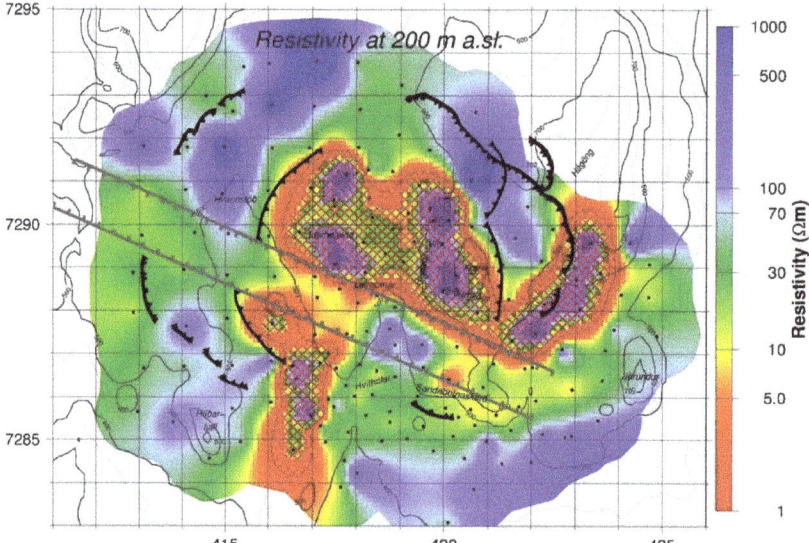

Figure 9. Resistivity at 200 m a.sl. (about 300 m depth) based on 1D inversion of TEM soundings. Areas where higher resistivity is observed below low resistivity are shown as red gridded (black dots are TEM soundings (reproduced from [30])). The thick black hachured lines show the 110 ka caldera. Thinner black hachured lines show buried inner caldera and grey lines mark a buried transverse structure inferred from gravity. Coordinates are UTM WGS84 zone 28 in km.

Figure 9 shows that high-temperature geothermal alteration is confined within the inner caldera, except in the western part of the southern arm of the fissure swarm and at the rim of the outer caldera in the southern end of Hágöng, east of Mt. Krafla. Both of these anomalies have been drilled into and geothermal alteration was found in accordance with the resistivity, but the present temperature is far below that responsible for the alteration. Therefore, cooling has occurred at these places outside the inner caldera. Figure 9 shows that no mature high-temperature geothermal system has developed outside the inner caldera except for those mentioned. It also shows that, at this depth, high-temperature alteration is absent in the transverse structure, which means a lower temperature.

To study the deeper resistivity structure, most of the TEM sounding sites (black dots in Figure 9) were later visited for MT soundings. Figure 10 shows slices through a resistivity model resulting from 3D inversion the MT soundings (static shift corrected by the TEM).

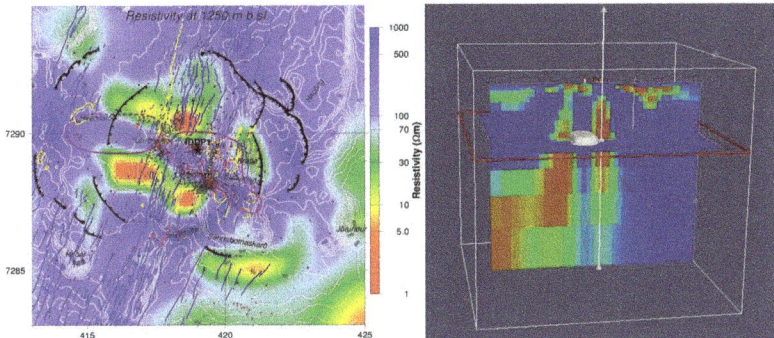

Figure 10. A resistivity model for Krafla from 3D inversion of MT data. Left: resistivity map at 1250 m bl. sl. The two calderas are shown as black hatched lines, purple lines mark the S-wave shadows, grey dots are earthquake epicentres and blue lines are faults and fissures. The location of the well IDDP-1 is shown as a black star. Right: 3D view from the west. The white surface shows the upper boundary of the western S-wave shadow at 3 km depth (reproduced from [31]). Coordinates are UTM WGS84 zone 28 in km.

The figure to the left, shows the resistivity at −1250 m a.sl. (about 1700–1800 m depth). It shows deep, low-resistivity bodies, mainly within the inner caldera except for a relatively low resistivity at the south-east rim of the outer caldera. The low-resistivity bodies within the inner caldera border an ESE-WNW high-resistivity structure, which roughly coincides with the S-wave shadows. This high-resistivity structure hosts most of the seismicity and the main production field is in its central and eastern part. The figure to the right shows the shallow conductive clay cap is fairly well in agreement with the 1D inversion of TEM and drilling, and that the two low-resistivity bodies extend down to about 6 km depth. Around that depth, the northern one seems to start to deviate to the north.

3.4. Seismic Studies

ÍSOR operates a local seismic monitoring network in Krafla for Landsvirkjun (National Power Company), which operates the Krafla power plant. Even though there have been no signs of volcanic unrest since the end of Krafla Fires, Krafla turns out to be very seismically active in the central part of the inner caldera. Figure 11 shows earthquakes located in Krafla from November 2015 to November 2016 [32], a rather typical annual seimicity. There are five main clusters, one just SW of Mt. Krafla, just north of the transverse structure, two at injection wells K-26 and IDDP-1 (probably mostly due to injection) and two at and north of Leirhnjúkur. The depth sections show that maximum depth of the earthquakes, i.e., the brittle/ductile boundary, which for basalt, is estimated in the range of 600–800 °C [33], is at about a 2–2.3 km depth north of the transverse structure. There is also a small cluster within the transverse structure south of Leirhnjúkur. The depth sections show that seismicity there extends much deeper than in the north or to about 3.5 km. This indicates lower upper crustal temperatures within the transverse structure. This is in agreement with the distribution of high-temperature alteration at a 300 m depth shown in Figure 9, i.e., a lack of high-temperature alteration within the transverse structure.

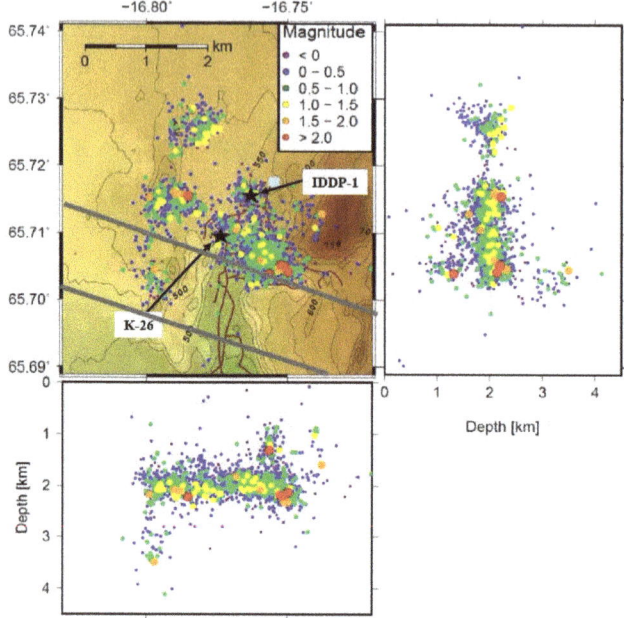

Figure 11. Earthquakes located in Krafla from November 2015 to November 2016. The map also shows the location of reinjection wells (K-26 and IDDP-1) and the transverse gravity low (grey lines) (modified from [32]).

Recent seismic tomography study of the Krafla area [34] shows some interesting results. Figure 12 (left) shows the estimated P-wave velocity (Vp) at 0.5 km bl.sl. (at about 1 km depth) and (right) at 2 km depth below sea level (about 2.5 km depth). At 0.5 km bl.sl., the tomography indicates low Vp within the transverse structure and higher Vp north of it. At 2 km bl.sl., the situation is the opposite, i.e., high Vp in the southern part of the calderas, but low Vp and low Vp/Vs ratio within the inner caldera north of the transverse structure and above the S-wave shadows. Schuler et al. (2015) suggest that this anomaly reflects rocks containing superheated steam.

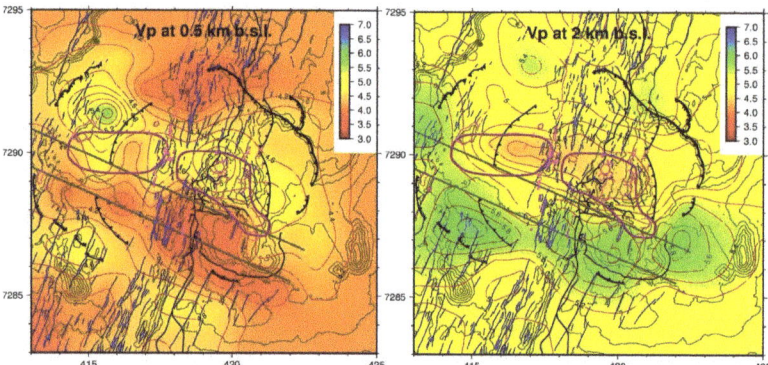

Figure 12. P-wave velocity (km/s) at 0.5 km bl.sl. (left) and at 2 km bl.sl. (right) according to recent seismic tomography (reproduced from [34]). Black hatched lines show the two calderas, grey lines mark the low-density transverse structure, blue lines show faults and fissures and the purple lines mark the S-wave shadows. Coordinates are UTM WGS84 zone 28 in km.

4. The Geothermal System(s)

Through the years, a total of 43 deep exploration and production wells have been drilled in Krafla, some of which are (or have been) used for injection. Figures 13 and 14 show their location and well tracks of deviated wells. The main well field is located in the eastern part of the inner caldera. Through drilling, the stratigraphic structure at depth has been outlined as well as the physical conditions of the geothermal system [11,12,35,36].

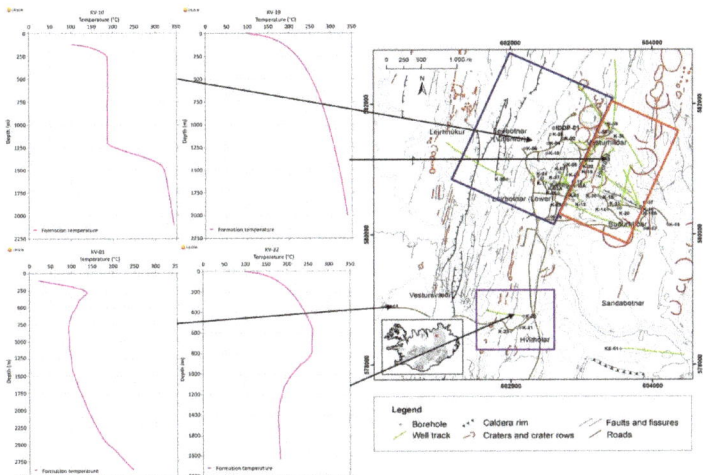

Figure 13. Drilled wells in Krafla (red dot on the little inlet map to the right) and the three main production fields (right) with different thermal character. The fields are: Vtísmór and Leribotnar (blue frame), southern and western slopes of Mt. Krafla (red frame) and Hvíthólar (purple frame). To the left are shown characteristic temperature profiles in wells in each production area and a well drilled in the fissure swarm west of Hvíthólar. For well KS-01, east of Hvíthólar, see text.

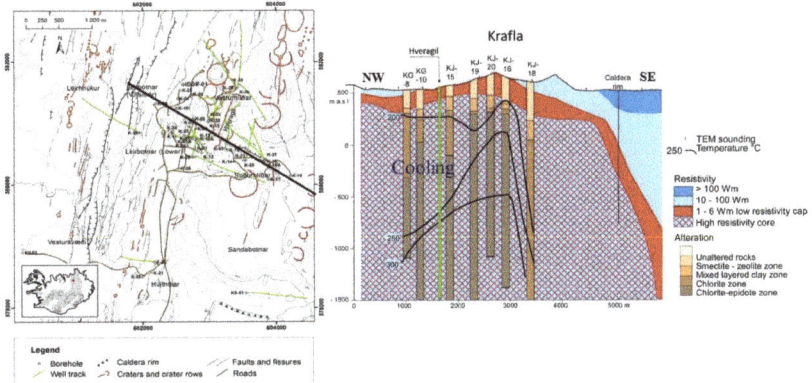

Figure 14. Left; The locations and tracks of wells in Krafla (green lines) and the location of the cross-section to the right (black line). Right; A resistivity cross-section with alteration mineralogy and formation temperature in wells.

The drilling has revealed that different "subfields" in Krafla have quite different thermo-hydraulic characteristics (Figure 13). In part of the production area, Leirbotnar and Vítismór (blue frame in Figure 13), the geothermal system has two different thermo-hydraulic characters at different depths.

Temperature profiles show that above 1000–1200 m in depth, there is a system with two phase (boiling) conditions close to the surface but then almost isothermal with depth, at about 200 °C. Below the isothermal part, the temperature rises almost linearly up to the boiling point curve (boiling point versus depth curve) in an underlying two-phase system. Wells in the southern and western slopes of Mt. Krafla (red frame in Figure 13) show two phase conditions from maximum drilled depth (2–2.5 km) to the surface and wells at Hvíthólar (purple frame in Figure 13) show a temperature inversion. Well KS-01 in Sandabotnaskarð, east of Hvíthólar, has a temperature profile (not shown) following the boiling point curve from the bottom (2500 m depth) and up to about 1000 m in depth, but temperatures below the boiling point curve at shallower depths (ÍSOR database). Finally, well KV-1, in the active fissure swarm west of Hvíthólar, shows extensive cooling. Resistivity surveys show high-temperature alteration there (Figure 9) and the alteration mineralogy of the well showed that temperatures there had been much higher than the current measured temperature.

Figure 14 shows a resistivity cross-section from Vítismór in the NW to the southern slopes of Mt. Krafla. Superimposed on the section is the alteration mineralogy and estimated formation temperature in wells close to the section. The figure shows an excellent correlation between resistivity and alteration, i.e., low resistivity at shallow depth in the smectite and zeolite alteration and increasing resistivity in the chlorite and chlorite/amphibole alteration [28,29]. The isotherms in the wells in the western part, west of Hveragil (a purple line in Figure 4), and also in the eastern most well (K-18), show much lower temperatures than the ones causing the alteration, so cooling has taken place.

5. New Conceptual Model

The complexity of the Krafla volcano and its geothermal system(s) has puzzled geoscientists for decades. Here, an attempt is made to shed some light on this complexity by putting it in context with the geological structures, tectonics and geological history discussed above. The first conceptual model of the geothermal system in Krafla was presented in 1977 [37]. It was based on limited data, mainly surface studies and the results for the first eleven wells in the area [35,38]. In that model, it was assumed that the shallow and the deep thermal regimes in the Vítismór-Leirbotnar area are separated by an impermeable cap-rock. It was assumed that there was a main up-flow from a deep two phase geothermal system through Hveragil (bold purple line in Figure 4) just west of Mt. Krafla and that the near-isothermal upper system in the west was due to a westward lateral flow from the up-flow under Hveragil. The difference in the contamination by volcanic gases during the Krafla Fires, discussed above, was considered to support this conceptual model [5]. This conceptual model has been considered valid until recently [39]. The author did, however, find this hypothesis unlikely and in the most recent review report, on the Krafla geothermal system [12], a quite different scenario was proposed. There is no obvious candidate for an impermeable cap-rock in the lithology in the wells in the Vítismór-Leirbotnar area, but the change in thermal character occurs at depths where intrusive rocks start to dominate. The area is within the presently active part of the fissure swarm and is likely to have very anisotropic permeability, i.e., high permeability along the fissure swarm but much lower permeability perpendicular to the faults and fissures. It is, therefore, unlikely that lateral flow, above a cap-rock from an up-flow in the east, would be to the west, perpendicular to the high permeability. An alternative explanation of the two thermo-hydraulic regimes is given below.

5.1. The Role of Permeability

Permeability plays a crucial role in the existence and development of volcanic high-temperature geothermal systems. Model calculations show that, in order for a magma intruded into the shallow crust to produce a two-phase geothermal system, permeability has to be within the rather narrow range of 0.5–5 mD (1 mD = 10^{-15} m^2) [40]. If the permeability is lower, the intrusion cools over a long time by heat conduction, but if the permeability is higher, the intrusion cools rapidly by vigorous one-phase (water) convection.

Model calculations show that if a magma intrusion is emplaced in rocks with permeability in the above range, a two-phase convective geothermal system is formed in time of the order of 1000 years and, if no further intrusions occur, the lifetime of such as system is of the order of 10 ka [40] (though depending on the size of the intrusion).

Figure 15 shows the results of two-dimensional (2D) numerical modelling using the HYDROTHERM programme [41,42]. The figure shows the thermo-hydraulic states of a geothermal system developed after the emplacement of an 800 m wide (infinitely long) "dike" intrusion with a temperature of 1100 °C, extending up to 2 km below the surface into rocks with a temperature gradient of 100 °C/km. The modelling takes into account the latent heat of the magma and the permeability created when the magma solidifies and contracts.

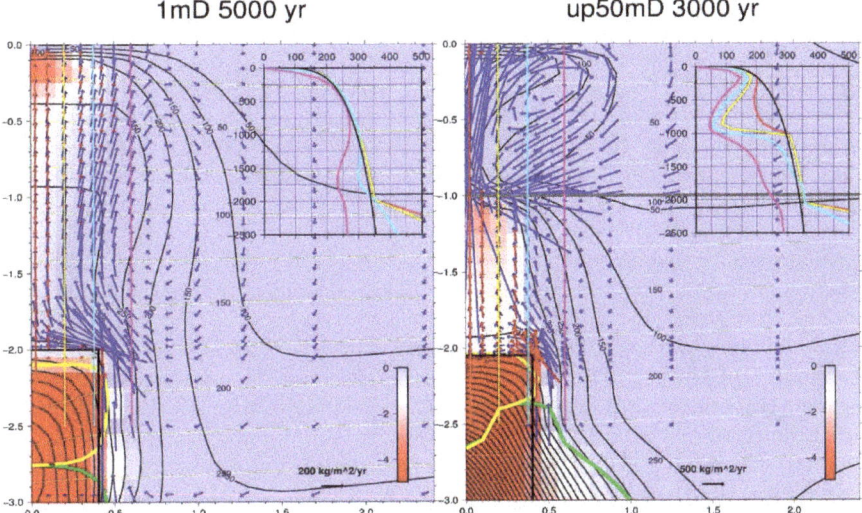

Figure 15. Two-dimensional modelling of thermo-hydraulic states of geothermal systems developed after the emplacement of an 800 m wide "dike" intrusion extending up to 2 km below surface into rocks with a temperature gradient of 100 °C/km (the scale of the horizontal and vertical axis is in km). To the left, is the system 5 ka after emplacement into rocks with the permeability 1 mD. The temperature (°C) is shown by black contours and the pressure (bars) by yellow contours. Phase conditions (water/steam) are shown by colours (light blue is pure water and dark red is almost pure steam, (actual scale 10 times the logarithm of water saturation). Mass flow (kg/m^2/year) is shown by arrows (scale at the bottom of the figures), blue for water phase and red for steam phase. The graph shows temperature profiles from hypothetical wells at different distances from the centre of the intrusion (related by colour) in the upper-right corners. The thick yellow line marks region of superheated steam and the green line a region with supercritical conditions. To the right, is shown a thermo-hydraulic situation 3 ka after an intrusion is placed in a host rock of permeability 1 mD, underlying a 1 km thick pile (thick grey line) of much higher permeability (50 mD in this case).

The figure to the left shows the developed geothermal system 5 ka after the dike injection into a host rock with a permeability of 1 mD. The figure shows a convecting two-phase geothermal system with up-flowing water and steam above the intrusion and a down flow of colder water to the sides down to the heat source. The panel in the upper-right corner shows calculated temperature profiles for hypothetical wells. Wells within the range of the dike (red and yellow) have temperature profiles corresponding to the boiling point curve (boiling point temperature vs. depth curve), while wells at the edge of or outside the dike (light blue and purple) roughly follow the boiling point curve at shallow depths but show temperature inversion at greater depths due to the in-flow of colder water towards the heat source.

Near the end of the lifetime of the system (of the order of 10 ka, if the heat source is not renewed), the heat source has lost most of its heat, and cold downwards flowing water can no longer gain enough heat to sustain two-phase conditions and the system cools down at depth. Two-phase conditions will, however, still prevail (for some hundreds to thousand years) at a shallower depth in a slowly upwards migrating "bubble", leading to temperature inversion with depth above the cooled heat source. Similar results, using different modelling software have been reported [43].

The right part of Figure 15 shows the thermo-hydraulic situation 3 ka after a similar dyke is emplaced in a host rock of permeability 1 mD, underlying a 1 km thick pile of a much higher permeability (50 mD in this case). The figure shows a two-phase system in the 1 mD host-rock but in the more permeable rocks above, a second and vigorous convection cell develops. The temperature profiles in the upper-right corner show that above the centre of the dike, the temperature follows the boiling point curve near the surface and then becomes almost isothermal until just above the permeability contrast, where it rises sharply towards the boiling point curve. It is, therefore, suggested that the two different thermal regimes in the Vítismór-Leirbotnar field are not due to cap-rock and lateral flow from east, but due to a much higher permeability in the fissure swarm above the intrusions at about 1 km depth. The model can also explain the difference in contamination by volcanic gases in the upper isothermal and the deeper two-phase part during the Krafla Fires, discussed above. The vigorous convection in the high permeability isothermal part above the deeper two-phase part would dilute the gases in fluids coming from below.

It may be questioned whether the simple 2D modelling shown here is justified in this case. However, it will be argued below that the main heat sources powering the geothermal system in Krafla are ESE-WNW trending intrusions and a dike complex in the inner caldera north of the low-density transverse structure, nearly perpendicular to the fissure swarm. Since permeability above the intrusions is mainly along faults and fissures perpendicular to the heat source(s), a 2D modelling is considered as a valid approximation.

A permeability of 50 mD in the upper 1000 m may seem unrealistically high. The Svartsengi geothermal system in the Reykjanes peninsula SW Iceland, has a wide spread confined nearly isothermal reservoir of about 235–240 °C from about 700 to 2000 m depth [44]. In numerical reservoir modelling of the system a permeability of about 220 mD is needed to simulate this isothermal system [45].

What, from resistivity, appears as one system in the northern part of the inner caldera is actually divided into two very different parts. East of Hveragil, in Vesturhlíðar and Suðurhlíðar, there are, generally speaking, two-phase conditions from the depth of drilling and to shallow depths, except in the easternmost well (K-18), which shows cooling in the upper parts although hot near the bottom. West of Hveragil (bold purple line in Figure 4), the geothermal system is divided into two very different parts, a deep two-phase system, below 1 km, and an upper, almost-isothermal, one-phase condition of about 200 °C, except for in the near-surface, where two-phase conditions are reached. Alteration shows, however, that sometime in the past, two-phase conditions existed in the upper part.

As discussed above, until about 8 ka ago, the rifting was in the presently active eastern part of the fissure swarm with considerable volcanism, mainly within the caldera (Mt. Krafla) and to the south. Then, the rifting was shifted to the west for about 5 ka. There seems to have been amazingly little extrusive volcanic activity during that period, the main event being a phreatic eruption from the explosion crater Hvannstóð, about 5 ka ago [7] and resistivity (Figures 9 and 10) shows that no pronounced geothermal system was formed in the western part of the 110 ka caldera.

Here, it is proposed that during the 5 ka, when the rifting was in the western part, intrusions continued to be emplaced in the eastern part and a two-phase geothermal system was formed and sustained there. About 3 ka ago, the spreading shifted back to the presently active part of the fissure swarm. This greatly increased the permeability in the upper part, west of Hveragil, leading to a vigorously convecting isothermal system. At greater depth, continuing intrusions maintained lower permeability with two-phase conditions. Fissures and faults east of Hveragil were not activated when the spreading moved back to the east and two-phase conditions still prevail there.

During the time when the spreading took place in the western part of the 110 ka caldera, intrusions did probably take place in the southwestern part of the inner caldera, south of the transverse low-density structure, and a geothermal system developed. When the rifting turned back east, the permeability increased drastically and cooling took place. Intrusions have not been frequent enough to maintain the heat source(s) and an active geothermal system there. The temperature profile of KV-1 (Figure 13) shows a temperature maximum at a shallow depth and lower temperatures below. This temperature maximum is likely due to a lateral flow in shallow permeable layers. It could be that a small geothermal system still exists further to the west inside the inner caldera and south of the transverse structure.

Hvíthólar (a purple frame in Figure 13) is a small production field with three deep wells, but only one of them is productive. All three wells have temperature profiles showing inversion with depth. Hvíthólar is a small geothermal system just east of the presently active fissure swarm. It is probably isolated from the main system(s) to the north, as indicated by stable isotope ratios (see below). It is suggested that the geothermal system in Hvíthólar is near the end of its lifetime with two-phase conditions at a shallow depth but lower temperature below because the heat sources have run out of heat. Similar conditions are found in the Krýsuvík geothermal area on the Reykjanes peninsula, SW Iceland (ÍSOR database). The geological settings are similar, i.e., hyaloclastite ridges from the end of glaciation. It is suggested that at the end of the glaciation, the rapid pressure drop in the crust did initiate extensive intrusions and eruptions, which generated geothermal systems still present today, but fading out, because their heat sources have not been maintained.

Finally, well KS-1 in Sandabotnaskarð, east of Hvíthólar, encountered a deep geothermal system. The well is deviated to the east, under a fossil alteration on the surface, showing that, recently, there has been geothermal surface activity there. Shallow holes indicate lower temperatures between KS-1 and Hvíthólar (ÍSOR database). Both 1D and 3D inversion MT soundings show a deep low resistivity there (Figure 10). The system encountered in KS-01 seems to be isolated from the systems to the north and Hvíthólar because its fluids have a quite different isotope signature than the others (see below).

5.2. Heat Sources, Magma Ascent and Volcanism

Figure 16 shows a zoom-in of the residual Bouguer gravity in and around the inner caldera. The figure shows a local gravity high within the general gravity low in the inner caldera, mainly north of the transverse structure, but also across it, from Leirbotnar and towards a gravity high at Hvíthólar. These gravity highs show the same general NNE-SSW trends as the fissure swarm. The gravity highs north of the transverse structure extend from Mt. Krafla in the east and to Leirhnjúkur in the west. They are confined within the area where high-temperature alteration is found north of the transverse structure (purple lines in Figure 16, from Figure 9). It is, therefore, natural to assume that they reflect intrusions which are (and have been) heat sources for the main geothermal system within the norther part of the inner caldera.

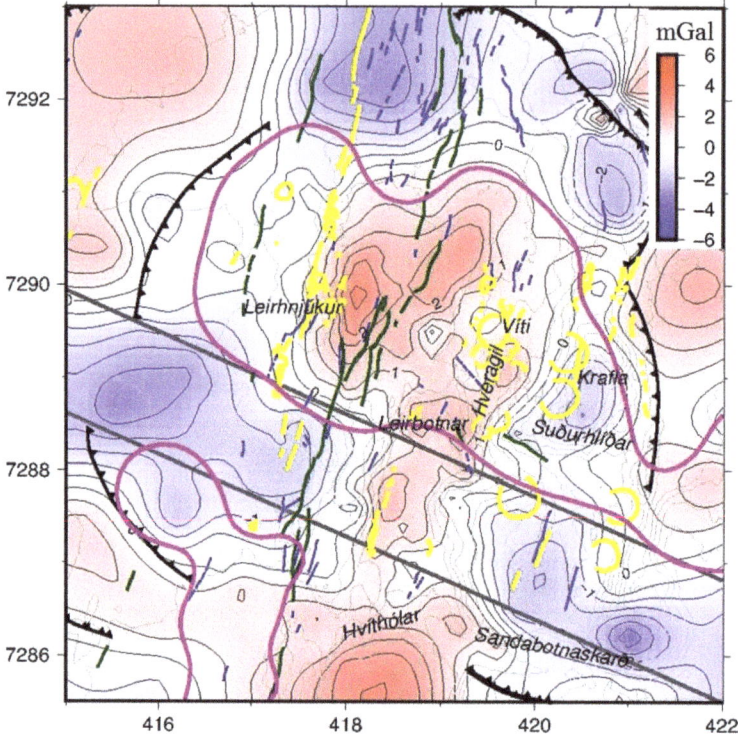

Figure 16. A zoom-in on residual Bouguer gravity in and around the inner caldera. Blue lines show faults and fissures, green lines mark the boundaries of the part of the fissure swarm that was active in the Krafla Fires and an ESE-WNW fault in the southern slopes of Mt. Krafla that moved in the Krafla Fires. Yellow arcs and lines show explosion craters and eruptive fissures. Note, the abundant explosion craters west and south of Mt. Krafla. Grey lines mark the transverse structure and purple lines mark the extent of high-temperature alteration at 200 m a.sl. (from Figure 9). Coordinates are UTM WGS84 zone 28 in km.

As mentioned above, there is a local gravity high through the transverse structure from Leirbotnar to Hvíthólar. It roughly coincides with an eruptive fissure from Daleldar, about 1.1 ka old [7]. Resistivity does, however, not show shallow high-temperature alteration there. This gravity high could be due to shallow dikes that have not managed to generate a high-temperature geothermal system. The high gravity crossing the transverse structure connects to a clear gravity high south of Hvíthólar. This gravity high is a part of a more regional gravity high SE of the 110 ka caldera (Figure 7). The gravity map in Figure 16 does not show any gravity anomalies that could be associated to heat sources in Sandabotnaskarð and in the SW part of the inner caldera.

The gravity highs north of the transverse structure reflect the heat sources of the main geothermal system there. These are probably dikes. The opening component discussed above (Figure 5) makes a pathway for basaltic magma in the deeper eastern part of the calderas from the mantle. At depth, these intrusions/dikes probably have an ESE-WNW orientation, but at shallower levels, some of the magma probably migrates into the faults and fissures between Mt. Krafla and Leirhnjúkur, resulting in the NNE-SSW pattern of the gravity highs seen in Figure 16. For some reason, dike injections to the south into the transverse structure seems limited except for the gravity high extending towards Hvíthólar, discussed above. The ESE-WNW dike complex seems to be delineated in the resistivity model from a 3D inversion of MT data (Figure 10). It appears as a high-resistivity zone bordered by low-resistivity, both to the south and north. The resistive central part of the dike complex probably reflects fully crystalized rocks, while the conductive margins might be a partial melt and hence explaining the S-wave shadows (see discussion below).

The main pathway of ascending basaltic magma from the mantle is probably offset to the east of the presently active spreading as manifested by the abundant explosive craters in the western slope and south of Mt. Krafla (Figure 16). In between rifting events, basaltic intrusions may enter east of the fissure swarm without reaching surface but triggering explosions/eruptions (Figure 16) ejecting silicic tephra (see discussion below). In major rifting events with long dike injections and fissure eruptions like the Mývatn and Krafla Fires, the magma flows up and to the west towards Leirhnjúkur in the presently active fissure swarm which, on surface, looked as the centre of activity. This idea gets support from the fact that the Mývatn Fires started with an explosive eruption in Víti, just west of Mt. Krafla and about 2 km east of Leirhnjúkur and the fact that at the beginning of the Krafla Fires, fumaroles SE of Víti had large increase in volcanic gases, even larger than fumaroles in Leirhnjúkur [11]. It also gets support from the fact that an ESE-WNW trending fault in the southern part of Mt. Krafla moved in the Krafla Fires (Figures 4 and 16). As the magma ascends, it starts to degas, causing the severe gas contamination in the deeper two-phase part of the geothermal system west of Mt. Krafla. As mentioned earlier, it has been suggested [18] that there are several magma reservoirs in the volcano, based on detailed crustal deformation during an eruption in Krafla Fires. These might be the magma plumbing system conveying magma from a deep up-flow in the east and towards the fissure swarm at Leirhnjúkur.

5.3. Origin of the Geothermal Fluids

The origin of the geothermal fluids in Krafla has been studied and debated for many years. The origin of geothermal fluids is normally estimated on the bases of the deviation of ratios of stable isotopes of oxygen, $\delta^{18}O$, and hydrogen, δD, from that of Standard Mean Oceanic Water (SMOW) [46] and comparison with local meteoric values [47]. There can be a significant oxygen isotope exchange due to water-rock interaction but δD is assumed to be better preserved and a better indicator of the original fluid.

The δD of local rainfall in the Krafla area is about −86‰ [48]. The δD of well discharge in the Vítismór-Leirbotnar area is about −87‰ but about −92‰ in the southern slopes of Mt. Krafla east of Hveragil and in Hvíthólar [47]. This is considered a significant difference. In the discharge from well KS-1 in Sandabotnaskarð, δD is even lower or about −115‰ (ÍSOR database), similar to wells in Námafjall geothermal system south of Krafla. In [47], it is concluded that the geothermal system under Vítismór-Leirbotnar is recharged by local meteoric water and state that the geothermal systems east of Hveragil and Hvíthólar seem to have "the same source of recharge" of water from higher elevation.

In a recent study [49], it is argued that the isotope composition of the discharged fluids from the high/excess enthalpy wells east of Hveragil is distorted from that of the actual reservoir fluid towards a lower δD due to phase segregation and that the reservoir fluids in the geothermal system(s) in the inner caldera, north of the transverse structure, are of the same origin, i.e., local rainfall. This supports the idea put forward above, that what looks like two different geothermal systems east and west of Hveragil, are actually one system, the only difference being different permeability in the upper part.

The low δD in Hvíthólar and in KS-01 in Sandabotnaskarð cannot be attributed to phase segregation like in the flanks of Mt. Krafla. The enthalpy there is about 2500 kJ/kg while the enthalpy in Hvíthólar is about 1200 kJ/kg [12] and about 1500 kJ/kg in KS-01 (ÍSOR database). The geothermal fluids in KS-01 seem to originate from the general groundwater flow from higher elevation in the south with a similar δD as the fluids in Námagjall and local groundwater south of Námafjall [47]. The geothermal fluids in Hvíthólar might be a mixture of local rainwater and groundwater from south. It, therefore, seems that there is no (or very little) hydrological connection across the transverse low-density structure between the geothermal system in the northern part of the inner caldera and Hvíthólar and the system encountered by KS-01 in the southern part. Furthermore, as mentioned above, geophysical data and the results of temperature measurements in boreholes indicate that the geothermal activity in Hvíthólar and Sandabotnaskarð are separate systems.

6. The Role of Melting

Two wells in Krafla have encountered silicic melt, K-39 and IDDP-1. K-39, which was drilled directionally towards east under the southern slopes of Krafla (for location see Figure 14), hit melt close to the bottom at about 2600 m below the surface [50]. The stuck drill string, cut loose by blasting, contained fresh rhyolite glass (69.0 to 78.8 wt.% silica). The bottom section of the hole was plugged with cement. The glass was interpreted to be quenched rhyolite magma derived by partial melting of geothermally altered basalt in a contact zone of a gabbroic intrusion [50].

IDDP-1, which was designed to be drilled vertically to 4 to 5 km in depth and into super critical fluids (for location see Figures 11 and 14), surprisingly hit silicic melt at the depth of 2104 m [51]. After drilling about 30 m with total loss of circulation, the penetration rate suddenly increased and the torque on the drilling assembly increased [52]. After pulling the drill string somewhat up and lowering it again, the drill bit got stuck at the depth of 2095 m, so the magma had been squeezed up into the borehole and a large amount of both brown and clear glass was recovered. The drilling was then halted and a cemented casing was installed to 1960 m depth. The well was cooled by injecting cold water into it for a month. After about seven months of heating up, discharge tests started. In the beginning of the flow-test, the well discharged dark water and steam (Figure 17). After the well had discharged for some time, it discharged superheated steam at a temperature of around 450 °C and a pressure of 140 bars at the well head. The maximum flow was around 45–50 kg/sec with an enthalpy of about 3200 kJ/kg giving about 200 MWt and 35 MWe [53].

Figure 17. The beginning of a flow test of IDDP-1 (photo G. Ó. Friðleifsson).

The fluid discharged from IDDP-1 was enriched in CO_2 and H_2S as well as chlorine (Cl), fluorine (F) and boron (B) relative to normal geothermal fluids in Krafla [54]. The black colour of the discharge in the beginning was due to corrosion of the casing by hydrochloric acid formed when the chlorine from the superheated steam dissolved in water droplets. Several other wells in Krafla had the same character when discharged, so-called "black death", and are thought to have hit superheated steam close to magma [55].

The results of the drilling and flow-test of IDDP-1 clearly showed that right above the molten magma there was a permeable layer with superheated steam. Since "black death" has been observed in several other wells, it seems that similar conditions are found to be widespread in the reservoir inside the inner caldera, north of the transverse structure. This is in agreement with the observation and interpretation of the low P-wave velocity around 2.5 km depth within the ESE-WNW dike complex north of the transverse structure [34].

Detailed analysis of the glass from IDDP-1 [52,56] showed that it was quenched, near-liquidus rhyolitic magma (~76.5 wt.% SiO_2) but with oxygen and hydrogen isotope values ($\delta^{18}O$ = 3.1‰; δ^2H = −118‰), depleted relative to mantle values. From this it was concluded that its composition is consistent with the formation by partial melting of basalt which had been hydrothermally altered by meteoric water. They estimate the in-situ temperature to be in the range of 850 to 920 °C and that it had degassed at a pressure of 40 MPa, which is higher than hydrostatic pressure but lower than lithostatic pressure expected at 2100 m depth. Based on crystal zoning, they further infer that the magma originated at a greater depth and migrated upward to its present position.

In a study of silicic rocks in Krafla and in the active volcanic zones in Iceland [13,57], it was concluded, based on chemical and oxygen isotopic composition, that they are derived from partial melting of older basaltic rocks that were hydrated by meteoric water. It was further concluded that the silicic magma is formed in intrusive complexes beneath the volcanoes rather than by differentiation in large long-lived magma chambers.

Krafla is not the only geothermal area where drilling into upper crustal magma has taken place. The first reported incidence was in 2005 in the Puna geothermal field, Big Island Hawaii [58], where a drilling hit dacite magma (67 wt.% SiO_2) at the depth of 2488 m. Furthermore, in at least three wells drilled in Menengai caldera in Kenya, trachitic magma (63 wt.% SiO_2) at the depth of 2040 to 2180 m was encountered [59]. A detailed petrological study has, so far, not been carried out on the quenched magma from Menengai.

All this suggests that rather shallow partial melting of old, hydrothermally altered volcanic rocks can be an important factor in volcanism and volcanic geothermal systems, at least in extensional tectonic environments. In evolved volcanos, the hydrothermal systems alter the host rocks, forming secondary hydrated minerals that lower the solidus temperature and produce more silicic early melt than the parent un-hydrated rocks. As the volcano builds up, the altered rocks subside to greater depths, particularly, in cases where a caldera structure develops. If a 1200 °C basaltic magma pools in this substrate, dehydration melting takes place, with partial melt segregating and buoyantly ascending. The melt is near liquidus and relatively rich in volatiles and hence with a low viscosity. As the melt migrates up, it cools, degases and the viscosity increases. Finally, it can be stagnated by minor structural obstructions, spreading out as sills. The occurrences discussed above indicate that for some reason, takes place at about 2–3 km depth.

If this is correct, re-melting of geothermally altered rocks and the ascent of the melt to shallower crustal levels can be a hitherto unrecognised but important heat transport mechanism from deep intrusions to the base of the geothermal system above. Therefore, as the volcano evolves, the geothermal system with the ascent of magma from below become a feedback system that facilitates heat flow from the mantle to the surface.

Geothermal drilling has shown that silicic intrusions are common in volcanic geothermal systems in Iceland. An exception is the Reykjanes peninsula, SW Iceland. In fact, no silicic rocks are found on surface west of the Hengill volcano SW-Iceland (this will be addressed below). In Krafla, silicic intrusions are more abundant in the eastern part of the wellfield, north of the transverse structure, where the rhyolitic explosion craters are found, than in the western part [12]. This supports the idea that basaltic intrusions at depth are more common in the eastern part of the inner caldera. Due to the nature of their formation, the silicic intrusions are probably in isolated pockets and, therefore, not detected/resolved by the MT.

IDDP-1 showed that there is a permeable zone with superheated steam above the magma pocket. The steam is probably trapped in ductile rocks. The low P-wave velocity anomaly at depth [34] (Figure 12 right), probably reflects the extent of superheated steam above recent silicic intrusions or magma pockets and could therefore be a proxy of their distribution.

Partial re-melting of hydrothermally altered basalt can explain the bimodal behaviour of evolved volcanos such as Krafla and Askja, i.e., occasionally having large and often phreatic silicic eruptions but erupting only basalt in-between. Over time, silicic intrusions and magma pockets accumulate at shallow crustal levels. When a substantial amount of silicic magma has accumulated, major basaltic intrusion(s) can mobilise the silicic magma triggering an eruption. After they have been erupted, the cycle starts again.

As mentioned above, no silicic rocks are found on the Reykjanes peninsula west of Hengill. The only occurrences are a basalt andesite intrusion encountered at a depth of around 800 m in well SG-12 in Svartsengi [60] and some felsite veins at the depth of 4634.5 m in IDDP-2 at the tip of the peninsula [61]. This is probably because the Reykjanes peninsula is a thin and young, basically oceanic, crust and different from crust further inland. It lacks the central volcanoes that have produced massive lava accumulations and caldera structures, where basalts are buried, hydrothermally altered, and subside to where they can melt.

The results of country-wide MT measurements show a deep, low-resistivity layer under most of Iceland [62], but this layer is absent under the Reykjanes peninsula. The nature of this low-resistivity layer has been a matter of discussion for a long time. It has been proposed [62] that it is partially molten rocks, but results of seismic studies, e.g., [63], show that this cannot be the case. An alternative explanation has been suggested [63] and supported by the author, namely that the deep conductive layer is due to dehydrating alteration minerals (chlorite, epidote and amphibole). Recent laboratory measurements [64,65] show an irreversible and substantial increase in electrical conductivity when rocks containing chlorite and epidote are heated above 600–700 °C under high pressure. This is explained by dehydration of the chlorite and epidote minerals, releasing hydrogen and water and resulting in the oxidation of Fe^{+2} to Fe^{+3} [66] and the minerals becoming electron/hole semi-conductors. If this is the correct explanation of the deep conducting layer under most of Iceland, it means that the crust is weathered/hydrothermally altered but the crust under the Reykjnes peninsula is not or at least to lesser extent. It, therefore, seems like the alteration of the crust somehow preconditions partial melting in central volcanoes, but this is a subject for a separate study.

7. The S-Wave Shadows

One thing remains to be addressed. The S-wave shadows observed during the Krafla Fires have been interpreted as due to a volume containing magma. They do appear as a high resistivity in the 3D MT model, but bordered by low resistivity, both to the south and north (Figure 10). Recent measurement of resistivity of molten basalt and rhyolite from Krafla [67] give 1.2–0.8 Ωm for rhyolite melt at temperatures from 900 to 1000 °C and 1.7–0.5 Ωm for basalt melt at temperatures from 1170 to 1250 °C. These are really low values and extended volumes of magma, as indicated by the S-wave shadows, should, therefore, be observed by MT as low-resistivity bodies. The most natural explanation of the absence of such a body is that the melt is in unconnected pockets. This seems, however, in contradiction with seismological observations. Seismologists say that in order to have a substantial attenuation of S-waves, connected magma/liquid volumes have to have dimensions of the order of the wavelength of the S-waves, which, in this case, is about 0.5–1 km (e.g., [68,69]). An anonymous reviewer suggests, however, that the "S-wave shadows" could be a manifestation of the radiation pattern and network distribution or that a sufficiently dense distribution of melt pockets may be "seen" by the S-waves as a single viscous body. This needs to be studied by numerical modelling.

Here, the ESE-WNW trending high-resistivity body in the S-wave shadows is considered to be a dike complex. One possible scenario is that the low resistivity bordering it is still-molten dikes. The low resistivity south of the complex could be a dike injected during the Mývatn Fires when diking into the fissure swarm and eruptions were mainly to the south. The low resistivity on the northern side could be due to a dike injected at the northern margin of the dike complex in the Krafla Fires, where dike injections into the fissures swarm and eruptions were manly to the north. This could explain the S-wave shadows, because the S-waves from which it is inferred are mainly traveling perpendicular to such dikes. The low-resistivity bordering the dike complex could also be due to dehydration of alteration minerals as discussed above. However, this contradicting evidence from seismic and resistivity studies still remains un-resolved.

8. Summary and Conclusions

This study indicates that the geological structure of the Krafla volcano is more complex than hitherto believed. The 110 ka caldera hosts, now buried, an 80 ka inner caldera and both calderas are bisected by an ESE-WNW transverse low-density structure. Resistivity surveys show that geothermal activity has mainly been within the inner caldera, but it is cut through by the ESE-WNW structure. A difference in the local crustal spreading directions south and north of the calderas leads to a N–S opening component, favouring the ascent of basaltic magma from the mantle and explaining why most of the volcanic activity is in the eastern part of the calderas.

The present thermo-hydraulic character of the main geothermal system can be understood by considering the geological and tectonic history. When crustal spreading moved back from the western part of the fissure swarm to the eastern part 3 ka ago, permeability above the deeper intrusions increased drastically, resulting in a vigorous almost isothermal convection in the part where faults and fractures where reactivated, but further to the east where fissures were not reactivated and two-phase conditions prevail there. Hydrogen isotope ratios in geothermal fluids in Krafla show that, north of the transverse structure, they are of a local meteoric origin, but the fluids in Sandabotnaskarð and Hvíthólar are from the general groundwater stream from higher altitude in the south. There seems to be little, if any, hydrological connection between the geothermal systems north and south of the low-density transverse structure.

The silicic magma encountered in K-39 and IDDP-1 indicates a hitherto overlooked heat transport mechanism in evolved volcanoes, i.e., an ascending silicic re-melt of altered basaltic rocks by basalt intrusions and getting stagnant magma pockets at shallow crustal levels, producing a superheated steam zone above them. Bimodal volcanic behaviour of evolved volcanos like Krafla and Askja, occasional silicic eruptions, often phreatic, but purely basaltic in-between, can be understood by considering the volcano and its geothermal system(s) as coupled systems, enhancing heat flow from the mantle. The geothermal system produces alteration for "distilling" out silicic magma. When a substantial amount of silicic intrusions/magma has accumulated, major basalt intrusion(s) may "ignite" them causing a silicic eruption and the cycle starts again.

An important conclusion from this work is that a holistic approach, considering different and independent datasets and information, can shed light on complex structures of volcanoes and associated geothermal systems. Of particular importance is to study and take into account tectonics and geological history. Geophysical surveys and monitoring, sensitive to different physical parameters, and geochemistry are also vital. These are inexpensive studies, but their derived models usually need to be checked by much more expensive drilling. Conceptual models, often based on limited data, should not get stagnant for decades like in the case of Krafla. Any new information should constantly be used to reconsider and update the conceptual models. This will make geothermal drilling and utilization more focused and cost-effective.

Funding: This research received no external funding.

Acknowledgments: The author wants to thank Thorbjörg Ágústdóttir and Halldór Ármannsson at ÍSOR and four anonymous reviewers for their valuable and constructive suggestions and corrections that greatly improved this paper.

Conflicts of Interest: The author declares no conflict of interest.

References

1. Guðmundsson, G.; Pálmason, G.; Grönvold, K.; Ragnarsson, K.; Sæmundsson, K.; Arnórsson, S.; Námafjall, K. *Áfangaskýrsla um Rannsókn Jarðhitasvæðanna (e. Námfjall-Krafla. An Interim Report on Exploration of the Geothermal Areas)*; Orkustofnun: Reykjavík, Iceland, 1971.
2. Karlsdóttir, R.; Johnsen, G.; Björnsson, A.; Sigurðsson, Ó.; Hauksson, E. *Jarðhitasvæðið við Kröflu. Áfangaskýrsla um Jarðeðlisfræðilegar Yfirborðsrannsóknir 1976–1978 (e. Krafla Geothermal Area. An Interim Report on Surface Geophysical Studies 1976–1978)*; Orkustofnun: Reykjavík, Iceland, 1978.
3. Gíslason, G.; Ármannsson, H.; Hauksson, T. *Krafla Hitaástand og Gastegundir í Jarðhitakerfinu (e. Krafla. Thermal Conditions and Gases in the Geothermal System)*; Orkustofnun: Reykjavík, Iceland, 1978.
4. Ármannsson, H.; Gíslason, G.; Hauksson, T. *Magmatic Gases in Well Fluids Aid. The Mapping of the Flow Pattern in a Geothermal System*; Orkustofnun: Reykjavík, Iceland, 1981.
5. Ármannsson, H.; Benjamínsson, J.; Jeffrey, A.W.A. Gas changes in the Krafla geothermal system, Iceland. *Chem. Geol.* **1989**, *76*, 175–196.
6. Árnadóttir Th Lund, B.; Jiang, W.; Geirsson, H.; Björnsson, H.; Einarsson, P. Glacial rebound and plate spreading: Results from the first countrywide GPS observations in Iceland. *Geophys. J. Int.* **2009**, *177*, 691–716. [CrossRef]
7. Sæmundsson, K. *Geology of Krafla Volcanic System*; Garðarsson, A., Einarsson, Á., Eds.; Náttúra Myávatns Hið Íslenska Náttúrufræðifélag: Reykjavík, Iceland, 1991; pp. 25–95.
8. Sæmundsson, K. *Krafla, Geological Map, 1:25,000*; Landsvirkjun: Reykjavík, Iceland, 2008.
9. Kennedy, B.M.; Holohan, E.P.; Stix, J.; Gravley, D.M.; Davidson, J.R.J.; Cole, J.W. Magma plumbing beneath collapse caldera volcanic systems. *Earth Sci. Rev.* **2018**, *177*, 404–424. [CrossRef]
10. Drouin, V.; Sigmundsson, F.; Ófeigsson, B.G.; Hreinsdóttir, S.; Sturkell, E.; Einarsson, P. Deformation in the Northern Volcanic Zone of Iceland 2008–2014: An interplay of tectonic, magmatic, and glacial isostatic deformation. *J. Geophys. Res. Solid Earth* **2017**, *122*, 3158–3178. [CrossRef]
11. Ármannsson, H.; Gudmundsson, Á.; Steingrímsson, B.S. Exploration and development of the Krafla geothermal area. *Jökull* **1987**, *37*, 13–30.
12. Weisenberger, T.B.; Axelsson, G.; Arnaldsson, A.; Blischke, A.; Óskarsson, F.; Ármannsson, H.; Blanck, H.; Helgadóttir, H.M.; Berthet, J.C.; Árnason, K.; et al. *Revision of the Conceptual Model of the Krafla Geotermal System*; ÍSOR: Reykjavík, Iceland, 2015; 111p.
13. Jónasson, K. Rhyolite volcanism in the Krafla central volcano, north-east Iceland. *Bull. Volcanol.* **1994**, *56*, 516–528. [CrossRef]
14. Einarsson, P. *Umbrotin við Kröflu 1975–1989*; Garðarsson, A., Einarsson, Á., Eds.; Náttúra Myávatns Hið íslenska náttúrufræðifélag: Reykjavík, Iceland, 1991; pp. 96–139.
15. Möller, D.; Ritter, B.; Wendt, K. Geodetic measurements of horizontal deformation in northeast Iceland. *Earth Evol. Sci.* **1982**, *2*, 149–154.
16. Kanngieser, E. Vertical component of ground motion in north Iceland. *Ann. Geophys.* **1983**, *1*, 321–328.
17. Tryggvason, E. Surface deformation at the Krafla volcano, North Iceland. 1982–1989. *Bull. Volcanol.* **1994**, *56*, 98–107. [CrossRef]
18. Tryggvason, E. Multiple magma reservoirs in a rift zone volcano. Ground deformation and magma transport during the September 1984 eruption of Krafla, Iceland. *J. Volanol. Res.* **1986**, *28*, 1–44. [CrossRef]
19. Árnadóttir Th Sigmundsso, F.; Delaney, P.T. Sources of crustal deformation associated with the Krafla, Iceland, eruption of September 1984. *Geophys. Res. Lett.* **1998**, *25*, 1043–1046. [CrossRef]
20. Buck, W.R.; Einarsson, P.; Brandsdóttir, B. Tectonic stress and magma chamber size as controls on dike propagation: Constraints from the 1975–1984 Krafla rifting episode. *J. Geophys. Res. Solid Earth* **2006**, *111*. [CrossRef]

21. Heimisson, E.R.; Einarsson, P.; Sigmundsson, F.; Brandsdóttir, B. Kilometer-scale Kaiser effect identified in Krafla volcano, Iceland. *Geophys. Res. Lett.* **2015**, *42*, 7958–7965. [CrossRef]
22. Einarsson, P. S-wave shadows in the Krafla caldera in NE-Iceland, evidence for a magma chamber in the crust. *Bull. Volcanol.* **1978**, *41*, 1–9. [CrossRef]
23. Sturkell, E.; Sigmundsson, F.; Geirsson, H.; Ólafsson, H.; Theodórsson, T. Multiple volcano deformation sources in a post-rifting period: 1989-2005 behaviour of Krafla, Iceland constrained by levelling and GPS observations. *J. Volcanol. Geotherm. Res.* **2008**, *177*, 405–417. [CrossRef]
24. Drouin, V.; (ÍSOR and Freistenn Sigmundsson, University of Iceland, Reykjavík, Iceland). Personal communication, 2019.
25. Johnsen, G.V. *Þyngdarkort af Kröflusvæði (e. GRAVITY Map of the Krafla Area)*; Hróarsson, B., Jónsson, D., Jónsson, S.S., Eds.; EYJAR Í ELDHAFI: Reykjavík, Iceland, 1995; pp. 93–100.
26. Árnadóttir Th Geirssom, H.; Jiang, W. Crustal deformation in Iceland: Plate spreading and earthquake deformation. *Jökull* **2008**, *58*, 59–74.
27. Sæmundsson, K.; (ÍSOR, Reykjavík, Iceland). Personal communication, 2016.
28. Árnason, K.; Haraldsson, G.I.; Johnsen, G.V.; Þorbergsson, G.; Hersir, G.P.; Sæmundsson, K.; Georgsson, L.S.; Rögnvaldsson, S.T.; Snorrason, S.P. *Nesjavellir-Ölkelduháls. Surface Exploration in 1986*; Orkustofnun Report: Reykjavík, Iceland, 1987; 112p.
29. Árnason, K.; Karlsdóttir, R.; Eysteinsson, H.; Flóvenz, Ó.G.; Guðlaugsson, S.T. The resistivity structure of high-temperature geothermal systems in Iceland. In Proceedings of the 2000 World Geothermal Congress, Toholu-Kyushu, Japan, 28 May–10 June 2000; pp. 923–928.
30. Árnason, K.; Magnússon, I.Þ. *Niðurstöður Viðnámsmælinga í Kröflu (e. Results of Resistivity Surveys in Krafla)*; Orkustofnun: Reykjavík, Iceland, 2001.
31. Rosenkjaer, G.K.; Gasperikova, E.; Newman, G.A.; Arnason, K.; Lindsey, N.J. Comparison of 3D MT inversions for geothermal exploration: Case studies for Krafla and Hengill geothermal systems in Iceland. *Geothermics* **2015**, *57*, 258–274. [CrossRef]
32. Blanck, H.; Ágústsson, K.; Gunnarsson, K. *Seismic Monitoring in Krafla*; ÍSOR: Reykjavík, Iceland, 2017.
33. Ágústsson, K.; Flóvenz, Ó.G. The thickness of the seismogenic crust in Iceland and its implications for geothermal systems in Iceland. In Proceedings of the World Geothermal Congress, Reykjavík, Iceland, 24–30 April 2005.
34. Schuler, J.T.; Greenfield, R.S.; White, S.W.; Roecker, B.; Brandsdóttir, J.M.; Stock, J.; Tarasewicz, H.; Martens, H.R.; Pugh, D. Seismic imaging of the shallow crust beneath the Krafla central volcano, NE Iceland. *J. Geophys. Res. Solid Earth* **2015**, *120*, 7156–7173. [CrossRef]
35. Stefánsson, V. The Krafla Geothermal Field Northeast ICELAND. In *Geothermal Systems: Principles and Case Histories*; Rybach, L., Muffler, L.J.P., Eds.; John Wiley and Sons Ltd.: Hoboken, NJ, USA, 1981; pp. 273–294.
36. Bodvarsson, G.S.; Pruess, K.; Stefansson, V.; Eliasson, E.T. The Krafla geothermal field, Iceland: 2. The natural state of the system. *Water Resour. Res.* **1984**, *20*, 1545–1559. [CrossRef]
37. Stefánsson, V.; Kristannsdóttir, H.; Gíslason, G. (Orkustofnun, Reykjavík, Iceland). *Holubréf nr. 7*, Orkustofnun; Unpublished work. 1977.
38. Stefánsson, V. Investigation on the Krafla high temperature geothermal field Náttúrufræðingurinn. *Natturufraedingurinn Reykjavik* **1980**, *50*, 333–359.
39. Mortensen, A.K.; Guðmundsson, Á.; Steingrímsson, B.; Sigmundsson, F.; Axelsson, G.; Ármannsson, H.; Björnsson, H.; Ágústsson, K.; Sæmundsson, K.; Ólafsson, M.; et al. *Jarðhitakerfið í Kröflu. Samantekt Rannsókna á Jarðhitakerfinu og Endurskoðað Hugmyndalíkan (e. The Geothermal System in Krafla. An Overview of Studies of the Geothermal System and Revised Conceptual Model)*; ÍSOR: Reykjavík, Iceland, 2009.
40. Haiba, D.O.; Ingebresen, S.E. Multiphase groundwater flow near cooling plutons. *J. Geophys. Res.* **1997**, *102*, 12235–12252. [CrossRef]
41. Haiba, D.O.; Ingebresen, S.E. *The Computer Model HYDROTHERM, a Tree-Dimensional Finite-Difference Model to Simulate Ground-Water Flow and Heat Transport in the Temperature range of 0 °C to 1200 °C*; USGS: Reston, VA, USA, 1994; pp. 94–4045.
42. Kipp, K.L.; Hsieh, P.A.; Charlton, S.C. *Revised Ground-Water Flow and Heat Transport Simulator: HYDROTHERM-Version 3*; USGS: Reston, VA, USA, 2008.
43. Scott, S.; Driesner, T.; Weis, P. The thermal structure and temporal evolution of high-enthalpy geothermal systems. *Geothermics* **2016**, *62*, 33–47. [CrossRef]

44. Björnsson, G.; Steingrímsson, B. *Temperature and Pressure in the Geothermal System in Svartsengi. Original Status and Changes Due to Production*; Orkustofnun: Reykjavík, Iceland, 1991; p. 6.
45. Arnalds, A.; (Vatnaskil ltd., Reykjavík, Iceland). Personal communication, 2018.
46. Homberg, G.M. New manuscript guidelines for the reporting of stable hydrogen, carbon, and oxygen isotope ratio data. *Water Resour. Res.* **1995**, *31*, 2895–2925.
47. Darling, W.G.; Ármannsson, H. Stable isotopic aspects of fluid flow in the Krafla, Námafjall and Theistareykir geothermal systems of Northeast Iceland. *Chem. Geol.* **1989**, *76*, 175–196. [CrossRef]
48. Árnason, B. Hydrothermal systems in Iceland traced by deuterium. *Geothermics* **1977**, *5*, 125–151. [CrossRef]
49. Pope, E.C.; Bird, D.K.; Arnórsson, S.; Giroud, N. Hydrology of the Krafla geothermal system, northeast Iceland. *Geofluids* **2016**, *16*, 175–197. [CrossRef]
50. Mortensen, A.K.; Grönvold, K.; Gudmundsson, Á.; Steingrímsson, B.; Egilson, T. Quenched Silicic Glass from Well KJ-39 in Krafla, North-East Iceland. In Proceedings of the World Geothermal Congress 2010, Bali, Indonesia, 25–29 April 2010.
51. Friðleifsson, G.Ó.; Pálsson, G.; Björnsson, F.; Albertsson, A.; Gunnlaugsson, E.; Ketilssonj, J.; Lamrache, R.; Andersen, P.E. Iceland Deep Drilling Project. The first IDDP Drill Hole Drilled and Completed in 2009. In Proceedings of the World Geothermal Congress 2010, Bali, Indonesia, 25–29 April 2010.
52. Elders, W.A.; Friðleifsson, G.Ó.; Zierenberg, R.A.; Pope, E.C.; Mortensem, A.K. Origin of a rhyolite that intruded a geothermal well while drilling at the Krafl a volcano, Iceland. *Geol. March* **2011**, *39*, 231–234. [CrossRef]
53. Markússon, S.H.; Einarsson, K.; Pálsson, P. *IDDP-1 Flow Tests 2010–2012*; Landsvirkjun: Reykjavík, Iceland, 2013.
54. Stefánsson, A. *Geochemical Assessment of the Utilization of IDDP-1, Krafla*; Landsvirkjun: Reykjavík, Iceland, 2014; 59p.
55. Einarsson, K.; Pálsson, B.; Gudmundsson, Á.; Hólmgeirsson, S.; Ingason, K.; Matthíasson, J.; Hauksson, T.; Ármannsson, H. Acid wells in the Krafla geothermal field. In Proceedings of the World Geothermal Congress 2010, Bali, Indonesia, 25–29 April 2010.
56. Zierenberg, R.A.; Schiffman, P.; Barfod, G.H.; Lesher, C.E.; Marks, N.E.; Lowenstern, J.B.; Mortensen, A.K.; Pope, E.C.; Fridleifsson, G.Ó.; Elders, W.A. Composition and origin of rhyolite melt intersected by drilling in the Krafla geothermal field, Iceland. *Contrib. Mineral. Petrol.* **2013**, *165*, 327–347. [CrossRef]
57. Jónasson, K. Silicic volcanism in Iceland: Composition and distribution within the active volcanic zones. *J. Geodin.* **2007**, *43*, 101–117. [CrossRef]
58. Teplow, W.; Marsh, B.; Hulen, J.; Spielman, P.; Kaleikini, M.; Fitch, D.; Rickard, W. Dacite melt at the Geothermal Venture Wellfield, Big Island of Hawaii. *GRC Trans.* **2009**, *33*, 989–1005.
59. Mibei, G.K. *Preliminary Report on Manengai Intrusive Mapping*; Geothermal Development Company: Nairobi, Kenya, 2014.
60. Franzson, H. *Svartsengi-Eldvörp, A Conceptual Model of the Geothermal Reservoir*; ÍSOR: Reykjavík, Iceland, 2017; 69p.
61. Friðleifsson, G.Ó.; Elders, W.A.; Zierenberg, R.A.; Stefánsson, A.; Fowler, A.P.G.; Weisenberger, T.B.; Harðarson, B.S.; Mesfin, K.G. The Iceland Deep Drilling Project 4.5 km deep well, IDDP-2, in the seawater-recharged Reykjanes geothermal field in SW Iceland has successfully reached its supercritical target. *Sci. Drill.* **2017**, *23*, 1–12.
62. Björnsson, A.; Eysteinsson, H.; Beblo, M. *Crustal Formation and Magma genesis Beneath Iceland: Magnetotelluric constraints, Plates, Plumes and Paradigms*; Geological Society of America: Boulder, CO, USA, 2005; pp. 665–686.
63. Vilhjálmsson, A.M.; Flóvenz, Ó.G. *Geothermal Implications from a Resistivity Survey in the Volcanic Rift Zone of NE-Icelandand Comparison with Seismic Data*; ÍSOR: Reykjavík, Iceland, 2017; 46p.
64. Manthilake, G.; Casanova, N.B.; Novella, D.; Mookherjee, M.; Andrault, D. Dehydration of chlorite explains anomalously high electrical conductivity in the mantle wedges. *Sci. Adv.* **2016**, *2*, e1501631. [CrossRef] [PubMed]
65. Nono, F.; Gibert, A.; Parat, F.; Loggia, D.; Cichy, S.B.; Violay, M. Electrical onductivity of Icelandic deep geothermal reservoirs up to supercritical conditions: Insight from laboratory experiments. *J. Volcanol. Geotherm. Res.* **2018**, in press. [CrossRef]
66. Wang, D.; Guo, Y.; Yu, Y.; Karato, S. Electrical Conductivity of Amphibole-Bearing Rocks: Influence of Dehydration. *Contrib. Mineral. Petrol.* **2012**, *164*, 17–25. [CrossRef]

67. Gibert, B.; Levy, L.; Sigmundsson, F.; Hersir, G.P.; Flovenz, Ó.G. Electrical Conductivity of Basaltic and Rhyolitic Melts from Krafla central Volcano, Iceland. In Proceedings of the IMAGE Final Conference, Akureyri, Iceland, 4–6 October 2017.
68. Jackson, D.D.; Anderson, D.L. Physical mechanisms of seismic-wave attenuation. *Rev. Geophys.* **1970**, *8*, 1–63. [CrossRef]
69. Einarsson, P.; (University of Iceland, Reykjavík, Iceland). Personal communication, 2018.

© 2020 by the author. Licensee MDPI, Basel, Switzerland. This article is an open access article distributed under the terms and conditions of the Creative Commons Attribution (CC BY) license (http://creativecommons.org/licenses/by/4.0/).

Article

Distribution and Transport of Thermal Energy within Magma–Hydrothermal Systems

John Eichelberger

International Arctic Research Center, University of Alaska Fairbanks, Fairbanks, AK 99775, USA; jceichelberger@alaska.edu

Received: 15 April 2020; Accepted: 25 May 2020; Published: 1 June 2020

Abstract: Proximity to magma bodies is generally acknowledged as providing the energy source for hot hydrothermal reservoirs. Hence, it is appropriate to think of a "magma–hydrothermal system" as an entity, rather than as separate systems. Repeated coring of Kilauea Iki lava lake on Kilauea Volcano, Hawaii, has provided evidence of an impermeable, conductive layer, or magma–hydrothermal boundary (MHB), between a hydrothermal system and molten rock. Crystallization on the lower face of the MHB and cracking by cooling on the upper face drive the zone downward while maintaining constant thickness, a Stefan problem of moving thermal boundaries with a phase change. Use of the observed thermal gradient in MHB of 84 °C/m yields a heat flux of 130 W/m^2. Equating this with the heat flux produced by crystallization and cooling of molten lava successfully predicts the growth rate of lava lake crust of 2 m/a, which is faster than simple conduction where crust thickens at $\sqrt{t}$ and heat flux declines with $1/\sqrt{t}$. However, a lava lake is not a magma chamber. Compared to erupted and degassed lava, magma at depth contains a significant amount of dissolved water that influences the magma's thermal, chemical, and mechanical behaviors. Also, a lava lake is rootless; it has no source of heat and mass, whereas there are probably few shallow, active magma bodies that are isolated from deeper sources. Drilling at Krafla Caldera, Iceland, showed the existence of a near-liquidus rhyolite magma body at 2.1 km depth capped by an MHB with a heat flux of ≥16 W/m^2. This would predict a crystallization rate of 0.6 m/a, yet no evidence of crystallization and the development of a mush zone at the base of MHB is observed. Instead, the lower face of MHB is undergoing partial melting. The explanation would appear to lie in vigorous convection of the hot rhyolite magma, delivering both heat and H$_2$O but not crystals to its ceiling. This challenges existing concepts of magma chambers and has important implications for use of magma as the ultimate geothermal power source. It also illuminates the possibility of directly monitoring magma beneath active volcanoes for eruption forecasting.

Keywords: magma energy; magma convection; hydrothermal system; heat flux; geothermal energy; eruption

1. Introduction

Given the tremendous difference between the rate of heat transport by conduction through solid rock and advection of heat by aqueous fluid through permeable rock, the strong control of hydrothermal activity on magma evolution cannot be questioned. It is therefore surprising that relatively little has been written on this topic. The reason is likely a separation of communities of practice. Until now, no one has investigated magmatic systems directly. Rather, experiments are conducted with micro synthetic or natural rock samples in the laboratory, field and petrologic studies explore "fossil" systems, remote (surface or supra-surface-based) sensing detects proxy signals of magma from active systems, and hydrodynamic models describe how magma could behave, subject to various assumptions. In contrast, hydrothermal systems are the domain of fluid geochemistry and alteration mineralogy, reservoir modeling involving porous flow, and most importantly, direct measurement of conditions and lithology from geothermal drilling, much of the data from which are proprietary.

There are notable exceptions to the inadequate treatment of the magma–hydrothermal coupling problem. For example, Lister [1] provided rigorous theoretical treatment of thermal cracking and crystallization, defining a moving magma–hydrothermal boundary (MHB), a Stefan problem, related to mid-ocean ridge volcanism. Hardee [2] applied this approach to analyzing temperature measurements from Kilauea Iki lava lake. Carrigan [3] drew an analogy between magma–hydrothermal systems and two resistors in series, where the largest resistance, the hydrothermal system, dominates energy flow. He previously argued for convection in magma bodies and large sills [4,5], which would keep the value of the magma resistor low, whereas convection in lava lakes would die out quickly. This is in contrast to, for example, Hort et al. [6], who asserted that a mush insulator grows at the top of the magma body and quickly shuts off convection. Recently, Lamy et al. [7] continued the view of simple inward-growing mush and stagnant magma, although their main point was the importance of a magmatic vapor phase (MVP) released during crystallization of intrusions. Hawkesworth et al. [8] made assumptions about heat loss from magma bodies through hydrothermal systems and concluded that the more vigorous hydrothermal circulation expected at shallow depth would force more rapid magmatic differentiation. Fournier [9] used advective heat discharge by rivers draining the Yellowstone Plateau to infer crystallization rates of the underlying magma. Scott et al. [10] modeled high-enthalpy fluid circulation above a heat source at magmatic temperature. Glazner [11], making arguments similar to those employed here, showed the importance of hydrothermal convection above magma and then further argued that an arid climate favors the development of large magma chambers.

The purpose of this paper is to discuss new implications beginning to emerge from accidental encounters with magma by geothermal drilling and the importance of understanding MHB. It is in part a review paper, gathering key observations where hidden bodies of magma have been accidently encountered. Complementing these are lessons learned from drilling into a lava lake as a magma chamber analogue. Possible explanations for the more surprising drilling results are proposed. The drilling data are sparse, however, so it is not possible to consider all allowable interpretations or to say that the postulates are uniquely constrained by the data. Rather, the intent is to stimulate thinking about the magma–hydrothermal connection, leading to direct observations through intentional scientific drilling into this critical zone [12]. In contrast, a lava lake is an imperfect analogue. It represents magma that has erupted and degassed to become lava, losing almost all of its dissolved water, which profoundly affects melt properties and phase relationships. A lava lake is also rootless, closed to introduction of heat and mass from below, whereas probably few shallow magma chambers exist in isolation from complex melt-bearing columns that extend from the mantle [13]. The only way to truly understand magma with its surrounding crust is to drill it.

2. Drilling into Molten Rock

It may come as a surprise that molten rock can be drilled [14]. Normal drilling practice is to circulate fluid down the drill stem, through the drill bit, and back up the annulus to the surface. The cold fluid is mostly water, usually with some additives. It serves the purposes of keeping the drill bit and the rock cool and returning rock, brittlely fractured in making the hole, to the surface. These fragments are in the form of sand-like cuttings. However, if a core bit and receiving core barrel are used, samples are retrieved as cylinders of rock. The core is of greater use scientifically because the exact depth of origin is known and the samples are intact, preserving a great deal of information about lithologic texture, interrelationships, structures such as veins, and temperature and chemical gradients.

Molten rock can be drilled if circulation is sufficiently fast and penetration sufficiently slow to quench melt to brittle drillable glass just ahead of the advancing drill bit. Magma, unerupted molten rock stored at depth, has never been cored, but in principle it could be. The molten rock in the interior of lava lakes has been cored many times for research purposes, most prominently in Hawaii where the occurrence of pit (collapse) craters frequently results in ponding of lava of sufficient depth to remain molten for years or decades. This reveals much about how magma crystallizes.

The most thoroughly studied lava lake is Kilauea Iki (Figure 1 [15]). Erupted near the summit of Kilauea Volcano in 1959 and filled to a depth of 130 m, it was cored many times during the next three decades. Thus, a thorough record of its cooling and crystallization was accumulated [16]. I am aware of three sites, the Puna site on Kilauea Volcano, Hawaii [17]; Menengai Caldera in the East African Rift, Kenya [18,19]; and Krafla Caldera, Iceland [20], where drilling has penetrated silicic magma. There may well be other incidents that have not been reported in the publicly accessible literature, or cases where return of glass cuttings was not recognized as evidence of magma. The most comprehensively studied case of actual magma is the Iceland Deep Drilling Program's IDDP-1 [21], drilled to a depth of 2104 m in Krafla Caldera, Iceland in 2009. The richness of the data is because IDDP-1 was a research and development well sponsored by a consortium of Icelandic geothermal companies, with extensive scientific participation under the aegis of the International Continental Scientific Drilling Program (ICDP). However, the objective of the well was to penetrate the domain of supercritical fluids at 4.5 km, where both temperature and fluid pressure were expected to be above the critical point of water. Instead, magma was encountered at 2.1 km, with a temperature well above the critical point but fluid pressure below it. IDDP-1 can now be viewed as opening the path to extensive exploration of the magma body it discovered, the objective of the Krafla Magma Testbed program (KMT [12]).

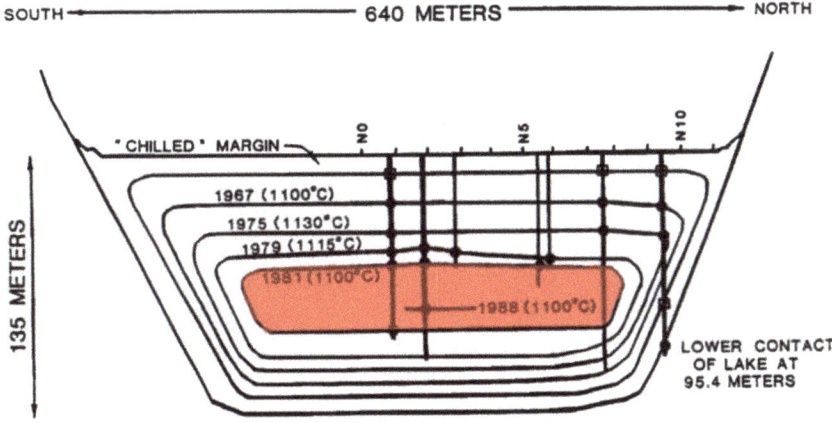

Figure 1. Cross section of Kilauea Iki lava lake showing some of the holes that have been cored and the position of the molten lava lens with time. 1981 (Figure 2A) is shown in red [15].

Table 1 is a compilation of pertinent data and calculation results for magma–hydrothermal systems. It makes sense that the first encounters with magma were shallow ones. However, are these oddities or are there many more such bodies to be encountered as drilling in quest of high-enthalpy fluids proceeds to greater depth? Superhot wells are expensive from a geothermal perspective, some USD $20 M for IDDP-1. One would infer that the reason why there is little scientific data from magma and near-magma encounters, other than from IDDP-1, is the economic pressure to make a well "pay" as soon as a viable hydrothermal resource is reached. It will take an alliance of scientists, engineers, and geothermal companies as pioneered by IDDP to really explore and put to use this final crustal frontier.

One striking feature of the compilation in Table 1, in addition to magma being unexpected, is that it is surprisingly shallow. Another is the absence of an upper crystallization or mush zone above the nearly pure melt magma. One could say that the magma intruded only just before drilling or that the upper mush zone, which would be gravitationally unstable, had only just collapsed. However, this is special pleading, and a general explanation would be that this is the way magma chambers work.

I will summarize the most pertinent lessons from Kilauea Iki, and then apply them, to the extent they can be, to the much sparser data set from Krafla and other systems. As a note on terminology, I use magma to refer to a material comprising silicate melt +/-crystals +/-vapor bubbles that is capable of flow, usually considered to require melt ≥50 vol%. [22]. At <50 vol%, I use mush to denote residual melt-bearing material that is the product of partial crystallization of magma, and partially melted rock (in the Krafla case felsite) to denote rock that is undergoing melting. Distinguishing between residual melt in mush and partial melt in melting rock is based on texture. Residual melt in mush is interstitial to generally euhedral crystals. Partial melt in partially melted rock is interstitial to crystals that are anhedral and contain embayments, indicative of resorption. In the former, crystals were growing from melt when quenched, in the latter they were dissolving into melt.

Table 1. Some modeling and drilling results pertaining to magma–hydrothermal systems.

Site	Lister Model	Kilauea Iki, Hawaii	Heimaey, Iceland	Grimsvotn Caldera, Iceland	Puna Venture, Hawaii	Menengai Caldera, Kenya	Krafla Caldera, Iceland
Setting	theory	Basalt lava lake	Basalt lava flow	Central volcano in plate rift	Magma in east Kilauea rift	Caldera in East African Rift	Central volcano in plate rift
Depth to magma/lava (m)	Diking at midocean ridge	0 @ 1959, 80@1988	—	>2000	2488	2080	2102
Roof rock	—	Own crust	Own crust	basalt	Diorite w/melt inclus.	syenite	Partially melted felsite
wt% SiO_2	Generic basalt	50 bulk, 75 last melt	48–50	51 basalt	67 dacite	63 trachyte	75 rhyolite
T (°C) where melt present	—	1000–1190	1030–1055	1100	1050	—	900
vol% Xtal	—	variable	—	—	5–8	<5	<1
Viscosity (Pa·s)	—	variable	—	10–2000	10^7	—	3×10^5
Flow up well (m)	—	0 in 1981	—	—	14	0	9
Last erupted silicic magma	—	—	—	—	never	8000 a	10^4 a; (trace in 1724)
ΔT/ΔZ (°C/m) in MHB	—	84	10,000	300	>5	>17	>16
MHB thickness (m)	—	11	0.1	2	—	—	<25
Heat flux W/m^2	—	130	40,000	120–600	—	—	>24
Thermal power output (MW)	—	14	300 (short term)	1000–5000	—	—	>100 from IDDP-1
Fracture penetration rate	30 m/a	2 m/a	1 m/day	5 m/a	—	—	—
Permeability therm. cracks	10 D	0.3	—	—	—	—	0.7
References	[1]	[2,15,16]	[23]	[23–25]	[17]	[18,19]	[20,26,27]

229

3. The Case of Kilauea Iki: A Stefan Problem

The results from drilling Kilauea Iki have been influential in thinking about magma chambers at depth, particularly results pertaining to the transition from the hydrothermal zone of the solid lake crust to the melt-rich zone [2]. The hydrothermal zone comprises basalt with fractures open to the surface and contains steam from vaporization of downward percolating rainwater. At the surface elevation of the lake of 1070 m, the boiling point is 96 °C, and the temperature profile is isothermal at that temperature to the base of the open system. The downward transition from the brittle, fractured, crystalline hydrothermal regime to the molten regime is a thin conductive, linear temperature gradient zone that has moved downward with time. The hydrothermal zone expanded downward by cooling the crust and cracking it towards the retreating melt zone (Figure 2). The result is that cooling and crystallization are approximately constant so that crustal thickness grows linearly with time (Figure 2C) conductive zone, essentially a growing layer of insulation. However, thermally, the transition zone, MHB, can be treated as bounded by two constant temperature surfaces bathed by hydrothermal fluid on the upper surface and melt-rich lava on the lower surface, producing a steady-state condition within the moving spatial coordinates of the zone [2]. The downward propagation of MHB is then a proxy for the thermal power output of the system.

Bjornsson et al. [23] suggested that the phenomena described above accounted for the high, constant thermal energy output of Grimsvotn Volcano, Iceland. They also noted the formation of columnar jointing (Figure 2B) within the lava flow at Heimaey, Iceland, where water from firehoses was used to stop a lava flow threatening to block the harbor. Fournier [9] likewise appealed to invasion of thermal cracking to enhance latent and sensible heat extraction from the very large magma body inferred to exist at Yellowstone, producing an advective thermal power output of 5 GW. Axelsson et al. [29] made a similar suggestion for IDDP-1 at Krafla, where an extended flow test yielded a thermal power output exceeding 100 MW. Thermal cracking was investigated experimentally by Lamur et al. [30], who found that it can begin during cooling as early as 100 °C below the solidus.

Kilauea Iki is the only case where the MHB concept can be quantitatively tested through a time series of temperature profiles and thicknesses of the crust. The temperature gradient in the conductive zone (using thermal parameters of Hardee [2] so as to be consistent) gives the heat flow in the z direction (F_z):

$$F_z = -k \, \Delta T / \Delta z \tag{1}$$

where thermal conductivity, k, is 1.57 J/m·s·°C and the average, linear (conductive) thermal gradient, $\Delta T/\Delta z$, is between the solidus at 1000 °C and the hydrothermal system at 100 °C using data from years 1962, 1967, 1979, and 1981 (data: [16]) is 84 °C/m. Therefore:

$$F_z = -130 \text{ W/m}^2$$

The total thermal power output (P_T) is the product of heat flux and area, A:

$$P_T = F_z \, A \tag{2}$$

Using melt lens radius r = 186 m from Figure 1, A is 1.6×10^5 m^2 and:

$$P_T = 14 \text{ MW}$$

The rate of growth of the volume of the crust of the lake is a proxy for power output. This is to say that the product of the rate of downward propagation of the solidus ($\Delta z_{1000}/\Delta t$) times the energy density of molten lava cooled to the solidus (ε_M), should equal the heat flux upward measured by the temperature gradient in the conductive zone (Figure 2A. To test this, we can solve for the predicted rate of thickening of the crust with heat flux, F_z, calculated with Equation (1).

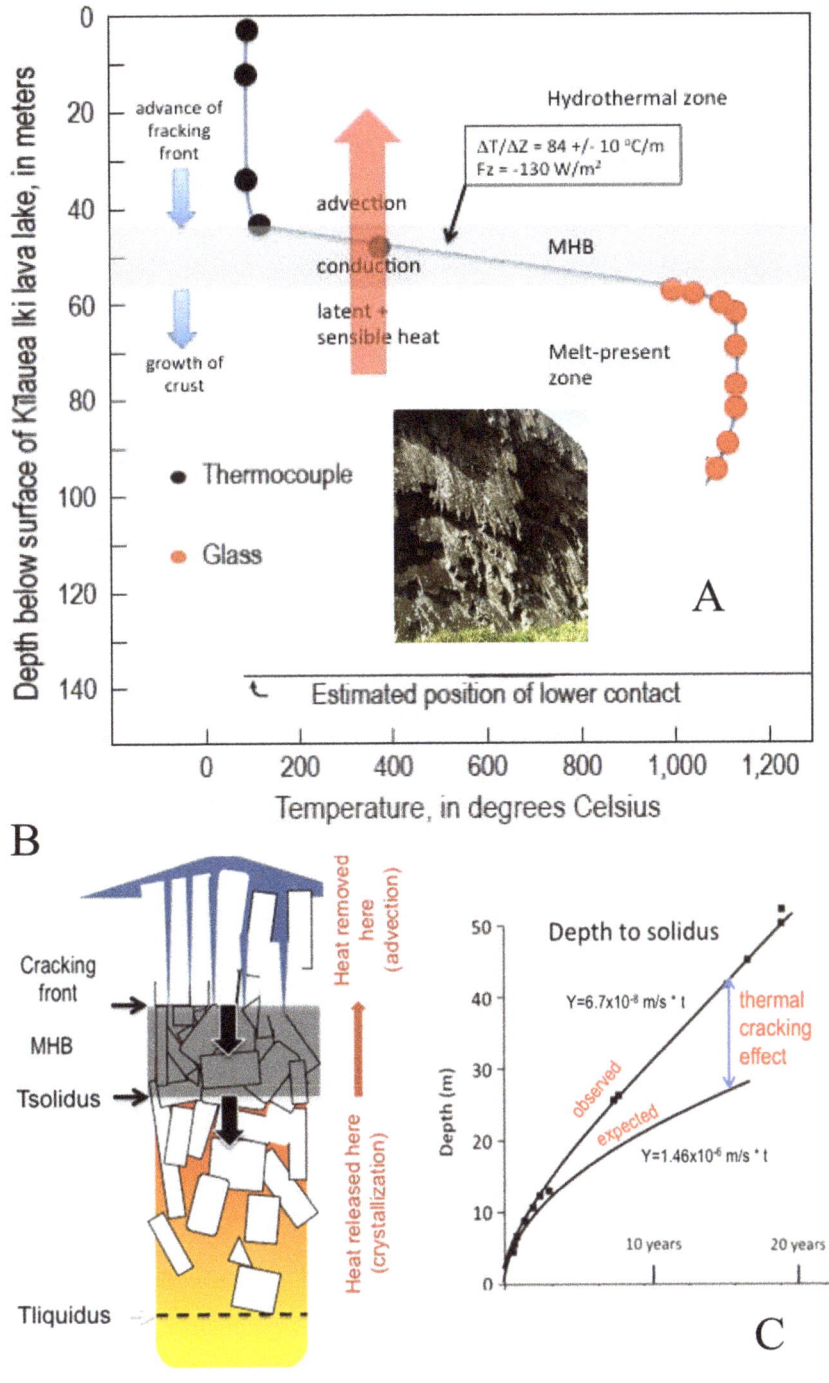

Figure 2. (**A**) A temperature profile from the surface into the melt lens in Kilauea Iki, 1981 (after

Helz [16]). This is one of many obtained over three decades after formation of the lake in 1959. A conductive MHB, moving downward by thermal cracking, divides a hydrothermal zone at local boiling temperature from the molten zone. Inset shows thermal contraction columns in a lava flow near Christ Church, NZ, the track left by a descending MHB (photo by author). (**B**) Cartoon displaying the processes envisioned for a cooling igneous body, based upon observations at Kilauea Iki. The tips of the contraction fractures define a thermal cracking front that propagates perpendicular to and following the retreating isotherms. The fractures define the familiar columnar jointing. (**C**) Core hole data on growth of crust over two decades show the effect of thermal cracking compared to that of simple conduction with crystallization (after Hardee [2] using the Carslaw and Jaeger thermal model [28]).

One can define an effective heat capacity of basalt lava, β_{Meff}, within the interval T_L to T_S:

$$\beta_{Meff} = \beta_M + \frac{L_M}{T_L - T_S} \tag{3}$$

where β_M = 1046 J/kg·°C, sensible heat capacity of molten basalt lava
L_M = 418,000 J/kg, latent heat of crystallization of molten basalt lava
T_L = 1150 °C, liquidus temperature and
T_S = 1000 °C, solidus temperature
$\Delta Z_{1000}/\Delta t$, downward velocity of solidus
ρ_M = 2700 kg/m^3, mass density of molten basalt
ρ_B = 2900 kg/m^3, mass density of basalt
Using the above values in Equation (3) gives:

$$\beta_{Meff} = 3800 \, J/kg°C$$

Energy density of magma, ε_M, i.e., energy per volume released by cooling from T_L to T_S, is:

$$\varepsilon_M = \rho_M \, \beta_{Meff} \, (T_L - T_S) \tag{4}$$

Yielding by volume (adjusted from molten density) of crystallized basalt:

$$\varepsilon_B = 1.7 \times 10^9 \, J/m^3$$

For the rate of movement of T_S downward ($\Delta z_{1000}/\Delta t$; growth of crust) to produce the heat flux upward:

$$\varepsilon_B \, (\Delta Z_{1000}/\Delta t) = Fz \tag{5}$$

Solving for the downward growth rate of the crust:

$$\Delta Z_{1000}/\Delta t = 7.6 \times 10^{-8} \, m/s \text{ or } 2.4 \, m/a$$

The observed value for 1962 to 1981 is 7.4×10^{-8} m/s or 2.3 m/a. This agreement is somewhat fortuitous because it is better than the uncertainties should provide and neglects that some latent heat of crystallization has already been lost. Nevertheless, if the lava lake is a good analogue for a magma chamber, then the conductive temperature gradient in the MHB can be used to determine the heat flux from magma to the hydrothermal system and to predict the rate of growth of a crystal layer at the roof of the magma chamber.

4. The Case of Krafla

Figure 3 is a plausible 3-D rendering of Krafla, centered at 65.71° N and 16.75° W. The encounter with magma by drilling was not anticipated through geophysical surveys. However, it was later shown that the Krafla rhyolite magma body coincided with a strong V_p/V_s anomaly [31] and its top with a seismic reflector [32].

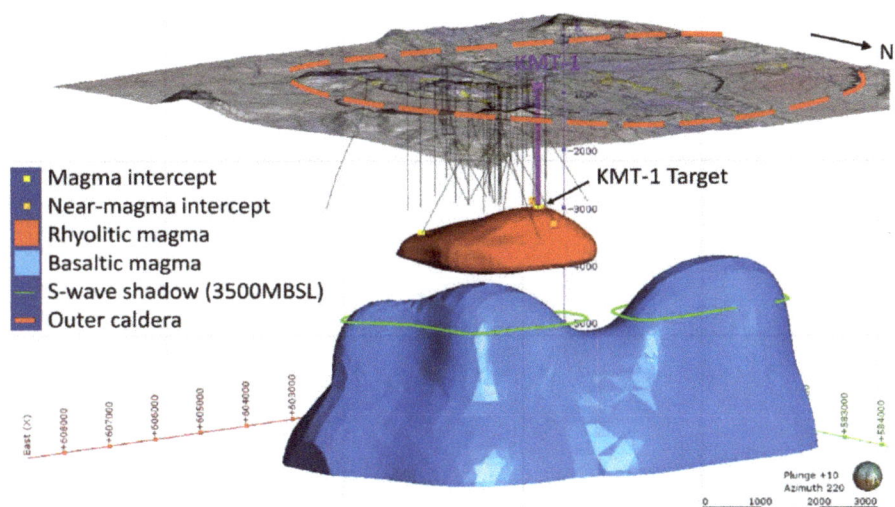

Figure 3. Idealized configuration of rhyolitic magma body (red) and deeper basaltic body (blue) under Krafla Caldera (red dashed line). Light gray lines are existing wells drilled by Landsvirkjun (National Power Company). The heavy line is planned KMT-1, the first well of KMT [12], closely following the path of IDDP-1 [21]. Credit: J.W. Catley. In addition to IDDP-1, two other wells have penetrated or come close to rhyolite magma. There is much debate about the configuration of the rhyolitic magma, ranging from small separate pockets or a larger, long-persistent unified body similar to what erupted from Askja in 1875 [33]. There is more general acceptance of the large persistent basaltic body shown in blue, which receives new basaltic injections, sends out N–S dikes during rifting "Fires", and is responsible for the two S-wave shadow zones and for most of the surface deformation.

The site for IDDP-1 was chosen to be near the center and hottest part of the caldera. It is also close to Viti Crater, site of a phreatomagmatic explosion during the Myvatn Fires in 1724 that ejected a small amount of rhyolite similar to magma of IDDP-1. As the most prominent example of drilling encountering rhyolite magma, results were presented in a special issue of Geothermics (2014) and in a number of separate publications; note especially [26,27,34].

In drilling IDDP-1, circulation was lost at 2070 m, and continued drilling encountered a soft formation at an average depth of 2102 m in the original well and two sidetracks. It is the second sidetrack (third penetration) that is of special interest here because magma flowed up the borehole, the bit became stuck, and for a brief period, restoration of circulation brought glass-bearing chips (Figure 4) to the surface, confirming that magma was present at the bottom of the well (Figure 5). Gamma logs indicate that lithology below 2070 m is predominantly felsite (A in Figure 5, fine-grained crystalline equivalent of rhyolite magma) through this zone [26]. The deepest temperature measurements are at 2077 m and through multiple measurements are extrapolated to an equilibrium formation temperature of 500 °C (Figures 5 and 6) [26]. Somewhere between that depth and the melt-present zone (B in Figure 5, checkered) should be a transition from brittle to ductile behavior in the felsite. This should be the lower limit of fracturing. However, high strain rates caused by cooling during drilling would have raised the temperature of the brittle/ductile boundary so that fractures could have propagated deeper. Only two kinds of melt-present (shown by quenched glass) chips were recovered: the partially melted felsite (B) and near-liquidus magma (D), although some chips of D contain micro-xenoliths of B (Figure 4).

Although the actual spatial relationships are unknown, it may be presumed that the high-temperature magma lies below the partially melting felsite. When the magma was penetrated, it rose 8 m from 2104 m to 2096 m within 9 min and then another meter where the bit subsequently became stuck. Lithology C is partially crystallized magma or mush, expected to occur between the melting felsite and the near-liquidus magma (Figure 5), but it is not represented in the chips recovered. 1 mm × 1 mm elemental electron microprobe maps of B and D are shown to the left of the borehole. SiO_2 contents are displayed as false colors indexed on the vertical bar. In B, interstitial rhyolite melt (red ~75 wt% SiO_2) lies between embayed andesine feldspar (lime green, ~56–58 wt% SiO_2) and quartz (white 100 wt% SiO_2). Black denotes cracks and voids. In D, the main component is rhyolite melt (red ~75 wt% SiO_2), with suspended euhedral crystals of andesine (lime green, ~56 wt% SiO_2) and clinopyroxene (dark green ~50 wt% SiO_2). Small black crystals are titanomagnetite (0 wt% SiO_2).

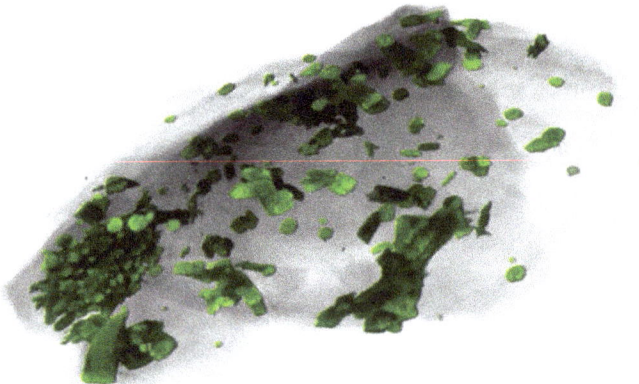

Figure 4. A 3-D perspective by X-ray tomography of a magma chip (2 mm long), from IDDP-1. False-color green crystals, mostly plagioclase, float in transparent gray melt. A clot of crystals at the lower left is partially melted felsite. The chip was quenched at about 2100 m depth and 900 °C. Credit: Sample provided by Landsvirkjun and A. Mortensen; image by F. Wadsworth, Munich U., Germany; E. Saubin, U. of Canterbury and C. I. Schipper, Victoria U., NZ.

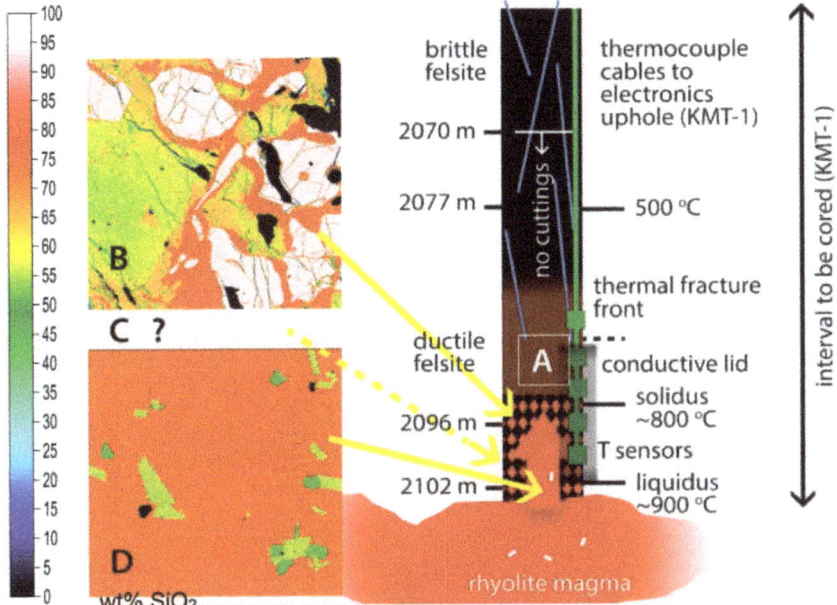

Figure 5. Important observations from the IDDP-1 borehole and 1 mm × 1mm Si element maps of the two lithologies represented in chips from the bottom of the well, color coded in wt% SiO$_2$ (bar at left). **B** is partially melted felsite, **C** is expected but missing mush, i.e., partially crystallized magma, and **D** is near-liquidus rhyolite magma. Although a layered sequence is shown, it is possible that lithology **B** occurs only as xenoliths (Figure 4) within the magma (**D**). Photo credit: N. Graham and P. Izbekov at University of Alaska Fairbanks's Advanced Instrumentation Laboratory. Also shown are plans for KMT-1 to obtain continuous core (black double-headed arrow) and a temperature profile from a thermocouple string (green line) through the same interval.

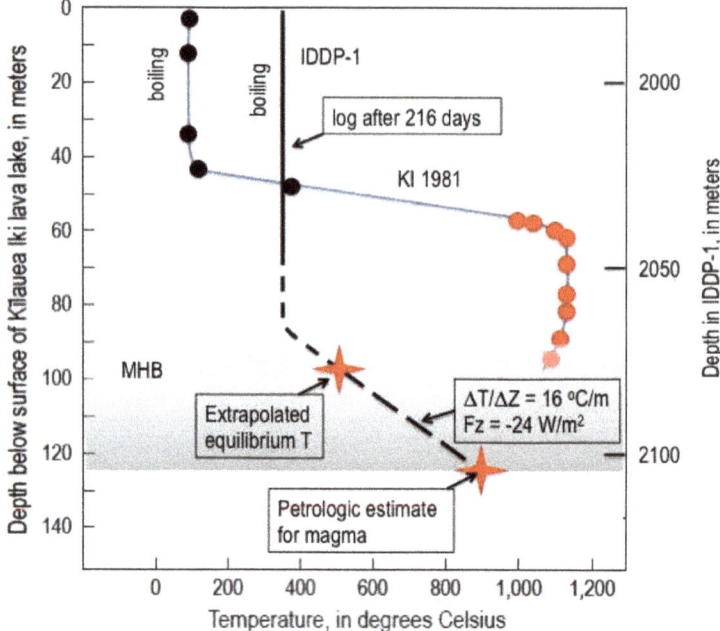

Figure 6. Temperature profile for IDDP-1 [26] compared with Kilauea Iki 1981 [16]. The depth and temperature scales are the same except that a much deeper interval is shown for IDDP-1; hence, the boiling-controlled portion of the profile is hotter. The molten portion in IDDP-1 is rhyolite rather than basalt, so that temperature is lower.

4.1. MHB in Krafla Compared to Kilauea Iki

IDDP-1 provides enough data to place some constraints on the Krafla magma–hydrothermal system. Allowing for the fact that the Krafla results are much deeper than for Kilauea Iki and therefore the hydrothermal temperatures much higher, there is a broad similarity in the temperature profiles. Again, we see an upper hydrothermal zone at the boiling point for the prevailing fluid pressure (Figure 6). Beneath that is a conductive zone with a steep thermal gradient, the MHB. The gradient is constrained by only two temperatures: (1) the time-extrapolated temperature to thermal equilibrium at 2077 m [26] and (2) petrologic estimates for the near-liquidus rhyolite magma, which can be regarded as about 900 +/−50 °C [27,34,35]. This yields a thermal gradient of 16 °C/m.

One of the great mysteries about the rhyolite magma is that it there: shallow yet undetected prior to drilling. Another is that there is no mush zone represented in the cuttings. There is, however, partially melted roof rock, formation of which would seem to require heat from partial crystallization of magma. Krafla is a dominantly basaltic central shield volcano with a summit caldera and essentially no rhyolite eruptions in the last 10^4 a. The very low crystal content would suggest the magma is "new", yet since the time of the Krafla Fires of 1975 to 1984, the volcano has been among the best monitored in the world [. It is unlikely that a substantial body of magma could have arrived at only 2.1 km beneath the center of the volcano during that time period without detection. However, if it were small enough to avoid detection, it should be entirely crystallized.

The existence of a conductive MHB with temperature constraints allows an approach at Krafla analogous to Kilauea Iki to predict how fast crystallization should be occurring in the magma below.

Using Equation (3) and values adopted by Axelsson [28]:

where β_M = 800 J/kg·°C, sensible heat capacity of rhyolite magma

L_M = 400,000 J/kg, latent heat of crystallization of rhyolite magma
T_L = 900 °C, liquidus temperature and
T_S = 800 °C, solidus temperature [36]
ρ_M = 2300 kg/m^3, mass density of rhyolite magma
ρ_R = 2700 kg/m^3, mass density of crystallized rhyolite (felsite)
yields:

$$\beta_{Meff} = 4,800 \, J/kg°C$$

Equation (4) yields (adjusted to the volume of crystallized rhyolite):

$$\varepsilon_M = 1.3 \times 10^9 \, J/m^3$$

We are neglecting here a contribution from the transfer of energy in any released magmatic vapor phase (MVP [7]), but the amount of vapor exsolved and the pressure and temperature drop it undergoes should be negligible at this local scale.

Applying this to Equation (5) gives:

$$\Delta Z_{800}/\Delta t = 1.9 \times 10^{-8} \, m/s \, or \, 0.6 \, m/a$$

This amount of crystallization expected at the roof of the rhyolite magma body over a period of three decades, before which emplacement of the magma might have gone unnoticed, would have been seen in chips from IDDP-1. It is hard to imagine randomly bringing up chips of magma and partially melted felsite without a substantial portion of mush, hypothetical lithology C in Figure 5, if C were present. The simplest interpretation, and one that can be tested by coring, is a stratigraphy comprised downward of felsite (A), partially melted felsite (B), and rhyolite magma (D) with no intervening mush (C). The hotter and higher enthalpy (more melt) rhyolite magma is actively melting its lid (less melt), which in turn means it is releasing heat to the MHB faster than the overlying hydrothermal system can take it away. Presumably, this would be moving MHB upward by effectively dissolving the lower face of MHB and closing fractures by thermal expansion at the cracking front at the upper face. In Kilauea Iki, we clearly see the progress of crystallization yielding mush and then solid crust at the roof of the molten lava lens, and its rate of downward growth is consistent with simple conduction through the MHB. Why not at Krafla? Likely because the magma of Krafla is convecting, and the molten lava of Kilauea Iki is stagnant.

One alternative to the convection of magma under Krafla is that the partially melted felsite is itself the source of the rhyolite magma. Rhyolitic melt could be percolating out of the walls of the bore hole or coming from preexisting segregation veins like those that flowed into the borehole at Kilauea Iki [14]. In this scenario, which aligns with quite a bit of thought about the source of near-liquidus rhyolite [37], the volume below MHB is mostly crystals with interstitial rhyolite melt. However, percolation is a slow process, all the more so with the walls of the borehole quenched. Segregation of melt into veins prior to sampling by drilling seems to be required [38]. However, the magma contains sparse euhedral phencrysts, totally unlike what is present in the melting felsite. Percolating these through the crystal network of felsite is a physical impossibility. The phenocrysts are larger than the pathways. However, growing them after segregation into veins is equally implausible. The magma and felsite are not in thermal equilibrium. Crystals in the felsite are melting and those in the magma are growing. Heat is being transferred. This could not be so in the segregation vein scenario where veins and their partially melting host should be at the same temperature. There must be flow within the magma relative to the felsite to maintain thermal disequilibrium.

It is worth noting that the calculated minimum heat flux of 24 W/m^2 in IDDP-1, if applied to the entire area of 3.5 km^2 that A. Mortensen (2012, unpublished data) suggested, which might be underlain by shallow rhyolitic magma, amounts to only 100 MW thermal power output, or less than the output of IDDP-1 during flow tests and less that the Krafla power plant uses to generate 60 MW electrical power. There are many uncertainties; for example, there may be other magmatic sources of heat in the caldera, the thermal gradient in MHB could be much higher, or there may be advective transport of heat by magmatic vapor phase through episodic rupturing of MHB [7,9] in addition to continuous conductive transport of heat across MHB. However, if the inferred values approximate reality, then realizing an order of continuous magnitude power output from this magma-sourced geothermal field likely requires penetrating and thermally fracturing MHB, as apparently occurred with IDDP-1.

4.2. MHB Represents a Discontinuity in the Stress Field as Well as in Temperature

Below the MHB is magma with properties of a liquid. As the interval between magma and thermal cracking, MHB is expected to be ductile and should not support a difference between the lithostatic load on it and pressure within the magma. Pressure within the magma can therefore be expected to be lithostatic and isotropic. Above MHB, stress in the rock is lithostatic and anisotropic, whereas fluid in cracks and connected pores is hydrostatic. Hydrostatic pressure in a borehole is a straightforward measurement or calculation. Neglecting additives to water in the drilling fluid (which for some boreholes are substantial) and density variations with temperature, the hydrostatic pressure (P_H) at the bottom of IDDP-1, 2100 m, with fluid standing 400 m below the surface and the density of water $\rho_w = 1,000$ kg/m^3, acceleration of gravity g ~ 10 m/s^2, and depth of water column z = 1700 m is:

$$P_H = \rho_w\, g\, z \tag{6}$$

$$P_H = 17\ \text{MPa}$$

Taking 2500 kg/m^3 as a common approximation for shallow, porous, and fractured volcanic rock, the lithostatic pressure, P_L, at z = 2100 m is:

$$P_L = 53\ \text{MPa}$$

Zierenberg et al. [27] concluded that the magma is at about 40 MPa, less than the lithostatic pressure of 53 MPa, though significantly higher than hydrostatic pressure at 17 MPa, because the volatile content of the quenched melt falls below (i.e., at lower volatile content than) the vapor saturation surface in $CO_2 + H_2O$ vs. P space. Taking this at face value, there are three ways to explain the discrepancy in inferred pressures (1) The rock envelope containing the magma is strong enough to protect it from the full weight of the overburden. This has already been suggested to be unlikely. (2) The magma is vapor undersaturated: i.e., $P_{CO2} + P_{H2O} < P_L$. This seems unlikely as well, because the magma appears to have been generated at greater depth by partial melting of hydrothermally altered basalt protolith [27,34,35]. The source rock contained chemically bound water that would be preferentially partitioned into the melt phase. Both partial melting and fractional crystallization concentrate water in melt just as they concentrate other incompatible elements. (3) The magma ascended to a higher level (lower pressure), degassed, and then descended to where it was sampled. This unlikely scenario would be necessary because degassing requires vapor saturation so that volatiles can escape as vapor.

A more likely conclusion is that the magma is vapor saturated at local lithospheric pressure, but the calculated vapor and/or lithostatic pressures are in error. There are multiple possible sources of errors: (1) The actual overburden pressure is unknown. Certainly, the density profile is not constant, and 2500 kg/m^3 is an arbitrary figure. (2) There are uncertainties in the experimentally determined solubility values. (3) Likewise, there are uncertainties in the analytical determination of the CO_2 and H_2O contents, as shown by error bars that overlap the vapor saturation surface at +1 σ [27]. (4) Finally, the solubilities of volatiles are dependent on melt temperature, and this is not known with confidence to within 50 °C. If the magma is hotter than 900 °C, the solubilities of H_2O and CO_2 will be lower and hence the pressure inferred from the volatile contents higher

There is another intriguing possibility and it arises because of active rifting, the regional stress field is anisotropic with the least principal stress, σ_3, oriented approximately east–west, causing dikes from the main basaltic magma system to propagate north–south. Or, if the caldera fill is decoupled from this field, σ_3 may be vertical within the caldera, resulting in the formation of sills. The stress field changes with time during rifting and magma intrusion events. This leads to further uncertainties, not only as to the current confining pressure on the magma but also the history of that pressure. It is possible that the magma degasses if it experiences a lower σ_3 during a rifting event and so becomes vapor undersaturated when pressure returns to equilibrium. The upper limit for pressure in the magma may be σ_3 plus the critical overpressure to initiate a magmatic hydraulic fracture or dike, thought to be < 10 MPa [39]. Therefore, we do not know the pressure on the magma accurately. With the progress presently underway in developing extreme sensors, direct measurement of magma pressure may become possible. This will provide a vast advance in understanding both the coupling of magma dynamics to tectonic events and changes in magma pressure that drive seismicity, surface deformation, and presage eruptions.

In any case, the magma is at or close to vapor saturation and at a much higher pressure than hydrostatic, as dramatically demonstrated by the magma flowing 8 m up IDDP-1 in 9 minutes or less after it was penetrated by drilling [26]. This provides a first direct measurement of magma pressure, although only a crude minimum. When the drill bit reached the magma, there was a weight loss from the drill string (hook load) of 50,000 kg. The magma pushed back with a force in N of (50,000 kg) × g. We can treat the drill bit as a piston, albeit an ill-fitting one (Figure 7). Net upward force (F) on the 12 $\frac{1}{4}$″ diameter drill bit is produced by ΔP acting on the area (A) of the bit face:

$$F = \Delta P\, A \tag{7}$$

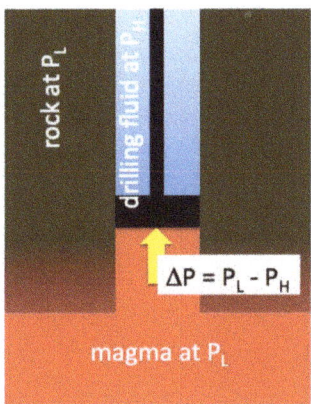

Figure 7. The difference between magma pressure beneath the drill bit and fluid pressure above it supports some of the weight of the drill string.

With a bit area of 0.076 m² (r = 0.15 m):

$$\Delta P = 7\ MPa$$

We know that P_H = 17 MPa, so this gives a minimum for P_M of 24 MPa. In fact, magma apparently flowed around and ended up on top of the drill bit at the same time it was pushing against the bit, so clearly the cylinder/piston leaked and P_M > 24 MPa. This leakage of magma and the fragments of partially melted borehole wall material it carried are the likely sources of the chips that were recovered. The pressure at the top of the rising magma column was also likely reduced by viscous drag as the magma cooled against the borehole walls and drilling fluid.

Note that the large change in the stress field across MHB in Krafla, or between the fluid-filled borehole and its wall, is in contrast to the lava lake case. In the latter, a shallow depth means that the difference between hydrostatic and lithostatic pressure is <2 MPa. The much larger pressure contrast at 2.1 km depth has implications for the stability of the borehole within the MHB at Krafla. Without extensive chilling or emplacement of casing to support ΔP of >30 MPa, the hottest portion of KMT-1 will close much faster than coreholes at Kilauea Iki. This can be used to advantage in understanding the MHB if a thermocouple string can be emplaced quickly (Figure 5) before the borehole closes on it.

4.3. Suppression of Upward Flow of Magma in the Borehole

There is always concern if the fluid in a well is overpressured. However, in general, "overpressured" is relative to hydrostatic. Here, we are dealing with a situation where the fluid, magma, could become overpressured with respect to lithostatic. If the Krafla magma were a simple liquid of $\rho = 2300$ kg/m^3, then $P_L = 53$ MPa at 2100 m would support a column of magma 2280 m high, 180 m higher than the surface. However, decompression during ascent would cause exsolution of the CO_2 and H_2O, forming a foam and then a dusty gas (because expansion of magmatic foam is limited by the strength of bubble walls to about 4 X), expanding about 200 X if the pre-eruption magma contains 2 wt% dissolved H_2O, e.g., [40]. This very low-density material within the borehole would reduce the pressure below it, establishing a condition similar to an airlift in a water well, that is, it becomes self-pumping, erupting at high velocity. An eruption, or in drilling terms a blowout, clearly did not happen, and from a safety standpoint in future exploration and use of magma the successful suppression of upward flow needs to be understood. Just as boiling in the drilling fluid column must be prevented, boiling, that is vesiculation, in the magma column, must be prevented as well.

Merely dropping the pressure on the magma to that prevailing at the bottom of the borehole, 17 MPa, results in significant foaming. Neglecting CO_2, which will not add much to the vapor volume, results from Zierenberg et al. [27] indicate that dropping the pressure of the magma to the hydrostatic pressure of 17 MPa will produce exsolution of about 0.5 wt% H_2O, or 0.005 kg of vapor in 1 kg of magma originally comprising melt (neglecting sparse crystals) with 1.8 wt% dissolved water. For magma density of 2300 kg/m^3 [29], the specific volume is $V_M = 4.3 \times 10^{-4}$ m^3/kg. The specific volume of vapor for ideal gas behavior is $V_v = (R \times T)/P$ where R = 456 J kg^{-1} °K^{-1}, T = 1173 °K, and P = 17 MPa, giving $V_v = 3.1 \times 10^{-2}$ m^3/kg (and consistent with extrapolation from tables of Burnham et al. [41]). The specific volume of the bubble-in-melt suspension is:

$$V_{mix} = X_V^{wt} V_V + X_M^{wt} V_M \qquad (8)$$

and porosity, Φ, is:

$$\Phi = V_V / V_{mix} \qquad (9)$$

This gives a porosity of 0.27, i.e., 27 vol% bubbles. The magma chips have less porosity than this, and many have no bubble content at all (Figure 4). This is clearly because the chips were effectively quenched by the drilling fluid before vapor exsolution and bubble growth due to decompression could occur. It was the cooling effect of the drilling fluid, rather than the pressure it exerted on the rising magma, that stopped the ascent.

The event described is essentially a small intrusion through an artificial perforation in the MHB. In the same way, a natural breach in MHB will cause quenching of magma unless it has sufficient volume and force to get through the formidable heat sink provided by the hydrothermal system and erupt.

4.4. Convection in Krafla Magma

To explain the absence of crystallization at the top of Krafla magma, Axelsson et al. [29] suggested, and I agree, that there must be convection within the magma body, continually sweeping away denser crystallizing magma and replacing it with uncooled magma at the roof zone (Figure 8).

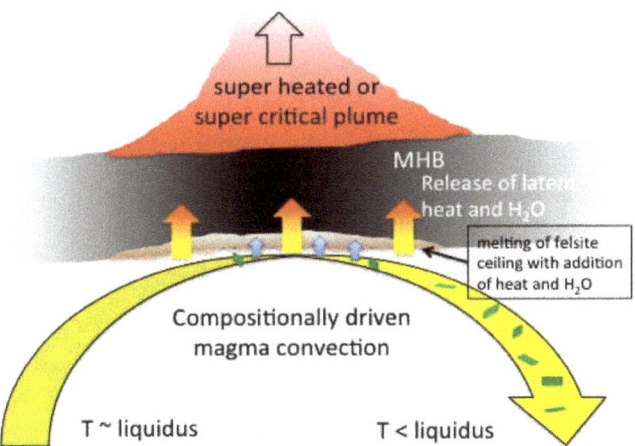

Figure 8. Thermally (density increase due to cooling) and compositionally (density increase due to crystal growth) driven circulation in magma could transfer heat and H$_2$O at the lower face of the MHB without coating it with crystals and, in the case of Krafla, partially melting it. This would explain the sparse crystal content of all shallow silicic magmas thus far encountered (Table 1).

An independent line of evidence supporting convection is the crystallization required to produce the observed melting in the roof at Krafla. In a static system, crystallization of magma should approximately balance melting in the roof because the latent heat of crystallization dominates the energy budget. Yet the meager crystallization observed at the top of the magma (~1 vol%) cannot account for the much more significant melting observed in some of the felsite (Figure 5). Convection of magma below the static roof would allow the crystallization that is driving the melting to be spread out over a much larger volume of magma, and so crystals would be less abundant in any volume element of magma than melt in a volume element of the roof.

But, melting of the roof requires delivery not only of heat but of water as well. The felsite contains no hydrous phases; it is anhydrous. Without water, the felsite solidus should be about 950 °C [36], well above that of the rhyolite magma with 1.8 wt% H$_2$O. Two mechanisms might accomplish this transfer of water. One is if the magmatic vapor phase exsolves in the upwelling magma because of decreasing pressure or increasing crystallization (second boiling) or both and enters the overlying felsite along grain boundaries. There it becomes a key though minor (in wt% but not mole%) ingredient in the melt phase as melting develops between quartz and feldspar crystals, which provide the other ingredients necessary to form the eutectic melt. Alternatively, this might be accomplished by diffusion of water within the melt phase. Given the presumed impermeability of MHB and the fact that the diffusivity of water in melt is five orders of magnitude smaller than for heat (10^{-11} m^2/s vs. 10^{-6} m^2/s), it is probably the transfer of water rather than heat to the felsite that is the rate-controlling step for melting. The extent of melting may therefore be quite limited, perhaps even to only occurring in fragments of roof rock that were engulfed by the magma. As a layer, the partially melted felsite would be a thin transient heat sink, blocking some of the upward heat flow as long as melting is occurring.

The convection hypothesis is consistent with the presence of an upper mush zone in Kilauea Iki and absence in magma bodies. Convection in the former may be limited in time and vigor due to the smaller size and to cooling from the bottom as well as the top in a lava lake, whereas in the magma chamber case, convection may be more vigorous and longer long-term [5].

The simplest form of convection occurs due to thermal contraction of magma on cooling and therefore negative buoyancy at the top of the cooling body. The coefficient of thermal expansion, a, is defined as:

$$a = \frac{1}{V}\frac{\Delta V}{\Delta T} \tag{10}$$

where V is specific volume, about 4×10^{-5} °C^{-1} for magma [5]. The Rayleigh number for thermal convection, Ra_T, which indicates whether a fluid body across which a vertical temperature drop exists will convect, is the dimensionless ratio of the time scale for heat transport by conduction ($_{cd}$) divided by the time scale for heat transport by convective flow ($_{cv}$).

$$Ra_T = {}_{cd}/{}_{cv} \tag{11}$$

The time scale for conduction (t_{cd}) can be approximated as d^2/K, where d is the thickness of the magma layer and K is thermal diffusivity of magma. The time scale for convection (t_{cv}) is d/u, where u is the terminal velocity of the sinking cooled fluid, dependent in turn on the density change with cooling (a) and magma viscosity (η). This gives:

$$Ra_T = \frac{\rho\,\alpha\,\Delta T\,d^3\,g}{K\,\eta} \tag{12}$$

Application of the viscosity model of [42] for the Krafla melt composition and T = 900 °C and P = 50 MPa gives viscosity $\eta = 3 \times 10^5$ Pa s. Taking $\Delta T = 100$ °C, d = 100 m, K = 10^{-6} m^2/s, and a as above gives:

$$Ra_T = 3 \times 10^8$$

That is well within the convective regime that begins above Ra ~ 2000.

The value of Ra_T is critically dependent upon the thickness of the magma layer, d. A 10-m thickness can be ruled out because the entire body would have crystallized since the Krafla Fires. A thickness of 1000 m is plausible for a long-lived rhyolite magma body. The 100-m thickness is chosen as being intermediate and conservative.

Another route to obtaining magma viscosity is the observation that magma quickly rose 8 m up the borehole after drilling penetrated magma on the second sidetrack [26]. After encountering magma, the drill stem was pulled back and then went back in, encountering magma 8 m higher in the well 9 minutes later. The magma continued to rise as the drill string was pulled back again, and 50,000 kg was lost from the hook load (see Section 4.2). The bit became stuck in magma 9.4 m above the previous level.

The Hagen–Poiseuille equation for fluid flow through a pipe gives a first approximation for this situation, although it applies to steady-state flow rather than abrupt entry of magma into the borehole, wherein the pressure gradient will start high but decrease with time and cooling of magma against the borehole wall and drilling fluid will cause the viscosity to rise.

The equation is:

$$\mu = \frac{\Delta P\,r^2}{8\,V\,L} \tag{13}$$

Using the same values as in Section 4.2 and velocity V = 0.015 m/s and L = 4 m, the midpoint in flow up the borehole to give an intermediate pressure gradient between the beginning and end of flow, gives:

$$\mu = 2 \times 10^6 \text{ Pa s}$$

This value is about an order of magnitude greater than that calculated from the laboratory measurement-based model of Giodano et al. [42] noted above. However, flow was being impeded by cooling of magma against the borehole walls and the drilling fluid, both raising its viscosity and constricting the radius (r) of flow. Zierenberg et al. [27] obtained quench temperatures averaging 850 °C based on speciation of water in melt. Dingwell et al. [43] found that a temperature decrease from 900 °C to 800 °C increases viscosity by more than an order of magnitude. So the estimate for magma viscosity from borehole flow is reasonable, but likely higher than that of the undisturbed magma.

There is an additional buoyancy force affecting convection due to crystallization of magma flowing below the roof. As is the case with heat capacity where the effect of heat from crystallization can be added to define an effective heat capacity, the effect of crystallization of crystals denser than melt can be added to obtain an effective thermal expansivity. Just as crystallization dominates heat transport through the release of latent heat, crystallization also dominates mass flow by increasing the bulk density of the melt + crystal mixture, magma.

We should first establish the validity of considering the bulk density of the melt+crystals suspension that is magma, wherein relative motion between melt and crystals can be neglected. The Stokes settling relationship, where buoyancy and drag forces are balanced for a sphere of r radius, is:

$$u = \frac{2 \, \Delta\rho \, g \, r^2}{9 \, \eta} \tag{14}$$

with $r = 1 \times 10^{-4}$ m (Figure 5), $\Delta\rho = 400$ kg/m^3 [29], and $\eta = 3 \times 10^5$ Pa·s [43].

$$u = 3 \times 10^{-11} \text{ m/s}$$

This works out to 1 mm/a, trivial even for convection in a lava lake [5].

The crystallization equivalent for the coefficient of normal thermal expansion, Equation (10), where we will assume linear crystallization with decreasing temperature (not accurate [36] but a useful approximation) from a weight fraction of $X^{wt}{}_C = 0$ at the liquidus of 900 °C to $X^{wt}{}_C = 1$ (so that $X^{wt}{}_L + X^{wt}{}_C = 1$) at the solidus of 800 °C, is:

$$a_{mix} = \frac{1}{V_{mix}} \frac{\Delta V_{mix}}{\Delta T} \tag{15}$$

where V_{mix} is the specific bulk density of the melt + crystal mixture, i.e., magma:

$$V_{mix} = \left(X^{wt}_L \times V_L\right) + \left(X^{wt}_C \times V_C\right) \tag{16}$$

Because $X^{wt}{}_L + X^{wt}{}_C = 1$, we can deal only with $X^{wt}{}_C$ as crystallinity, eliminating $X^{wt}{}_L$ as $(1 - X^{wt}{}_C)$:

$$V_{mix} = V_L + X^{wt}_C \times (V_C - V_L)$$

Because of the relatively small temperature interval, we can treat V_C and V_L as constant with T. Differentiating with respect to T:

$$dV_{mix}/dT = \left(dX^{wt}_C/dT\right) \times (V_c - V_L) \tag{17}$$

Using $V_L = 4.35 \times 10^{-4}$ m^3/kg, $V_C = 3.70 \times 10^{-4}$ m^3/kg, and $dX^{wt}_C/dT = 1/100$ we get:

$$dV_{mix}/dT = 0.65 \times 10^{-6} \text{ m}^3/\text{kg}/°C$$

Converting to the form of Equation (15) and using a 90:10 liquid to crystal mixture for Vmix:

$$a_{mix} = 1.5 \times 10^{-3} \, °C^{-1}$$

This is almost two orders of magnitude larger than the thermal expansivity of melt, so it will obviously dominate convection if crystallization is occurring, as it must. The compositional Rayleigh number, as it is called, analogous to Equation (12), is then:

$$Ra_C = 4 \times 10^{10}$$

This means that convection will occur in a thinner magma body or at a lower thermal gradient than in the melt-only case.

During convection, there is another process that helps to ensure that the hottest magma is delivered to the base of MHB. That is, the preferential migration of lower viscosity fluid within a fluid of variable viscosity to regions of high shear during flow. This has been observed in the chemical zonation of a volcanic conduit [44] and in pipes in an industrial setting [45]. In convecting magma, the highest shear will be at the base of MHB [45], driving—in addition to the buoyancy forces at work—the hottest magma with least crystallinity there.

However, some caveats should be recognized. One is that crystal growth must be sufficiently rapid to release the latent heat of fusion as the magma is flowing under the roof. If there is a time lag so that crystallization occurs when the magma has already been returned by thermally driven convection to deep in the magma chamber, then it will not contribute to convection as a whole. In fact, it may retard it. Also, in a vapor-saturated system, crystallization of anhydrous phases, as all the crystalline phases in Krafla magma are, will induce vapor exsolution, that is, bubble growth with the vapor fraction dependent upon depth (pressure). This would tend to compensate for the negative buoyancy effect of growing crystals. These complications have been treated in a number of sophisticated models, e.g., [46] and are beyond the scope of this simple analysis. The conclusion here is that the release of latent heat and forcing convection by crystallization are linked and can contribute greatly to heat transport from magma into the suprajacent hydrothermal system. Arguments against convection of magma, both experimental and theoretical, rely on growth crystals at the roof of the magma body in response to heat lost through MHB. Thus far, drilling encounters with magma indicate that this is not the case. A difference in views is whether convection is limited to the earliest stage of a magma body and therefore can be ignored in its overall cooling history [47] or whether it can continue until flow is locked as the crystal content approaches 50 vol% [4].

To get an idea of how effective convection is in delivering heat to MHB, we can ask how large the upwelling limb of magma convection cell needs to be to continuously deliver 1 GW of thermal energy. Assumptions are a vertical velocity in cross sectional area A, rising at vertical velocity u, say 0.01 m/s, and releasing heat to the roof by cooling 10 °C.

$$P_T = X_C^{wr} \beta_{Meff} A u \qquad (18)$$

Using the same parameters as before and solving for the radius of the heat pipe:

$$r = 17 \text{ m}$$

There may be time-discontinuous mechanisms of heat transport as well, for example, magma breaching the MHB and cooling within the hydrothermal system, transferring its energy directly. Based upon observations at Yellowstone, including inflation and deflation cycles and migrating seismic swarms, Fournier [9] postulated that build up of vapor pressure in a magma body due to crystallization would periodically rupture the MHB, injecting the magmatic vapor phase into the hydrothermal system (and occasionally causing phreatomagmatic eruptions).

4.5. Conceptual Model of Krafla as a Closely Coupled Magma–Hydrothermal System

A view of the Krafla system that is consistent with, but not unique to, most of the observations cited is presented in Figure 9. Like its nearby companion, Askja Caldera [33], Krafla is an essentially basaltic system lying astride Iceland's northern rift zone. Although one might be tempted to think of rift magmatism as uniformly distributed along the spreading plate boundary, these central volcanoes are clearly foci of basaltic magma upwelling that then feed dikes laterally at shallow depth along the rift structure. If there is a deeper zone of uniform along-strike magma generation, the magma must gather into plumes or heat pipes, which in this case then feeds into chambers or dike/sill complexes that in turn redisperse magma laterally at shallower depth. In a way, this is like subduction zone volcanism where existing models suggest uniform, along-strike magma generation above the subducting slab that coalesces upward to focus into discrete, long-lived centers rooted in the crust.

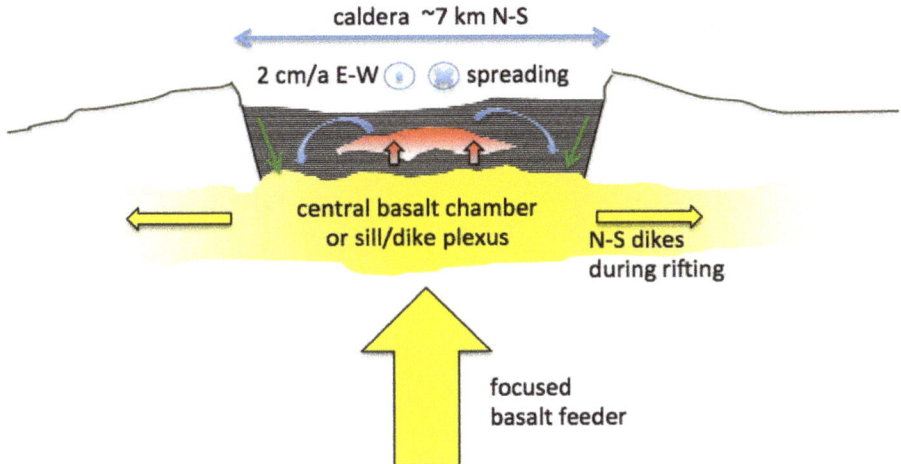

Figure 9. Suggested concept for Krafla shield and caldera. A basaltic magma plume beneath the rift is responsible for building the shield and maintaining a large and relatively shallow basaltic magma chamber. This chamber may be a plexus of dykes and sills rather than an oblate bladder. During rifting events, the magma accumulation feeds dikes that travel tens of kilometers to the north and south, deflating and contributing to caldera subsidence in the process. Central eruptions and intrusions of basalt and more silicic magmas comprise the caldera fill, which is hydrothermally altered in this crucible and gradually fed downward to the basaltic hearth. Partial melting ensues, probably continuously dribbling rhyolitic melt upward to its level of neutral buoyancy at about 2 km depth. If the input is insufficient, only felsite sills are produced, but if more vigorous, a homogeneous convecting rhyolite magma body is established. It is stable for long periods of time, but can be expelled if a massive influx of basaltic magma occurs below it during a major rifting event. Undisturbed, the shallow rhyolite contributes only a portion of the thermal output of the caldera, but it represents a massive concentration of latent heat of crystallization at shallow depth that can be accessed if the MHB above it is penetrated by drilling, allowing fluid to invade and fracture the magma and near-magma region.

Sustained, focused magmatism produces impressive edifices, but less so in rifts because the structures are being torn apart as they are built. Focused magmatism also produces calderas that are the surface expression of shallow magma chambers. As such, calderas reflect subsidence above shallow magma bodies because over time the magma entering the chamber is more than balanced by magma dispersed beyond the immediate volcanic center. The two processes that do this are lateral diking that drains the central chamber in the subsurface and large volume explosive eruptions that disperse its contents widely on the surface. The former is more frequent and the latter more catastrophic. Both processes appear to have contributed to the topographically rather subtle Krafla Caldera.

The association of rhyolitic magmas with an isotopic signature suggestive of a hydrothermally altered basalt protolith [27] fits with this view. Intense hydrothermal circulation is concentrated within these caldera structures, and the gradual or episodic subsidence feeds hydrothermally altered basaltic caldera fill towards the hot plate of focused basaltic magmatism. Partial melting ensues, generating rhyolitic melt that accumulates at shallow depth, perhaps at its level of neutral buoyancy, as a coherent body. This may be a continuous process and requires only local vertical rearrangement of materials. Thus it produces little by way of surface deformation or gravity signals. Being viscous and at neutral buoyancy, the magma accumulation remains stable over millennia, requiring a massive "kick" by a basaltic intrusion as occurred at Askja [33] in 1875 or on a smaller scale at Krafla's Viti Crater in 1724 [20]. To remain near liquidus, these rhyolitic accumulations must themselves be continuously heated by the basaltic furnace, either through their own intervening conductive boundary zone, or directly, because the strong contrast in density and viscosity and relatively low water content inhibit mixing [48]. Indeed, Iceland is one of the classic sites exhibiting bimodal volcanism [49]. This is in contrast to the subduction zone case where high water contents in the basalt cause foaming and mixing due to the temperature contrast at the boundary [50].

This is an admittedly simplistic picture, and three variants in the magma–hydrothermal relationship seem obvious. One is where the shallow magma body maintains an open pathway to the surface. This prevents the hydrothermal system from forming as a cap above the magma, but allows it to develop as a lateral fluid outflow zone. Mutnovsky Volcano, Kamchatka, Russia, is an example [51]. The direct connection between Mutnovsky and its adjacent geothermal system is supported both by oxygen isotope composition of fluids and evidence that changes in geothermal energy extraction appear to modulate the behavior of the volcano itself (A. Kiryukhin, unpublished data, 2019). Another is the rift setting, such as Reykjanes or in submarine midocean ridges, where frequent unfocused diking events maintain high temperature fluid circulation in the crust. The development of a magma chamber requires frequent, focused input [11]. Finally, as the end stage of a silicic system, the melt may disappear and with hydrothermal circulation driven by residual heat. However, in this case, shallow heat is not replenished by the combination of the latent heat of crystallization and magma convection.

5. Magma as an Energy Source for Geothermal Power Production

It is useful to consider energy density as it varies within a magma–hydrothermal system. This reveals where the energy is stored and what volume must be accessed to produce a thermal power output over a given period of time. Ultimately, whether by natural process or enhanced to extract geothermal energy, heat transport is some combination of short path-length conduction and long-path length convection. The enthalpy of hydrothermal fluid is important in that it determines how rapidly energy can be transported within a hydrothermal system and out of it for geothermal power production. However, because the fluid occupies only pores and fractures within a hydrothermal reservoir, it does not contribute much to the total energy in the system.

For magma as the source of energy, we need to consider energy per liquid volume (2300 kg/m^3) rather than its crystallized product (2700 kg/m^3) calculated previously. This yields:

$$\varepsilon_R = 1.1 \times 10^9 \text{ J}/m^3$$

In this case, I include the enthalpy of the 2 wt% H_2O that will be released as the magma crystallizes and eventually adds to power output at the surface. For water at 50 MPa and 800 °C [41], the enthalpy is 4×10^9 J/kg. Added to the magma, this yields a total energy density of:

$$\varepsilon_R = 1.3 \times 10^9 \text{ J}/m^3$$

There are processes not considered here that reduces the total energy in magma. One is the heat of vesiculation, which is opposite in sign from heat of crystallization. Another is any work, $P\Delta V$, done by expansion of the system against pressure as it exsolves the magmatic vapor phase. However, Sahagian and Proussevich [52] have shown that these energy terms can be neglected.

The Krafla magma is shallow and fairly dry. For magmas with the more typically estimated 4–6 wt%, the additional enthalpy in expelled magmatic vapor is more significant and may play an important role in mass and heat transfer, particularly during the latter half of the crystallization history of a magma body [7].

A useful reference amount of energy, say Q_{30}, is an amount required to produce 1 GW of thermal power output for 1 Gs, or about 30 a, a standard business horizon and also about 10^9 s. The volume of Krafla-like magma required is 0.8 km³. For visualization purposes, the face of the monolith El Capitan, Yosemite National Park, California, USA, is roughly 1 km², so a cube extending 800 m into it would be 0.8 km³. That is not to say that all the energy would be extractable, especially since convection would cease before 50% crystallization is reached, and heat transport would then be limited to invasion of the hydrothermal system by thermal cracking towards the hot center.

By comparison, for the same temperature drop in solid rock with 5 vol% superheated fluid in pores at 50 MPa and 800 °C, the contribution of superheated fluid in pores, even at 4×10^6 J/kg, is insignificant (0.2 wt%) and (Figure 10):

$$\varepsilon = 2.2 \times 10^8 \, J/m^3$$

$$V = 4.5 \, km^3$$

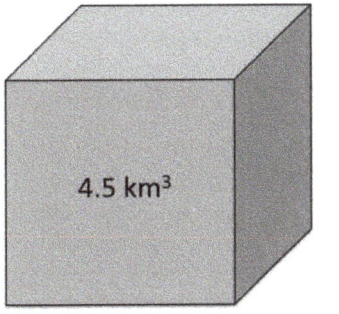
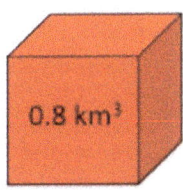

Figure 10. The volume of rock (gray) or magma (red) required to release 10^9 GJ of thermal energy during cooling of $\Delta T = 100$ °C. For the magma, the temperature drop is chosen to be liquidus to solidus, thereby releasing all the latent heat of crystallization. For the rock, where the energy content is entirely sensible heat, only the size of the temperature interval is important.

Accepting the argument that conduction through the MHB is the rate-controlling step for heat transport, then it also controls the response time over which a perturbation in the magma reservoir or hydrothermal reservoir will affect the other part of the system. Such a perturbation might be due to a new influx of very hot magma heating the magma reservoir, as, for example, with the Krafla Fires of 1975–1984 [20], or initiation of geothermal power production, cooling the hydrothermal reservoir as at Mutnovsky [51].

The characteristic time, τ, for thermal diffusion where ΔZ is the thickness of the MHB and K is thermal diffusivity is:

$$\tau = \Delta Z^2 / K \tag{19}$$

Taking 25 m as the upper limit of ΔZ at Krafla and $K = 10^{-6}$ m²/s,

$$\tau = 6.3 \times 10^8 s \text{ or about 20 a}$$

This is on the threshold of geothermal production by humans influencing magmatic behavior. If the actual value is $\Delta Z = 10$ m, then $\tau = 3$ a, which is to say that the power plant is actually tapping magma energy. Because the geothermal energy extraction is increasing cooling at the top of the magma body, the response may be to accelerate convection, thereby pulling thermal energy from deeper in the body. On the other hand, if the magmatic heat source is distant, say 500 m, then $\tau = 10^4$ a and the two reservoirs are effectively separate. Put another way, exploitation of the hydrothermal system would be mining old magmatic heat.

6. Applications

The picture of the Krafla system painted here, if a general case, would imply that magma-related geothermal systems contain far more energy, with higher enthalpy fluids, and are more sustainable than would otherwise be expected. The key may be to penetrate the MHB, by drilling and thermal cracking, so that hydrothermal fluid gains access to the magma itself. Convection within the magma, supplying latent heat of crystallization, enthalpy of magmatic vapor phase from a voluminous source, combined with continued thermal cracking due to energy extraction, could maintain thermal power output without or with lesser need of makeup wells. We need to test this scenario with space-continuous samples and temperature measurements through the MHB at Krafla, and at multiple points in time. We also need to improve geophysical techniques for imaging magma. Such prospecting tools have been tested against reality and improved through much drilling experience in the case of oil and gas reservoirs, but this was never, until now, possible for testing the imaging of magma reservoirs.

There are also large challenges in drilling engineering and technology, because for superhot geothermal power generation, production wells must be sustainable over a long period. This challenge encompasses casing alloys, accommodation of thermally induced stresses in casing, cementing at high pressure and temperature, corrosion and precipitation by and from acid fluids, and extreme sensor technology. The objection that truly high-grade geothermal systems are too far from the customer base to be viable is diminished by recent applications of High Voltage Direct Current (HVDC) technology that enables transmission of electrical power over great distances, including under the ocean, without losses due to the conductor skin effect and external inductive field of AC.

Development of extreme sensor technology, now underway for combustion engine and planetary exploration applications, opens another use for which there is a compelling need: monitoring of restless volcanoes and forecasting of eruptive events [12,53]. All the sensors used to date for volcano monitoring, which measure thermal and gas emission, surface deformation, and seismicity, are in essence remote sensing techniques that are interpreted but not tested against ground truth to reflect processes in the source magma that may or may not lead to eruption. Direct measurement of conditions within or at least proximal to magma bodies would test and improve these interpretations and in the case of very high risk volcanoes might partially supplant labor-intensive surface-based multisensor networks themselves. Monitoring temperature in magma is already within our grasp. Inflation, increased CO_2 and thermal emission, and increased seismicity are often interpreted as the rise of new magma into a shallow magma chamber, portending an eruption. This should quickly cause a temperature rise in the convecting magma body being intruded and therefore detected and quantified by in situ sensors. Another, and perhaps simpler question that could be answered, and always arises during periods of volcano unrest, is: are the geophysical and geochemical signals a consequence of underlying magmatic processes or merely reflecting of some structural reconfiguration within the hydrothermal system?

Accidental drilling encounters with magma are beginning to provide new insights into the relationship between magma and hydrothermal systems. Encounters planned for scientific and engineering objectives, as, for example, proposed for the Krafla Magma Testbed (KMT) [12,54], have the potential to lead to a new, much need clean bedload electric energy source and a reliable means to warn populations at risk of impending eruptions The scientific, geothermal energy, and volcanic hazard needs are compelling. It seems likely therefore that planned explorations of this new frontier are both essential and inevitable.

7. Conclusions

Existing in situ data from magma bodies are few and without precedent. Data from Krafla confirm the existence of an MHB such as found in lava lake drilling and predicted from theoretical considerations. The MHB at Krafla is remarkably thin. However, with such sparse data, it is not possible to say how representative Krafla is of magma–hydrothermal systems in general. Combining Krafla results with insights from the two other sites of magma encounters and results from lava lake drilling, it is suggested that:

- The MHB is a moving boundary. It can ascend by melting of the magma ceiling if the magma heats up or descend by thermal cracking, ultimately to the center of the magma body as it cools.
- The MHB is not only a thermal boundary but also a stress field boundary because of the upward transition from fluid to ductile to brittle behavior. It forms a barrier between magma where pressure is isotropic and lithostatic, and above which stresses are anisotropic and lithostatic within rock, and isotropic and hydrostatic in fluid-field fractures. Understanding this is essential to understanding the processes of volcano unrest and eruption.
- Although heat transport through the MHB is impressive, it is of the same order as conventional geothermal power extraction. The order of magnitude increase in thermal power output of near-magma wells likely requires penetrating and thermally cracking the MHB. Once done, crystallization-driven convection within the magma body, release of the magmatic vapor phase, and continued thermal fracturing by induced fluid circulation may make much of the energy contained within magma accessible for use.
- First-order data needed to understand magma–hydrothermal coupling are continuous petrological (core) and temperature profiles across the MHB, from solid, brittle rock to near-liquidus magma. In view of the apparent difficulty of magma breaching the MHB, using pressure sensors to detect an increase in magma pressure may be an effective way to predict eruptions. These advances, ultimately leading to exploitation of magma energy and reliable eruption forecasting, will require a testbed approach of multiple boreholes and engineering and scientific experiments over an extended period of time.

Funding: I benefited from workshops on the Krafla Magma Testbed supported by Landsvirkjun National Power Company, International Continental Scientific Drilling Program (ICDP), and the US National Science Foundation (NSF). In addition, I received travel support from Landsvirkjun, Iceland; the Geothermal Research Group (GEORG), Iceland; and the British Geological Survey.

Acknowledgments: Thanks are due for discussions with many colleagues involved in the Krafla Magma Testbed Project and for the pioneering work of IDDP. I am especially grateful to Charles Carrigan, Yan Lavallee, Allen Glazner, and Pavel Izbekov for stimulating insights, many of which are cited here, but that is not to say that they agree with everything herein. Charles Carrigan and Pavel Izbekov contributed to magma viscosity calculations.

Conflicts of Interest: I declare that I have no competing financial interests or personal relationships that could have influenced the work reported in this paper.

References

1. Lister, C.R.B. On the Penetration of Water into Hot Rock. *Geophys. J. Int.* **1974**, *39*, 465–509. [CrossRef]
2. Hardee, H.C. Solidification in Kilauea Iki lava lake. *J. Volcanol. Geotherm. Res.* **1980**, *7*, 211–223. [CrossRef]
3. Carrigan, C.R. Biot number and thermos bottle effect: Implications for magma-chamber convection. *Geology* **1988**, *16*, 771–774. [CrossRef]
4. Carrigan, C.R. Time and temperature dependent convection models of cooling reservoirs: Application to volcanic sills. *Geophys. Res. Lett.* **1984**, *11*, 693–696. [CrossRef]
5. Carrigan, C.R. The magmatic Rayleigh number and time dependent convection in cooling lava lakes. *Geophys. Res. Lett.* **1987**, *14*, 915–918. [CrossRef]

6. Hort, M.; Marsh, B.D.; Resmini, R.G.; Smith, M.K. Convection and Crystallization in a Liquid Cooled from above: An Experimental and Theoretical Study. *J. Petrol.* **1999**, *40*, 1271–1300. [CrossRef]
7. Lamy-Chappuis, B.; Heinrich, C.A.; Driesner, T.; Weis, B. Mechanisms and patterns of magmatic fluid transport in cooling hydrous intrusions. *Earth Planet. Sci. Lett.* **2020**, *535*, 116111. [CrossRef]
8. Hawkesworth, C.J.; Blake, S.; Evans, P.; Hughes, R.; Macdonald, R.; Thomas, L.E.; Turner, S.P.; Zellmer, G. Time Scales of Crystal Fractionation in Magma Chambers—Integrating Physical, Isotopic and Geochemical Perspectives. *J. Petrol.* **2000**, *41*, 991–1006. [CrossRef]
9. Fournier, R.O. Geochemistry and Dynamics of the Yellowstone National Park Hydrothermal System. *Annu. Rev. Earth Planet. Sci.* **1989**, *17*, 13–53. [CrossRef]
10. Scott, S.; Driesner, T.; Weis, P. Boiling and condensation of saline geothermal fluids above magmatic intrusions. *Geophys. Res. Lett.* **2017**, *44*, 1696–1705. [CrossRef]
11. Glazner, A.F. Climate and the Development of Magma Chambers. *Geosciences* **2020**, *10*, 93. [CrossRef]
12. Eichelberger, J. Magma: A journey to inner space. *EOS* **2019**, *100*, 26–31.
13. Cashman, K.V.; Sparks, R.S.J.; Blundy, J.D. Vertically extensive and unstable magmatic systems: A unified view of igneous processes. *Science* **2017**, *355*, eaag3055. [CrossRef] [PubMed]
14. Colp, J.L. *"Final Report-Magma Energy Research Project"*; Sand82-2377; Sandia National Laboratories: Albuquerque, NM, USA, 1982; pp. 1–36.
15. Barth, G.A.; Kleinrock, M.C.; Helz, R.T. The magma body at Kilauea Iki lava lake: Potential insights into mid-ocean ridge magma chambers. *J. Geophys. Res.* **1994**, *99*, 7199–7217. [CrossRef]
16. Helz, R.T.; Clague, D.A.; Sisson, T.W.; Thornber, C.R. Petrologic insights into basaltic volcanism at historically active Hawaiian volcanoes. *Prof. Pap.* **2014**, 237–292. Available online: https://www.researchgate.net/profile/Rosalind_Helz/publication/271829142_Petrologic_Insights_into_Basaltic_Volcanism_at_Historically_Active_Hawaiian_Volcanoes/links/54d2698b0cf25017917dfedc.pdf (accessed on 26 May 2020).
17. Teplow, W.; Marsh, B.; Hulen, J.; Spielman, P.; Kaleikini, M.; Fitch, D.; Rickard, W. Dacite melt at the Puna geothermal venture wellfield, Big Island of Hawaii. *Geotherm. Resour. Counc. Trans.* **2009**, *33*, 989–994.
18. Mbia, P.K.; Mortensen, A.K.; Oskarsson, N.; Hardarson, B. Sub-surface geology, petrology and hydrothermal alteration of the Menengai geothermal field, Kenya: Case study of wells MW-02, MW-04, MW-06 and MW-07. In Proceedings of the World Geothermal Congress, Melbourne, Australia, 19–25 April 2015; p. 20.
19. Mibei, G.; Lagat, J. Structural controls in Menengai geothermal field. In Proceedings of the Kenyatta International Conference Centre, Nairobi, Kenya, 21–22 November 2011; p. 21.
20. Árnason, K. New Conceptual Model for the Magma–hydrothermal-Tectonic System of Krafla, NE Iceland. *Geosci. J.* **2020**, *10*, 34. [CrossRef]
21. Elders, W.A.; Friðleifsson, G.Ó.; Albertsson, A. Drilling into magma and the implications of the Iceland Deep Drilling Project (IDDP) for high-temperature geothermal systems worldwide. *Geothermics* **2014**, *49*, 111–118. [CrossRef]
22. Glazner, A.F.; Bartley, J.M.; Coleman, D.S. We need a new definition for "magma". *EOS* **2016**, *97*, 12–13. [CrossRef]
23. Björnsson, H.; Björnsson, S.; Sigurgeirsson, T. Penetration of water into hot rock boundaries of magma at Grímsvötn. *Nature* **1982**, *295*, 580–581. [CrossRef]
24. Reynolds, H.I.; Gudmundsson, M.T.; Högnadóttir, T.; Pálsson, F. Thermal power of Grímsvötn, Iceland, from 1998 to 2016: Quantifying the effects of volcanic activity and geothermal anomalies. *J. Volcanol. Geotherm. Res.* **2018**, *358*, 184–193. [CrossRef]
25. Haddadi, B.; Sigmarsson, O.; Devidal, J.-L. Determining intensive parameters through clinopyroxene-liquid equilibrium in Grímsvötn 2011 and Bárðarbunga 2014 basalts. In Proceedings of the Geophysical Research Abstracts, Vienna, Austria, 24–25 April 2015; Volume 17. EGU General Assembly.
26. Mortensen, A.K.; Egilson, Þ.; Gautason, B.; Árnadóttir, S.; Guðmundsson, Á. Stratigraphy, alteration mineralogy, permeability and temperature conditions of well IDDP-1, Krafla, NE-Iceland. *Geothermics* **2014**, *49*, 31–41. [CrossRef]
27. Zierenberg, R.A.; Schiffman, P.; Barfod, G.H.; Lesher, C.E.; Marks, N.E.; Lowenstern, J.B.; Mortensen, A.K.; Pope, E.C.; Bird, D.K.; Reed, M.H.; et al. Composition and origin of rhyolite melt intersected by drilling in the Krafla geothermal field, Iceland. *Contrib. Mineral. Petrol.* **2013**, *165*, 327–347. [CrossRef]
28. Carslaw, H.S.; Jaeger, J.C. *Conduction of Heat in Solids*, 2nd ed.; Clarendon Press: Oxford, UK, 1959.

29. Axelsson, G.; Egilson, T.; Gylfadóttir, S.S. Modelling of temperature conditions near the bottom of well IDDP-1 in Krafla, Northeast Iceland. *Geothermics* **2014**, *49*, 49–57. [CrossRef]
30. Lamur, A.; Lavallée, Y.; Iddon, F.E.; Hornby, A.J.; Kendrick, J.E.; von Aulock, F.W.; Wadsworth, F.B. Disclosing the temperature of columnar jointing in lavas. *Nat. Commun.* **2018**, *9*, 1–7. [CrossRef] [PubMed]
31. Schuler, J.; Greenfield, T.; White, R.S.; Roecker, S.W.; Brandsdóttir, B.; Stock, J.M.; Tarasewicz, J.; Martens, H.R.; Pugh, D. Seismic imaging of the shallow crust beneath the Krafla central volcano, NE Iceland. *J. Geophys. Res.* **2015**, *120*, 7156–7173. [CrossRef]
32. Kim, D.; Brown, L.D.; Árnason, K.; Ágústsson, K.; Blanck, H. Magma reflection imaging in Krafla, Iceland, using microearthquake sources. *J. Geophys. Res.* **2017**, *122*, 5228–5242. [CrossRef]
33. Sigurdsson, H.; Sparks, R.S.J. Petrology of Rhyolitic and Mixed Magma Ejecta from the 1875 Eruption of Askja, Iceland. *J. Petrol.* **1981**, *22*, 41–84. [CrossRef]
34. Schiffman, P.; Zierenberg, R.A.; Mortensen, A.K.; Friðleifsson, G.Ó.; Elders, W.A. High temperature metamorphism in the conductive boundary layer adjacent to a rhyolite intrusion in the Krafla geothermal system, Iceland. *Geothermics* **2014**, *49*, 42–48. [CrossRef]
35. Masotta, M.; Mollo, S.; Nazzari, M.; Tecchiato, V.; Scarlato, P.; Papale, P.; Bachmann, O. Crystallization and partial melting of rhyolite and felsite rocks at Krafla volcano: A comparative approach based on mineral and glass chemistry of natural and experimental products. *Chem. Geol.* **2018**, *483*, 603–618. [CrossRef]
36. Hammer, J.E.; Rutherford, M.J. An experimental study of the kinetics of decompression-induced crystallization in silicic melt. *J. Geophys. Res.* **2002**, *107*, ECV-8. [CrossRef]
37. Bachmann, O.; Bergantz, G.W. On the origin of crystal-poor rhyolites: Extracted from batholithic crystal mushes. *J. Petrol.* **2004**, *45*, 1565–1582. [CrossRef]
38. Helz, R.T. Crystallization history of Kilauea Iki lava lake as seen in drill core recovered in 1967–1979. *Bull. Volcanol.* **1980**, *43*, 675–701. [CrossRef]
39. Rubin, A.M. Tensile fracture of rock at high confining pressure: Implications for dike propagation. *J. Geophys. Res.* **1993**, *98*, 15919. [CrossRef]
40. Eichelberger, J.C.; Carrigan, C.R.; Westrich, H.R.; Price, R.H. Non-explosive silicic volcanism. *Nature* **1986**, *323*, 598–602. [CrossRef]
41. Burnham, C.W.; Holloway, J.R.; Davis, N.F. Thermodynamic Properties of Water to 1,0000 C and 10,000 Bars. *Geol. Soc. Am.* **1969**, *132*, 96.
42. Giordano, D.; Russell, J.K.; Dingwell, D.B. Viscosity of magmatic liquids: A model. *Earth Planet. Sci. Lett.* **2008**, *271*, 123–134. [CrossRef]
43. Dingwell, D.B.; Romano, C.; Hess, K.-U. The effect of water on the viscosity of a haplogranitic melt under PTX conditions relevant to silicic volcanism. *Contrib. Mineral. Petrol.* **1996**, *124*, 19–28. [CrossRef]
44. Carrigan, C.R.; Eichelberger, J.C. Zoning of magmas by viscosity in volcanic conduits. *Nature* **1990**, *343*, 248–251. [CrossRef]
45. Stockman, H.W.; Cygan, R.T.; Carrigan, C.R. Modelling viscous segregation in immiscible fluids using lattice-gas automata. *Nature* **1990**, *348*, 523–525. [CrossRef]
46. Cardoso, S.S.S.; Woods, A.W. On convection in a volatile-saturated magma. *Earth Planet. Sci. Lett.* **1999**, *168*, 301–310. [CrossRef]
47. Brandeis, G.; Marsh, B.D. The convective liquidus in a solidifying magma chamber: A fluid dynamic investigation. *Nature* **1989**, *339*, 613–616. [CrossRef]
48. Eichelberger, J.C.; Gooley, R. Evolution of silicic magma chambers and their relationship to basaltic volcanism. In *The Earth's Crust*; IntechOpen: London, UK, 1977; pp. 57–77.
49. Eichelberger, J.C. Andesitic volcanism and crustal evolution. *Nature* **1978**, *275*, 21–27. [CrossRef]
50. Eichelberger, J.C. Vesiculation of mafic magma during replenishment of silicic magma reservoirs. *Nature* **1980**, *288*, 446–450. [CrossRef]
51. Kiryukhin, A.; Chernykh, E.; Polyakov, A.; Solomatin, A. Magma Fracking Beneath Active Volcanoes Based on Seismic Data and Hydrothermal Activity Observations. *Geosci. J.* **2020**, *10*, 52. [CrossRef]
52. Sahagian, D.L.; Proussevitch, A.A. Thermal effects of magma degassing. *J. Volcanol. Geotherm. Res.* **1996**, *74*, 19–38. [CrossRef]

53. Lowenstern, J.B.; Sisson, T.W.; Hurwitz, S. Probing magma reservoirs to improve volcano forecasts. *EOS* **2017**, *98*. Available online: https://www.researchgate.net/profile/Jacob_Lowenstern/publication/320588486_Probing_Magma_ReReservoi_to_Improve_Volcano_Forecasts/links/59f937f3aca272607e2f71a4/Probing-Magma-Reservoirs-to-Improve-Volcano-Forecasts.pdf (accessed on 26 May 2020). [CrossRef]
54. Eichelberger, J.C.; Carrigan, C.R.; Ingolfsson, H.P.; Lavallee, Y.; Ludden, J.; Markusson, S.; Mortensen, A.; Papale, P.; Sigmundsson, F.; Saubin, E.; et al. Magma-sourced geothermal energy and plans for krafla magma testbed, iceland. In Proceedings of the World Geothermal Congress, Reykjavik, Iceland, 27 April–1 May 2020.

© 2020 by the author. Licensee MDPI, Basel, Switzerland. This article is an open access article distributed under the terms and conditions of the Creative Commons Attribution (CC BY) license (http://creativecommons.org/licenses/by/4.0/).

MDPI
St. Alban-Anlage 66
4052 Basel
Switzerland
Tel. +41 61 683 77 34
Fax +41 61 302 89 18
www.mdpi.com

Geosciences Editorial Office
E-mail: geosciences@mdpi.com
www.mdpi.com/journal/geosciences

www.ingramcontent.com/pod-product-compliance
Lightning Source LLC
LaVergne TN
LVHW070454100526
838202LV00014B/1724